Electricity for Agricultural Applications

Electricity for Agricultural Applications

Carl J. Bern • Dean I. Olson

Carl J. Bern, P.E., Ph.D., Agricultural and Biosystems Engineering Department, Iowa State University, Ames.

Dean I. Olson, Ph.D., Agricultural Engineering Technology Department, University of Wisconsin–River Falls.

Iowa State Press
2121 State Avenue, Ames, Iowa 50014

Orders: 1-800-862-6657
Office: 1-515-292-0140
Fax: 1-515-292-3348
Web site: www.iowastatepress.com

Word processing and drawings: Kristine Wilson, Agricultural and Biosystems Engineering Department, Iowa State University, Ames; drawings: Nicholas Krueger, student in Agricultural Engineering, Iowa State University, Ames.

∞ Printed on acid-free paper in the United States of America
First edition, 2002

Library of Congress Cataloging-in-Publication Data

Bern, Carl J.
Electricity for agricultural applications/Carl J. Bern, Dean I. Olson; word processing and drawings, Kristine Wilson.—1st ed.
p. cm.
Includes bibliographical references and index.
ISBN 0-8138-2199-1 (alk. paper)
1. Electricity in agriculture. I. Olson, Dean I. II. Title.

TK4018.B46 2002
631.3′71—dc21 2002003496

The last digit is the print number: 9 8 7 6 5 4 3 2 1

CONTENTS

PREFACE

Can you envision agriculture without electric motors, electric lights, or electric controls? In 1920 when 2% of U.S. farms had line power, this was the reality of American agriculture. Visionaries of that age saw the need and the possibilities of electricity, and through their efforts almost every American farm was electrified within 35 years. This transformed life for the farm family.

Since that date, the uses and usage of electricity in agriculture have increased dramatically, and today, an understanding of electricity is a prerequisite for anyone who wants to work in technical agriculture or in a supporting industry. This text was written to help you, the reader, to attain this understanding.

The study of electricity is necessarily mathematical. This text uses algebra throughout and an understanding of algebra is needed by the reader. Trigonometry and vectors are used in chapters 2, 4, and 10, and Boolean algebra is used in chapter 13. However, these uses are all explained thoroughly and prior knowledge of trigonometry, vectors, and Boolean algebra is not necessary for those using the text. Readers with math backgrounds beyond algebra will be able to skip over some of the explanations and problems.

This text includes a significant emphasis on safety, especially in chapter 3 (humans), chapter 6 (humans and equipment), and chapter 14 (animals). The authors consider this emphasis to be necessary because of the inherent dangers of electricity, and see this emphasis as a strong point of this text.

This text is not intended as an instruction manual for electrical wiring practice. However, topics covered here allow the reader to understand good wiring practice, and knowledge gained here will help the reader to understand instructions in a how-to wiring manual, and to do wiring more quickly and more confidently.

The chapters and problems have been developed over several years as we have taught electrification courses for many technology and engineering students. We are indebted to these students for asking questions, making suggestions, and pointing out errors. Student input has helped make this a quality text.

Electricity for Agricultural Applications

Chapter 1

Basic Electricity

The Nature of Electricity

Electricity is a physical phenomenon that involves movement of electrons. Figure 1.1 illustrates how this movement can occur. The dots represent atoms of a material such as copper with many loosely held electrons (a good conductor). If by some means an abundance of electrons is created at one end of this material and a scarcity of electrons is created at the other end, electrons will flow, atom to atom, from the point where they are abundant to the point where they are scarce. The relative abundance and scarcity can be created by electromechanical, electrochemical, thermoelectrical, photoelectrical, or some other means. As electrons flow, they can be made to do something useful, such as turning the shaft of an electric motor, heating the filament in a light bulb to incandescence, transmitting data to a computer, or illuminating a video screen.

Circuit Parameters

Three related parameters (current, voltage, resistance) are used to quantify the flow of electrons.

Current (symbol: I)

Current is a measure of the rate of electron flow. I is the symbol for current. The unit of current is the ampere or amp (A). Here is a definition of the amp:

$$1\text{ A} = 6.3 \times 10^{18}\,\frac{\text{electrons}}{\text{second}}\text{ moving past a point}$$

This number is so large that it is almost incomprehensible. The units show that current is a flow and is somewhat analogous to the flow of a fluid like water expressed in liters per second or gallons per minute.

Voltage (symbol: E)

Voltage is a measure of the driving force or pressure that can cause the electrons to flow. The unit of voltage is the volt and E or V are common symbols. Voltage, or electrical pressure, is analogous to Pa (pascals) or lb/in^2 pressure in a fluid system. It is important to note that, like fluid pressure, voltage is never *at* a point, but is always *between* two points. For example, a 9-volt battery has a voltage difference of 9 volts between its terminals.

Resistance (symbol: R)

Resistance is a measure of the difficulty encountered by the electrons as they flow through the conductor. The resistance is a property of the conductor material. For most materials, resistance increases as material temperature increases. The unit of resistance is the ohm, often expressed as Ω (omega). R is the symbol for resistance.

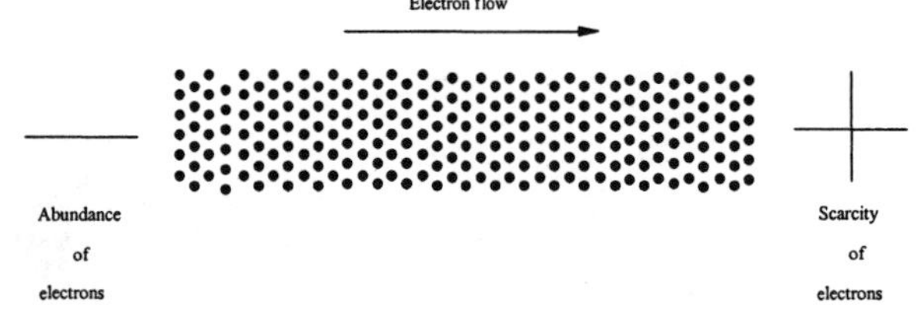

Fig. 1.1. Electron flow in a conductor.

Direction of Current Flow

Although electrons flow from − to + as illustrated in Figure 1.1, the conventional direction assumed for current flow in electrical circuit analysis is from + to − as illustrated in Figure 1.2. This discrepancy came about because the + to − current flow direction was assumed and used long before electric current was discovered to be a flow of electrons. Some texts use electron flow for discussion purposes because some concepts are more easily understood using electron flow analysis. Mathematical relationships presented in this text are valid regardless of which direction is assumed.

Direct Current and Alternating Current

Electrical circuits can be conveniently classified according to the manner in which the current varies with time. A direct current (DC) system employs a power source (such as a battery or a DC generator) that causes current to continually flow in one direction. Figure 1.3 shows a DC current which is constant with time.

An alternating current (AC) system employs a power source (such as an AC generator) which produces a current that continually varies with time and periodically reverses direction (Figure 1.4). Usually, in AC systems, the current varies with time the way the trigonometric sine varies with angle. Hence, the wave form is said to be sinusoidal. AC line power systems in the USA operate at 60 Hz where current and voltage complete 60 cycles per second. Most discussion in this text will involve 60 Hz AC because of its wide use.

Example 1.1

How many minutes does it take for 60-Hz power to complete 1 million cycles?

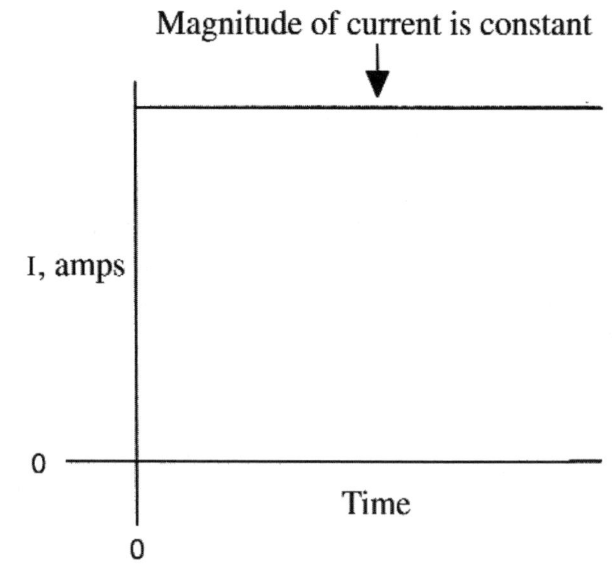

Fig. 1.3. Current–time relationship in a DC system.

$$\left(\frac{1{,}000{,}000 \text{ cycles}}{}\right)\left(\frac{1 \text{ second}}{60 \text{ cycles}}\right)\left(\frac{1 \text{ min.}}{60 \text{ seconds}}\right) = 277.8 \text{ min.}$$

Circuit Analysis

Ohm's Law

In 1823, Simon Ohm set down this relationship among the factors I, E, and R:

$$I = \frac{E}{R} \tag{1.1}$$

This says that the current flowing in a circuit is proportional to the voltage between the points and inversely proportional to the resistance of the current path between the points.

Placing I, E, and R in a triangle as shown in Figure 1.5 is a convenient aid to memory in manipulating Ohm's Law to solve for various parameters. If the parameter to be solved for is covered, the remaining two parameters appear in the correct positions for solution. For example, covering R leaves E

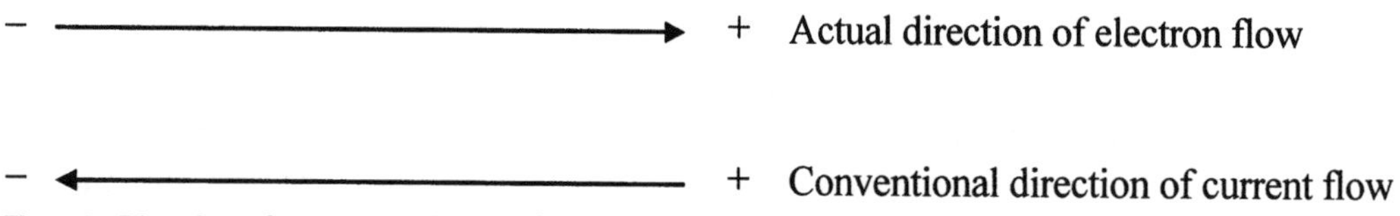

Fig. 1.2. Direction of current or electron flow.

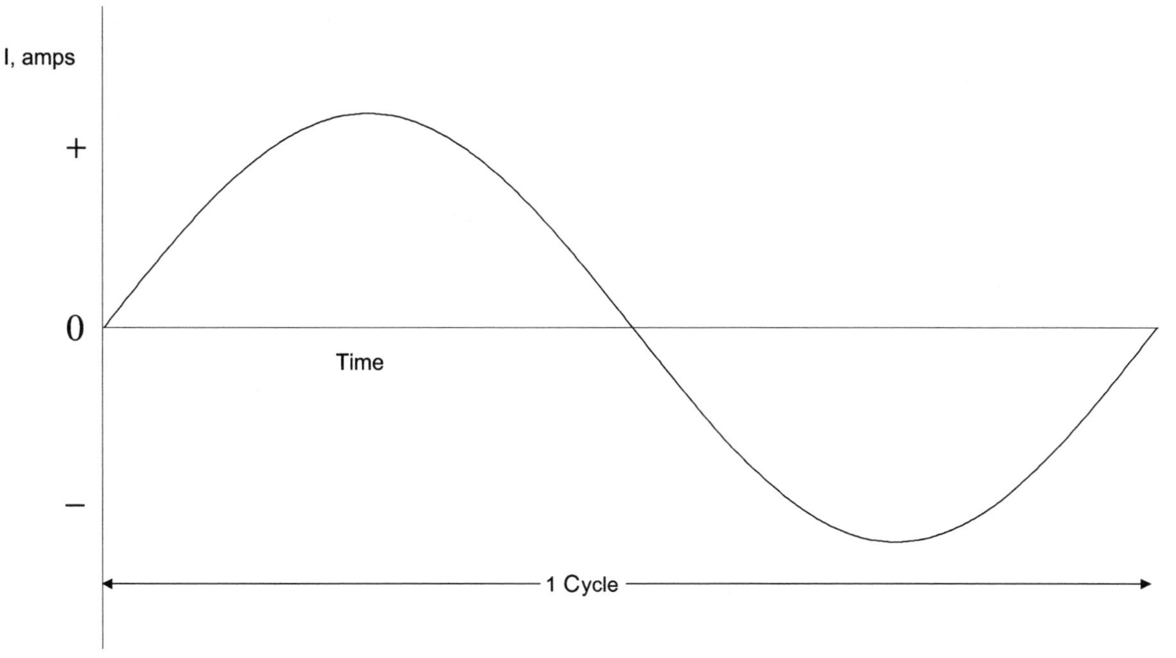

Fig. 1.4. Current–time relationship in an AC system.

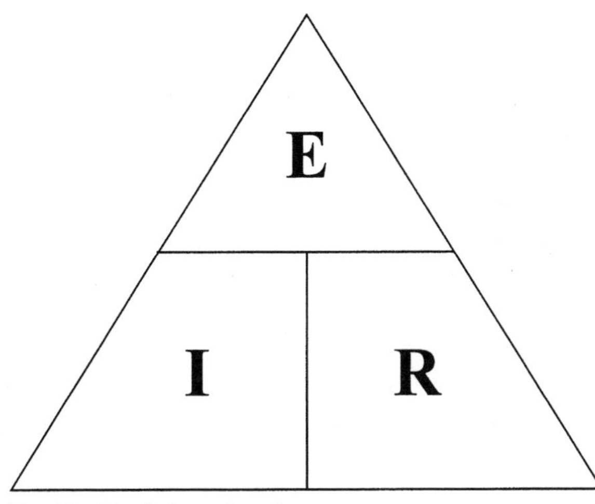

Fig. 1.5. Ohm's Law triangle.

over I, showing that:

$$R = \frac{E}{I} \tag{1.2}$$

Simple Circuit Analysis

It is very useful to model electrical systems using lines and symbols drawn on paper. System behavior can be predicted without actual construction of the system. Figure 1.6 is a model of a simple DC circuit. Examine this figure carefully and note descriptions of the components. Figure 1.6 is an *open circuit.*

The switch is open and hence no current can flow. If the switch is closed as shown in Figure 1.7, a *closed circuit* exists and electrons flow through the

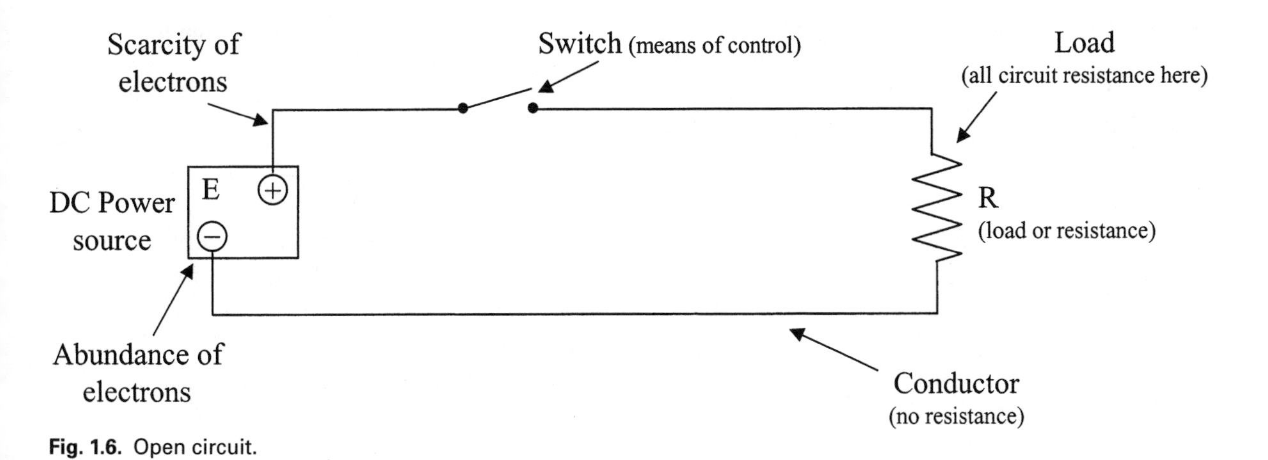

Fig. 1.6. Open circuit.

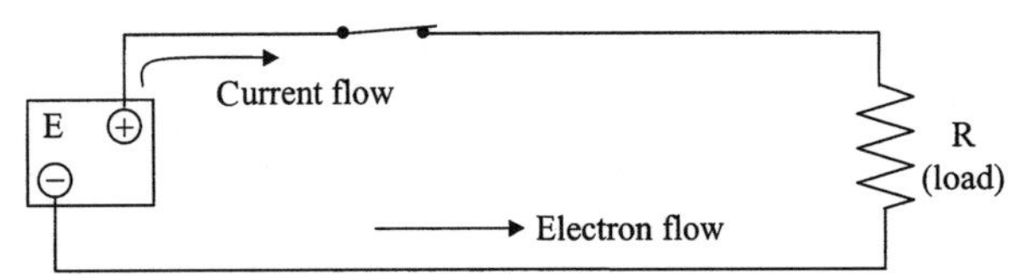

Fig. 1.7. Closed circuit.

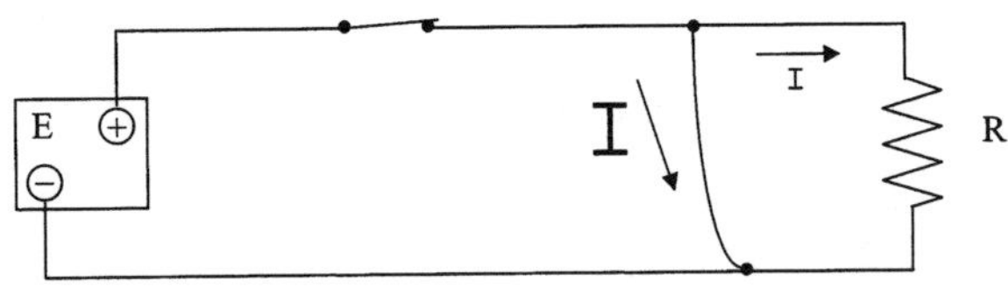

Fig. 1.8. Short circuit.

circuit at a flow rate dependent on the voltage of the power source and the resistance of the load. This text will, from now on, assume + to − current flow.

Note that in fluid systems, the opposite convention is observed. An open water valve allows water to flow, while a closed valve prevents flow.

Another type of closed circuit is the *short circuit* as shown in Figure 1.8. Here, a low resistance path has been placed around the load. This allows most of the current to bypass the load, which normally limits the current flow. An abnormally high current flows from the power source and if the circuit is not equipped with an overcurrent protective device, some circuit elements may be damaged. Current through the short circuit path is limited by the power source or the very low resistance of conductors and connections. Most (but not all) of this high current bypasses the load since an alternative or parallel path has been established.

Using Ohm's Law

Example 1.2

Using Ohm's Law, a quantitative circuit analysis can be made. In the circuit of Figure 1.9, the power source has 9 V and is causing 18 A to flow through a resistor. The value of the resistance can be calculated

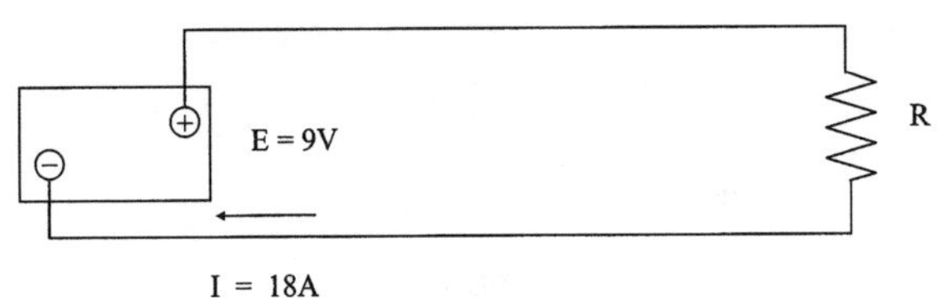

Fig. 1.9. Simple circuit analysis.

TABLE 1.1 Power and energy units

Power units		Energy units	
horsepower	hp	horsepower · hour	hp h
foot pounds per minute	ft lb/min	foot pounds	ft lb
joule per second	J/s	joule	J
Watt	W	watthour	Wh
kilowatt	kW	kilowatt hour	kWh
British thermal unit per hour	Btu/h	British thermal unit	Btu

Conversions 1 hp = 33,000 ft lb/min; 1 hp = 746 W; 1 J/s = 1 W; 1 kWh = 3413 Btu; 1 kW = 3413 Btu/h; kilowatt 1 kW = 1,000 W = 10^3 W; megawatt 1 MW = 1,000,000 W = 10^6 W; gigawatt 1 GW = 1,000,000,000 W = 10^9 W

by solving Ohm's Law for R, substituting for I and E, and computing an answer:

$$R = \frac{E}{I} = \frac{9}{18} = 0.5\ \Omega \qquad \textbf{(1.3)}$$

Power and Energy

Power is a rate of doing work, while energy is a quantity of work done. Table 1.1 lists some common units of power and energy. Electric power is usually expressed in watts or kilowatts and electric energy is usually expressed in kilowatt hours.

Notice that in each case, the energy unit is obtained by multiplying the power unit by a time unit. For example,

$$\underset{\text{(power)}}{\text{kW}} \times \underset{\text{(time)}}{\text{h}} = \underset{\text{(energy)}}{\text{kWh}} \qquad \textbf{(1.4)}$$

DC Power

In all DC circuits, and in AC circuits which have loads made up only of resistance (such as heaters and incandescent lamps), power used by the load can be computed using Equation 1.5:

$$P = IE \qquad \textbf{(1.5)}$$

where: P = power, watts
I = current through the load, amps
E = voltage across the load, volts

Example 1.3

Compute the power used by the resistor in Figure 1.9.

$$P = (18\ A)\ (9\ V) = 162\ W \tag{1.6}$$

How much energy will this load use if it is left on for 100 hours? Recalling that energy is power × time:

$$\begin{aligned} \text{Energy} &= (162 \text{ watts})\ (100 \text{ hours}) = 16{,}200 \text{ watthours} \\ &= 16.2 \text{ kilowatthours} \\ &= 16.2 \text{ kWh} \end{aligned} \tag{1.7}$$

If the electrical energy costs \$0.09/kWh, the cost to operate this load for 100 hours will be:

$$\frac{(16.2\ \not{\text{kWh}})(\$0.09)}{\not{\text{kWh}}} = \$1.46 \tag{1.8}$$

AC Power

For AC circuits having loads containing reactance such as motors or fluorescent lights, we must consider the power factor and:

$$P = IE(PF) \tag{1.9}$$

Where: P = power used by load, watts
I = current through the load, amps
E = voltage across the load, volts
PF = power factor of the load, decimal
IE = total power carried by conductors, volt amps

The power factor (PF) is a characteristic of the load and is always between 0 and 1. It is the fraction of the total power carried by the conductors to the load which is used by the load. (The power not used is momentarily stored by the load and then passed back to the generator.) If the load is resistive, PF = 1 and Equation 1.9 reduces to Equation 1.5 (P = IE). Then watts (power used) equals volt amps (total power).

Figure 1.10 will help to explain the concept of power factor. Both voltage and current vary with time in alternating current. If the load is a resistor, voltage and current peaks occur at the same time, as shown in the upper wave diagram.

Since peaks occur at the same time, instantaneous voltage and instantaneous current always have the same sign (positive or negative) and their product, p (the instantaneous power), is always positive. Positive power is power flowing from generator to load. Note in the upper figure that the p curve is never less than zero.

A load like a motor causes current to lag voltage (bottom wave diagram) because a magnetic field is produced by the windings of the motor. The building and collapsing of this field 120 times per second (twice per cycle) is what causes current to lag voltage. When this happens, the voltage and current peaks do not occur at the same time. Also, during a portion of the cycle, instantaneous voltage and instantaneous current have different signs. Since the product of two numbers of different signs is negative, this means p is negative, indicating that power is flowing from the load toward the generator during this portion of the cycle. During periods of negative power, energy stored in the magnetic field is flowing back toward the generator.

A power factor less than 1.0 indicates that more current is flowing to the load than is required to supply the actual power used by the load. (The watthour meter registers, and the customer pays for, only the actual power.) The power supplier must have in place generators, wires, transformers, and other equipment heavy enough to carry the total power. Therefore, the power supplier benefits from a high power factor.

If the power factor of a load is improved—that is, increased (with no other change)—the power used by the load stays the same, but the current to the load is reduced. This subject will come up again in chapter 2.

Example 1.4

What is the power factor of a load which draws 975 W and 14.5 A when connected across 115 V?

Solving Equation 1.9 for PF, we get:

$$PF = \frac{P}{IE}$$

$$\text{then} \quad PF = \frac{975}{(14.5)(115)} = 0.58$$

Efficiency

To a consumer of electrical energy, an important quantity is efficiency. Efficiency indicates how effective a machine is at converting electrical power to some other form:

$$\%\ \text{Efficiency} = \frac{\text{output power}}{\text{input power}}(100) \tag{1.10}$$

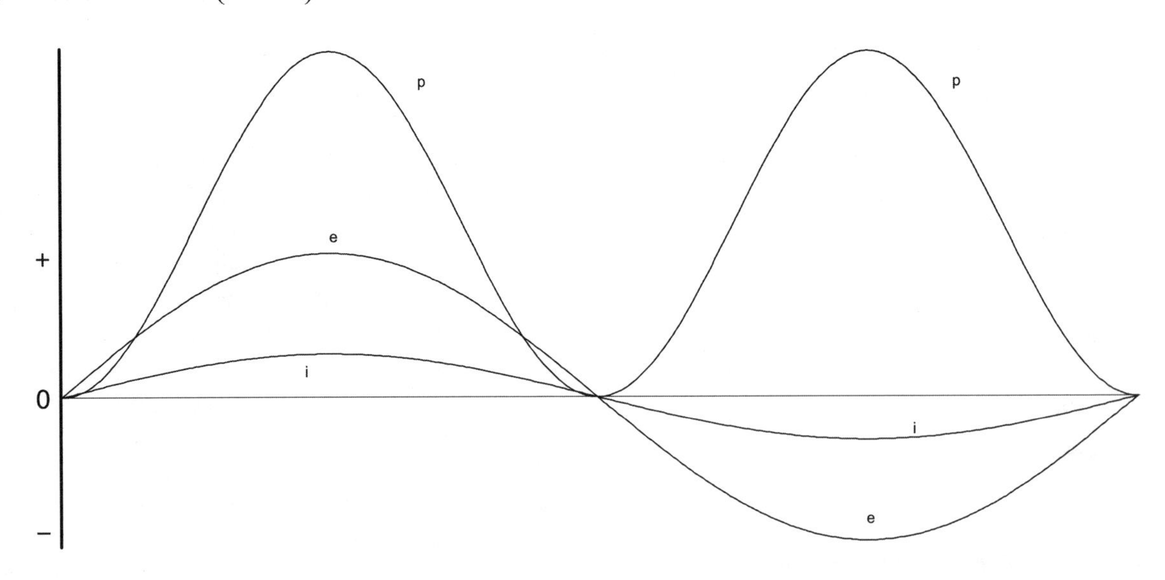

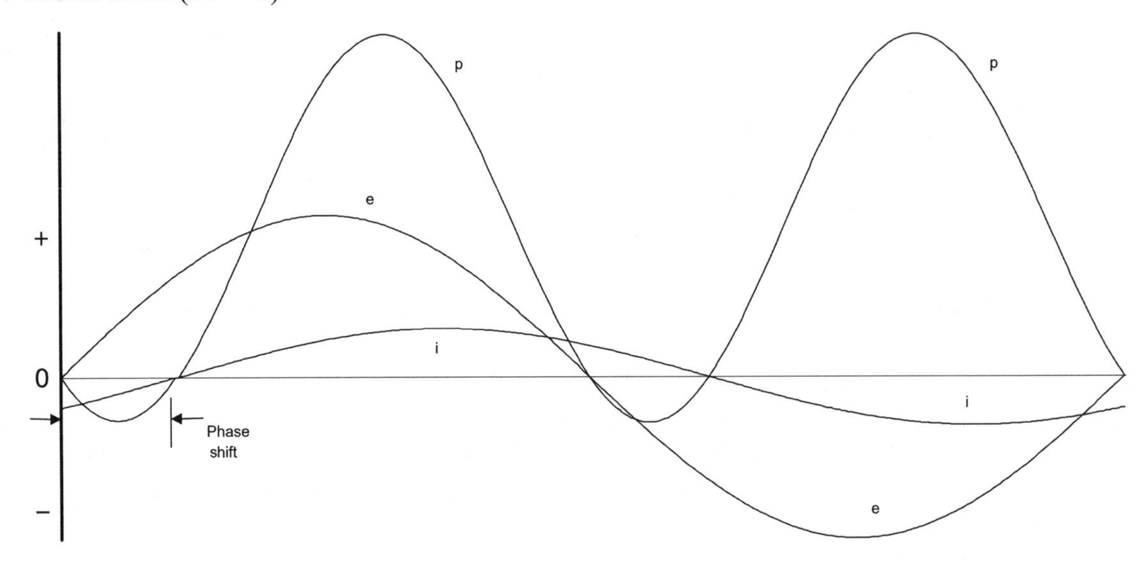

i = instantaneous current
e = instantaneous voltage
p = instantaneous power = ie

Fig. 1.10. AC time relations.

At full load, a common type of electric motor is about 75% efficient. An incandescent light bulb is about 8% efficient at converting electric power to visible radiation. Electrical resistance heaters are 100% efficient. Efficiency is not dependent on power factor.

Appendix B lists some typical electrical equipment data. Examples will illustrate use of this data.

Example 1.5

How much energy will a hog waterer use in one month if the heater element is on 10% of the time?

From Table B.1: A hog waterer uses 200 W.

$$\text{Energy} = \frac{(0.2\ \text{kW})(24\ \text{h})(30\ \cancel{\text{day}})(0.1)}{\cancel{\text{day}}\ \text{month}} = \frac{14.4\ \text{kWh}}{\text{month}}$$

Example 1.6

Calculate actual power, current, and total power, for a fully loaded 1.5-hp, 230-V electric motor. (*Fully loaded* means motor output equals the rated power output stated on the nameplate.)

From Table B.1: Power use = 1000 W/hp, PF = 0.75.

(Footnote 2 of Table B.1 explains why 1000 W/hp is used.)

$$\text{actual power} = (1.5\ \cancel{\text{hp}})\left(\frac{1000\ \text{W}}{\cancel{\text{hp}}}\right) = 1500\ \text{W}$$

$$I = \frac{P}{E(PF)} = \frac{1500\ \text{W}}{(230\ \text{V})(0.75)} = 8.7\ \text{A}$$

$$\text{Total power} = IE = (8.7\ \text{A})(230\ \text{V}) = 2000\ \text{VA}$$

Buying Electrical Energy

Electric power companies sell electrical energy to their customers. The kilowatthours of electrical energy used by a customer are measured by the power company's watthour meter which is installed in the line ahead of all the customer's loads. The change in the reading on the meter during the billing period is the kWh of electrical energy used during that period. The electric power company may also bill customers for kW of electrical demand. This is explained in chapter 14.

Example 1.7 shows the computation of a residential electrical energy bill.

Example 1.7

What is the charge for 2200 kWh of electrical energy by a residential customer in Ames during a summer month and the average cost per kWh? The 2000 summer rate structure is as follows:

Service Charge		$ 3.50
First 400 kWh	@7.70¢/kWh	30.80
Next 600 kWh	@7.30¢/kWh	43.80
All over 1000 kWh	@6.70¢/kWh	80.40
		$158.50

$$\text{Average electrical energy cost} = \frac{\$158.50}{2200\ \text{kWh}} = 7.20¢/\text{kWh}$$

Circuit Analysis

The circuit of Figure 1.6 has only one load. In many applications, several loads are connected in a circuit. Series, parallel, and combination connections of resistances will be considered here.

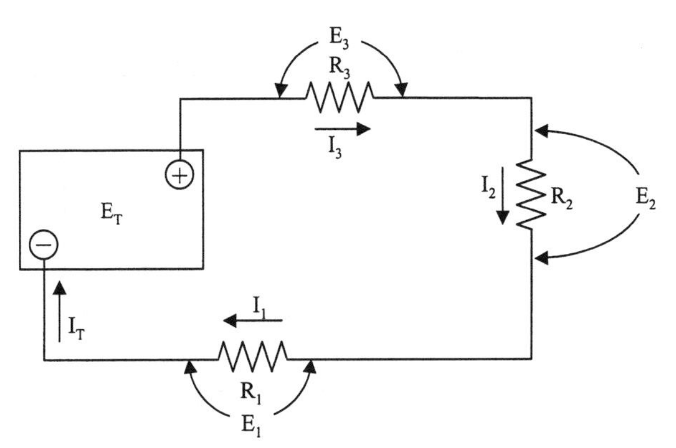

Fig. 1.11. Series circuit.

Series Circuits

Figure 1.11 shows a closed circuit composed of three resistances connected in series to a DC power source having a voltage of E_T across its terminals. (The analysis would also hold if the power source was AC since all the loads are resistive.)

In this analysis, we want to establish relationship among the voltages, currents, and resistances existing on this circuit. Our tool in this analysis is Ohm's Law (Equation 1.1) which applies for each resistor.

We can, by sight, establish the relationship among the currents. Because no parallel paths exist, current leaving the −terminal must *pass through each resistor in series* before reaching the +terminal. Thus the current is everywhere the same and:

$$I_T = I_1 = I_2 = I_3 \quad \textbf{(1.11)}$$

Some thought about a fluid analogy may be useful in accepting the validity of Equation 1.11. If an incompressible fluid such as oil is pumped through a series circuit of tubing, a flow meter placed at any point in the circuit will read the same flow rate, since fluid is not stored or produced anywhere along the tubing.

Current must pass successively through each resistance. Thus the total resistance encountered by an electron will be the sum of the individual resistances (conductor resistance is assumed zero):

$$R_T = R_1 + R_2 + R_3 \quad \textbf{(1.12)}$$

A single resistor having R_T ohms would look exactly the same to the power source as R_1, R_2, and R_3 in series. That is, the circuit of Figure 1.12 looks exactly the same to the power source as the circuit of Figure 1.11.

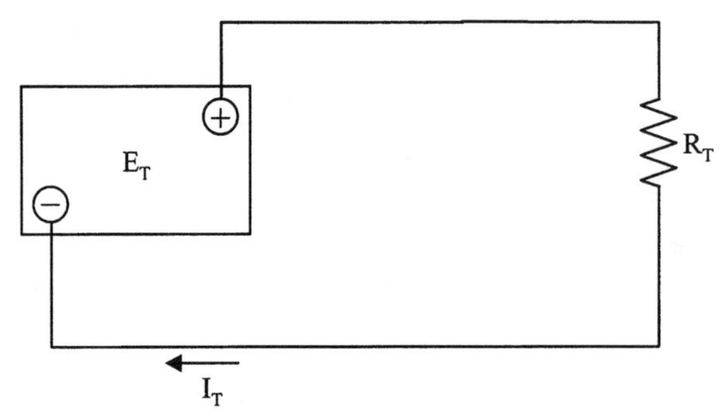

Fig. 1.12. Circuit equivalent to Figure 1.11.

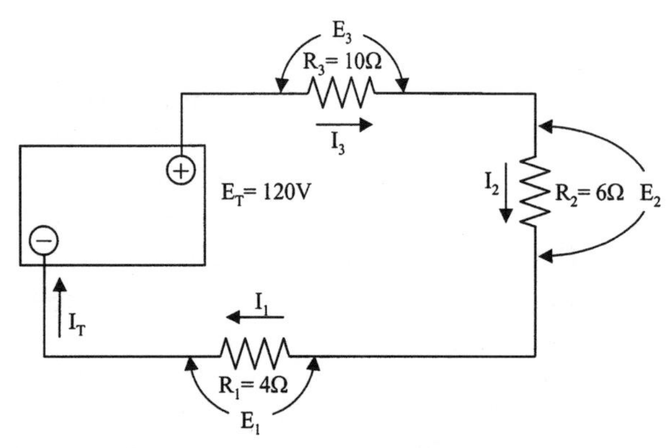

Fig. 1.13. Series circuit with numerical parameter values.

There is a voltage drop across each resistor in a series circuit. To analyze this voltage drop, numerical values are placed on the parameters of the circuit of Figure 1.11 (Figure 1.13). E_1, E_2, and E_3 are the voltage drops across R_1, R_2, and R_3, respectively:

Using Equation 1.12: $R_T = 4 + 6 + 10$

$$R_T = 20\ \Omega$$

Since we know R_T and E_T, we can apply Ohm's Law to R_T in Figure 1.12:

$$I_T = \frac{E_T}{R_T}$$

$$I_T = \frac{120}{20}$$

$$I_T = 6\ \text{A}$$

Because all currents are the same (Equation 1.11) we know the current and resistance of each load and we can calculate the voltage across each load using Ohm's Law.

Solving Ohm's Law for E:

$$\begin{aligned} E_1 &= I_1R_1 = (6)(4) = 24\ \text{V} \\ E_2 &= I_2R_2 = (6)(6) = 36\ \text{V} \\ E_3 &= I_3R_3 = (6)(10) = 60\ \text{V} \\ E_T &\qquad\qquad 120\ \text{V} \end{aligned}$$

If all of the voltage drops around the circuit are added, the total voltage is obtained. Thus:

$$E_T = E_1 + E_2 + E_3 \tag{1.13}$$

This analysis has been for a circuit of three series resistances. However, the same relationships apply regardless of the number of series resistances. Here are the equations which describe the operation of a series circuit consisting of n resistors arranged in series (n can be any positive number):

$$E_T = E_1 + E_2 + E_3 + \ldots + E_n \tag{1.14}$$

$$I_T = I_1 = I_2 = I_3 = \ldots = I_n \tag{1.15}$$

$$R_T = R_1 + R_2 + R_3 + \ldots + R_n \tag{1.16}$$

Loads are seldom arranged in series on power circuits in actual applications because of problems encountered:

- Control — If the circuit is opened to shut off one load, all the circuit loads are shut off.
- Load voltage — The voltage drop across any resistance (load) in a series circuit (and hence the power used by that load) is dependent on the size (resistance) of all the other loads in the circuit. Placing an additional load in a series circuit will lower the current through, the voltage across, and the power used by, all the other loads on the circuit. Removing a load will have the opposite effect.

Some Christmas tree light strings are series configurations because this arrangement reduces the length of conductor needed for a certain length of tree light string.

Example 1.8

Four resistors are connected in a circuit across 230 V as shown in Figure 1.14.

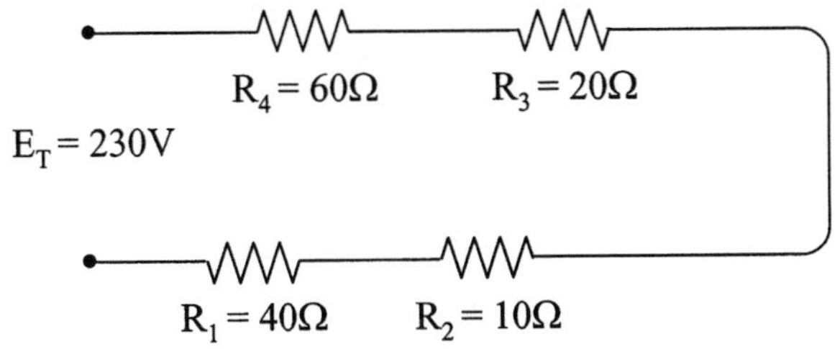

$R_4 = 60\Omega$ $R_3 = 20\Omega$

$E_T = 230V$

$R_1 = 40\Omega$ $R_2 = 10\Omega$

Figure 1.14. Series circuit for Example 1.8.

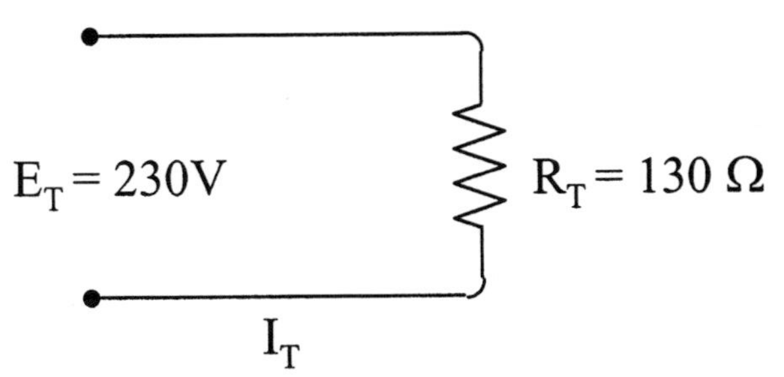

Figure 1.15. Equivalent circuit for Example 1.8.

a. What is the total resistance of this series circuit?
Substitute into Equation 1.16, with n = 4:

$$R_T = R_1 + R_2 + R_3 + R_4$$
$$R_T = 40 + 10 + 20 + 60$$
$$R_T = 130\ \Omega$$

b. What is the total circuit current?
Knowing R_T, we can draw the equivalent circuit (Figure 1.15). Substituting into the Ohm's Law equation (Equation 1.1):

$$I_T = \frac{E_T}{R_T} = \frac{230\ \text{V}}{130\ \Omega} = 1.77\ \text{A}$$

c. What is the voltage drop across each resistor?
Note from Equation 1.15 that all currents in a series circuit are equal:

$$I_T = I_1 = I_2 = I_3 = I_4 = 1.77\ \text{A}$$

Now we know the current and resistance of each resistor, and we can use Ohm's Law to compute voltage:

$$E_1 = I_1R_1 = (1.77)(40) = 70.8\ \text{V}$$
$$E_2 = I_2R_2 = (1.77)(10) = 17.7\ \text{V}$$
$$E_3 = I_3R_3 = (1.77)(20) = 35.4\ \text{V}$$
$$E_4 = I_4R_4 = (1.77)(60) = 106.2\ \text{V}$$

As a check on our work, we can substitute into Equation 1.14 and verify that the sum of the voltage drops equals the applied voltage:

$$E_T = E_1 + E_2 + E_3 + E_4 = 70.8 + 17.7 + 35.4 + 106.2 = 230.1\ \text{V}$$

Rounding errors cause the calculated total to exceed the applied voltage by 0.1 V.

d. What power is used by resistor 3?
The power equation for a resistive load is Equation 1.9, with PF = 1 (or Equation 1.5): P = IE, where I is the current through resister 3, and E is the voltage across resistor 3. Thus,

$$P_3 = I_3E_3 = (1.77)(35.4) = 62.7\ \text{W}$$

Noting that E = IR, and substituting into the power equation,

$$P = IE = I(IR) = I^2R \quad \text{or}$$
$$P = IE = \left(\frac{E}{R}\right)E = \frac{E^2}{R} \qquad \textbf{(1.17)}$$

Substituting into this equation,

$$P = (1.77)^2(20) = 62.7\ \text{W} \quad \text{or}$$
$$P = \frac{(35.4)^2}{20} = 62.7\ \text{W}$$

Parallel Circuits

Figure 1.16 is a parallel circuit.

By sight we see that an electron never passes through more than one resistance in its journey from − to +, since there are parallel paths available. Each resistor has the total voltage, E_T, across it, since there is a solid conductor path between each resistor end and each battery terminal. In equation form:

$$E_T = E_1 = E_2 = E_3 \qquad \textbf{(1.18)}$$

To find the current and resistance relationship, it is convenient to assign numerical values for E_T and all the loads (Figure 1.17).

Now we want to find the value of a single resistance (R_T) which looks exactly the same to the power source as the parallel circuit of R_1, R_2, and R_3 (Figure 1.18).

Equation 1.18 says each resistance has 120 volts across it. Knowing the value of each resistance and

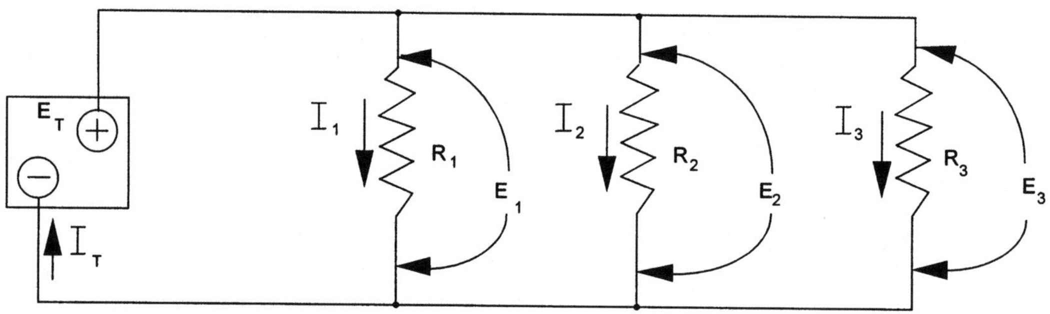

Fig. 1.16. Parallel circuit.

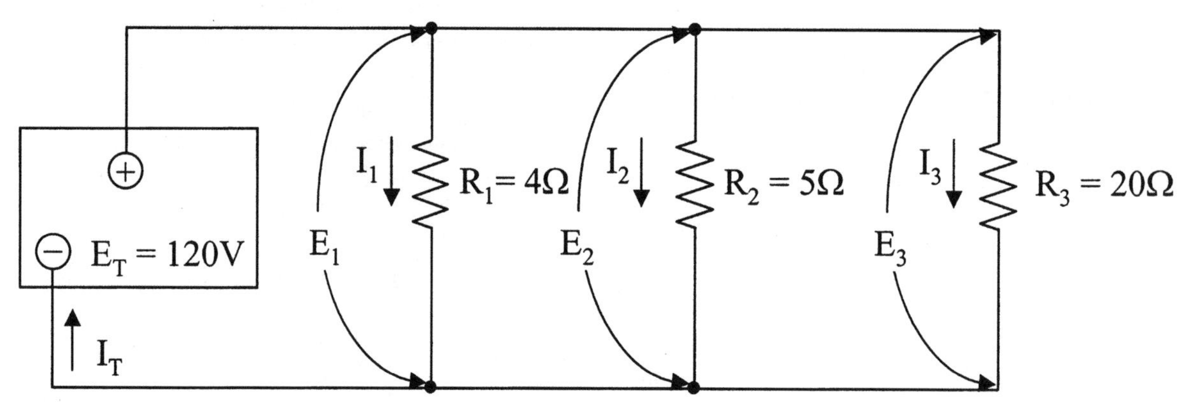

Fig. 1.17. Parallel circuit with numerical parameter values.

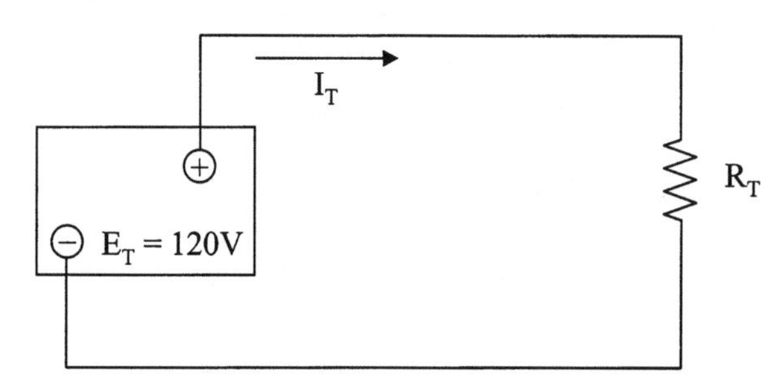

Fig. 1.18. Equivalent circuit of Figure 1.16.

the voltage across it, we can apply Ohm's Law to each resistor:

$$I_1 = \frac{E_1}{R_1} = \frac{120}{4} = 30\text{ A}$$

$$I_2 = \frac{E_2}{R_2} = \frac{120}{5} = 24\text{ A}$$

$$I_3 = \frac{E_3}{R_3} = \frac{120}{20} = 6\text{ A}$$

$$\overline{I_T} \qquad = 60\text{ A}$$

Because all the current must come from the power source, the total current must be the sum of the individual load currents:

$$I_T = I_1 + I_2 + I_3 \tag{1.19}$$

Since Equation 1.18 holds, we can write this equation another way:

$$I_T = \frac{E_T}{R_1} + \frac{E_T}{R_2} + \frac{E_T}{R_3} \tag{1.20}$$

Factoring out E_T:

$$I_T = E_T\left(\frac{1}{R_1} + \frac{1}{R_2} + \frac{1}{R_3}\right) \tag{1.21}$$

Ohm's Law will also apply across R_T in Figure 1.18:

$$I_T = \frac{E_T}{R_T}$$

or

$$I_T = E_T\left(\frac{1}{R_T}\right) \tag{1.22}$$

Notice that Equations 1.21 and 1.22 are identical except for the form of writing the resistance term:

$$I_T = E_T\left(\frac{1}{R_T}\right) = E_T\left(\frac{1}{R_1} + \frac{1}{R_2} + \frac{1}{R_3}\right)$$

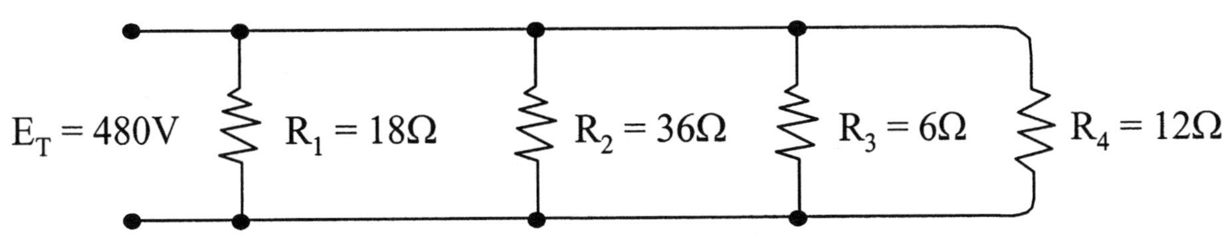

Fig. 1.19. Parallel resistive load.

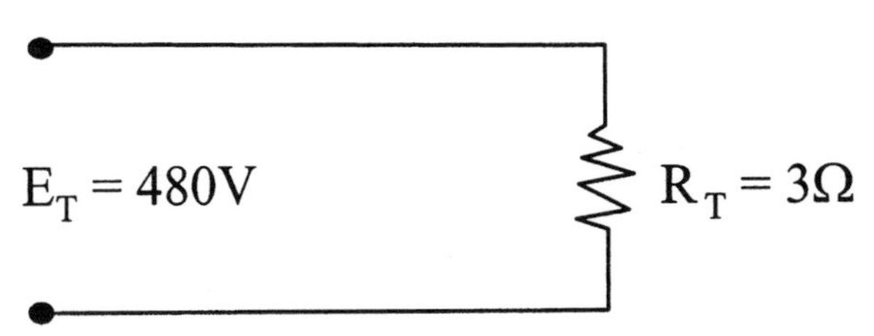

Fig. 1.20. Equivalent circuit of Figure 1.19.

We conclude that:

$$\frac{1}{R_T} = \frac{1}{R_1} + \frac{1}{R_2} + \frac{1}{R_3} \quad \textbf{(1.23)}$$

Equation 1.23 says that in a parallel circuit, the reciprocal of the total resistance is the sum of the reciprocals of the individual resistances. Caution: This equation cannot be turned upside down to get Equation 1.12!

The validity of Equation 1.23 can be checked by computing R_T and then I_T and comparing the answer with our previous figure for I_T of 60 amps:

$$\frac{1}{R_T} = \frac{1}{4} + \frac{1}{5} + \frac{1}{20} = \frac{5 + 4 + 1}{20} = \frac{10}{20}$$

$$R_T = \frac{20}{10} = 2\ \Omega$$

or

$$\frac{1}{R_T} = \frac{1}{4} + \frac{1}{5} + \frac{1}{20} = 0.25 + 0.20 + 0.05 = 0.50$$

$$R_T = \frac{1}{0.50} = 2\ \Omega$$

Applying Ohm's Law to Figure 1.18:

$$I_T = \frac{E_T}{R_T} = \frac{120}{2.0} = 60\ \text{A}$$

This is seen to be the same as the sum of the individual currents.

This analysis used a circuit with three parallel loads. The same relationships hold regardless of the number of parallel loads. For review, here are the parallel circuit equations for a parallel circuit of n resistors (n can be any positive number):

$$E_T = E_1 = E_2 = E_3 = \ldots = E_n \quad \textbf{(1.24)}$$

$$I_T = I_1 + I_2 + I_3 + \ldots + I_n \quad \textbf{(1.25)}$$

$$\frac{1}{R_T} = \frac{1}{R_1} + \frac{1}{R_2} + \frac{1}{R_3} + \ldots + \frac{1}{R_n} \quad \textbf{(1.26)}$$

Electrical loads are almost always connected in parallel because individual loads on a parallel circuit can be turned on or off without affecting the voltage across, the current through, or the power used by any other loads on the circuit:

Example 1.9

Four resistors are connected in a circuit across 480 volts, as shown in Figure 1.19:

a. What is the equivalent resistance of the circuit? Substituting into Equation 1.26, with n = 4 resistors:

$$\frac{1}{R_T} = \frac{1}{18} + \frac{1}{36} + \frac{1}{6} + \frac{1}{12} = \frac{2 + 1 + 6 + 3}{36} = \frac{12}{36}$$

$$R_T = \frac{36}{12} = 3\ \Omega$$

b. What is the total current flowing to the four loads?

$$I_T = \frac{E_T}{R_T} = \frac{480}{3} = 160\ \text{A}$$

c. What current flows through each of the loads? From Equation 1.24 (with n = 4),

$$E_T = E_1 = E_2 = E_3 = E_4 = 480\ \text{V}$$

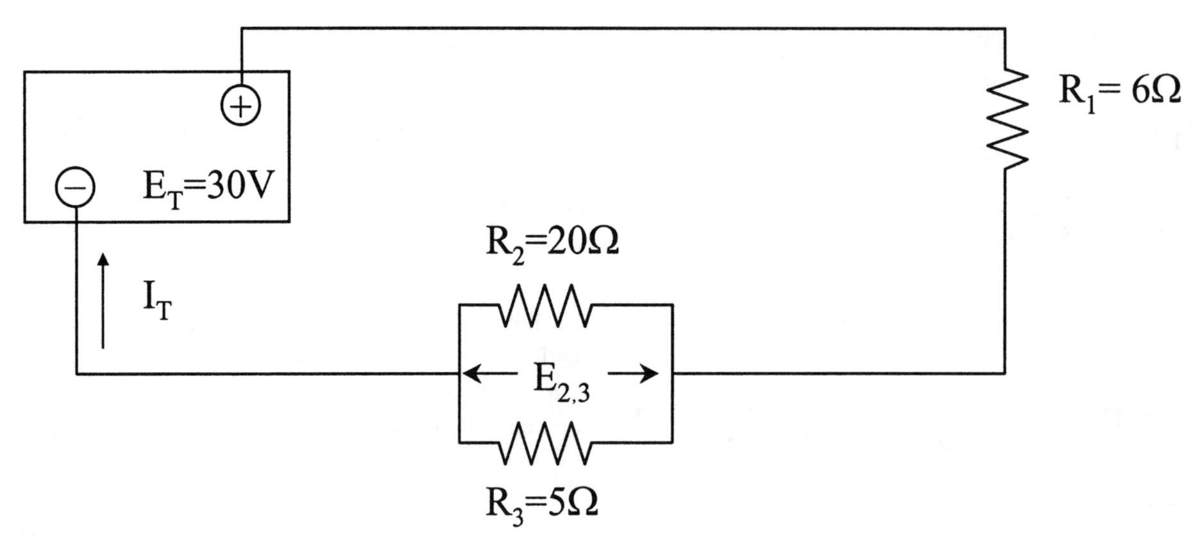

Fig. 1.21. Combination circuit.

Knowing I and R for each resistor, we can apply Ohm's Law:

$$I_1 = \frac{E_1}{R_1} = \frac{480}{18} = 26.7 \text{ A}$$

$$I_2 = \frac{E_2}{R_2} = \frac{480}{36} = 13.3 \text{ A}$$

$$I_3 = \frac{E_3}{R_3} = \frac{480}{6} = 80.0 \text{ A}$$

$$I_4 = \frac{E_4}{R_4} = \frac{480}{12} = 40.0 \text{ A}$$

$$I_T = 160 \text{ A}$$

We can check our work by noting from Equation 1.25 that the sum of the currents for a parallel circuit equals the total current. In part b, this was computed to be 160 A, so our work is correct.

d. What power does R_3 use?
Applying Equation 1.9, with PF = 1 for this resistive load,

$$P_3 = I_3E_3(PF) = (80)(480)(1) = 38{,}400 \text{ W}$$

Combination Circuits

In a combination circuit, some resistors are in series and others are in parallel. Analysis is accomplished by use of Ohm's Law, and the series and parallel equation sets to repeatedly find simpler equivalent circuits until a single equivalent resistor remains. The procedure is illustrated by Example 1.10.

Example 1.10

Figure 1.21 is a circuit composed of a combination of series and parallel loads. Parallel resistors R_2 and R_3 are in series with R_1.

What is the voltage across, the current through, and the power used by each load? First we simplify the parallel set of R_2 and R_3:

$$\frac{1}{R_{2,3}} = \frac{1}{R_2} + \frac{1}{R_3}$$

$$\frac{1}{R_{2,3}} = \frac{1}{5} + \frac{1}{20} = \frac{4+1}{20} = \frac{5}{20}$$

$$\frac{1}{R_{2,3}} = \frac{20}{5} = 4\ \Omega$$

Placing $R_{2,3}$ in the original circuit (Figure 1.22), we now have a series circuit, so:

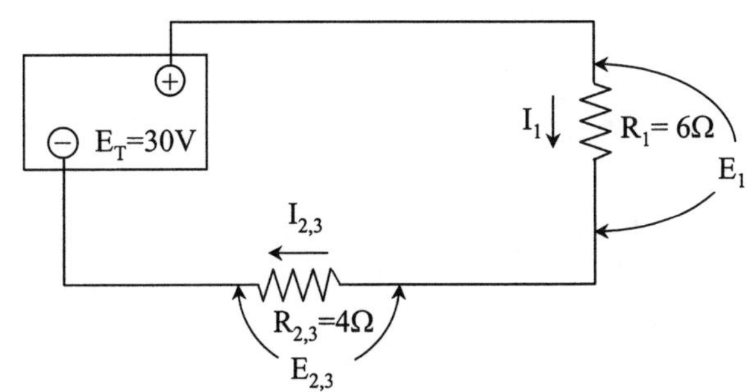

Fig. 1.22. Simplified Figure 1.21 circuit.

$$R_T = R_{2,3} + R_1 = 4 + 6 = 10\ \Omega$$

The circuit is now reduced to one resistance (Figure 1.23).

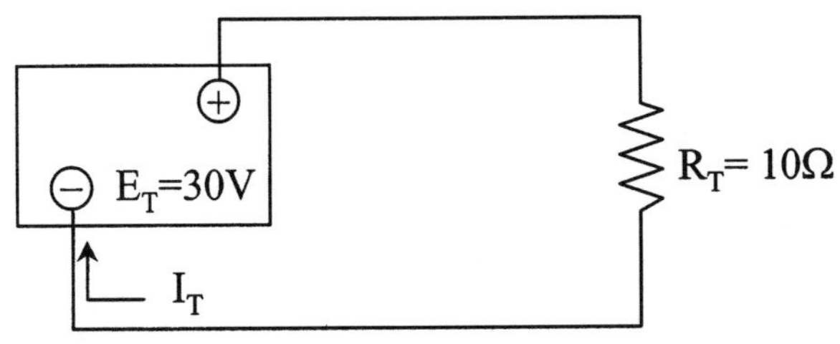

Fig. 1.23. Equivalent circuit of Figure 1.21.

Applying Ohm's Law:

$$I_T = \frac{E_T}{R_T} = \frac{30}{10} = 3\ A$$

Referring to Figure 1.22:

$$I_1 = I_T = 3\ A$$
$$E_1 = I_1R_1 = (3)(6) = 18\ V$$

Applying Equation 1.5:

$$P_1 = I_1E_1 = (3)(18) = 54\ W$$
$$E_{2,3} = I_{2,3}R_{2,3} = 3(4) = 12\ V$$

Referring to Figure 1.21, we see R_2 and R_3 both have $E_{2,3}$ across them:

$$I_2 = \frac{E_{2,3}}{R_2} = \frac{12}{20} = 0.6\ A$$
$$P_2 = I_2E_{2,3} = (0.6)(12) = 7.2\ W$$
$$I_3 = \frac{E_{2,3}}{R_3} = \frac{12}{5} = 2.4\ A$$
$$P_3 = I_3E_{2,3} = (2.4)(12) = 28.8\ W$$

Line Power

Transformers

Most electric power is produced as AC. One reason for this is the ease with which AC voltages can be changed up or down by means of a transformer. A transformer is a device with no moving parts which transfers energy from one circuit to another by electromagnetic induction. Figure 1.24 shows a simple transformer circuit. The winding connected to the AC power source is called the *primary winding,* and the winding which delivers power to a load is called the *secondary winding.* The windings are electrically insulated, so there is no current conducted to the iron core or from one winding to another. When an AC power source is connected to the primary, I_P, the alternating current in this winding sets up an alternating magnetic flux in the iron core, which links with the secondary winding and induces an alternating current, I_S, to flow in the secondary, and AC voltage E_S to appear across the secondary winding. The ratio of turns of conductor around the primary and secondary windings determines the ratio of primary and secondary voltages:

$$\frac{E_1}{E_2} = \frac{N_1}{N_2} \qquad \textbf{(1.27)}$$

In practice, power transformers usually operate with energy efficiencies of 97% to 99%. We will consider here only an ideal transformer where power in = power out:

$$I_P E_P = I_S E_S \qquad \textbf{(1.28)}$$

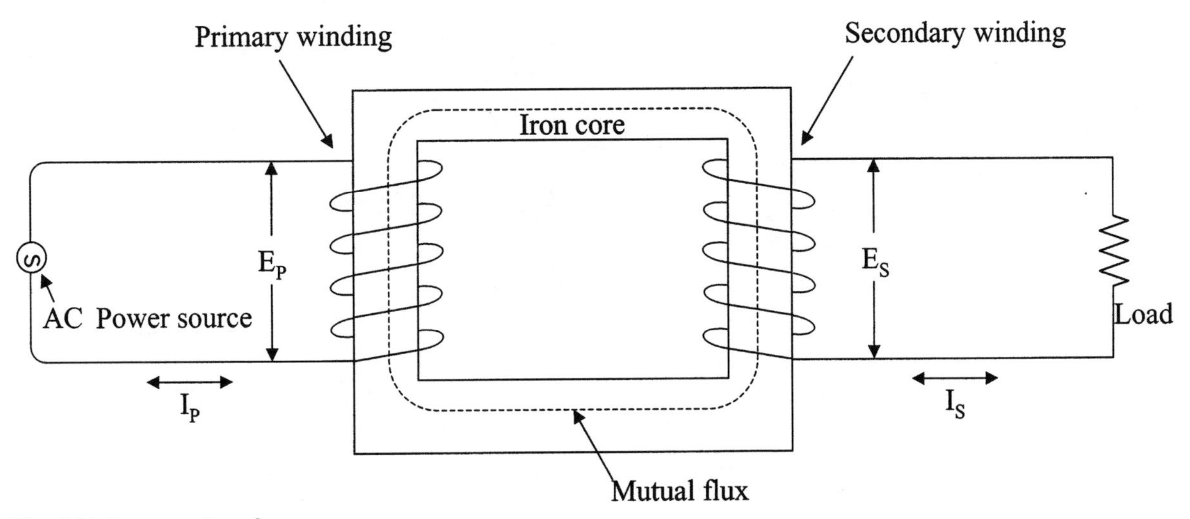

Fig. 1.24. Iron core transformer.

Example 1.11

A transformer is needed to raise 120 VAC to 240 VAC to allow use of a 240-V machine. If the primary winding has 200 turns, how many turns should the secondary have?
Solving Equation 1.27 for N_2:

$$N_2 = \frac{N_1 E_2}{E_1} = \frac{(200)(240)}{(120)} = 400 \text{ turns}$$

What is the primary current when the load connected on the secondary is 744 VA?
First we solve Equation 1.5 for I:

$$I_S = \frac{P}{E_S} = \frac{744 \text{ VA}}{240 \text{ V}} = 3.1 \text{ A}$$

Solving Equation 1.28 for I_P, we get:

$$I_P = \frac{I_S E_S}{E_P} = \frac{(3.1)(240)}{(120)} = 6.2 \text{ A}$$

Several other characteristics of transformers should be noted:

AC only—Since transformer action depends on a continually changing mutual flux caused by alternating primary current, transformers will change only AC voltages. Transformers cannot be used to change a DC voltage to a different DC voltage.

Frequency—Secondary frequency is the same as primary frequency. The waveforms, however, will have different amplitudes.

Isolation—Unless there is a conductor added to the circuit, primary windings and secondary windings are isolated from each other. This means there is no voltage defined between any point on the primary windings and any point on the secondary windings.

Self regulation—Transformers are self regulating. If the load connected to the secondary changes, secondary current and primary current both change, but stay in balance so that power in = power out, as specified in Equation 1.28.

Three-Wire Circuits

The most common type of electrical service provided by electric power companies in the United States is the 3-wire, 120/240-V type. The circuits are illustrated in Figure 1.25. The service transformer is shown in schematic form. It operates like the machine shown pictorially in Figure 1.24. Note from the figure that:

- The service originates from a distribution line, which typically operates with 7200 volts to ground.
- The service transformer cuts this voltage to 240 volts.
- A lead called a center tap is brought out halfway between the ends of the 240-V side of the transformer, cutting this voltage in half (120 V and 120 V).

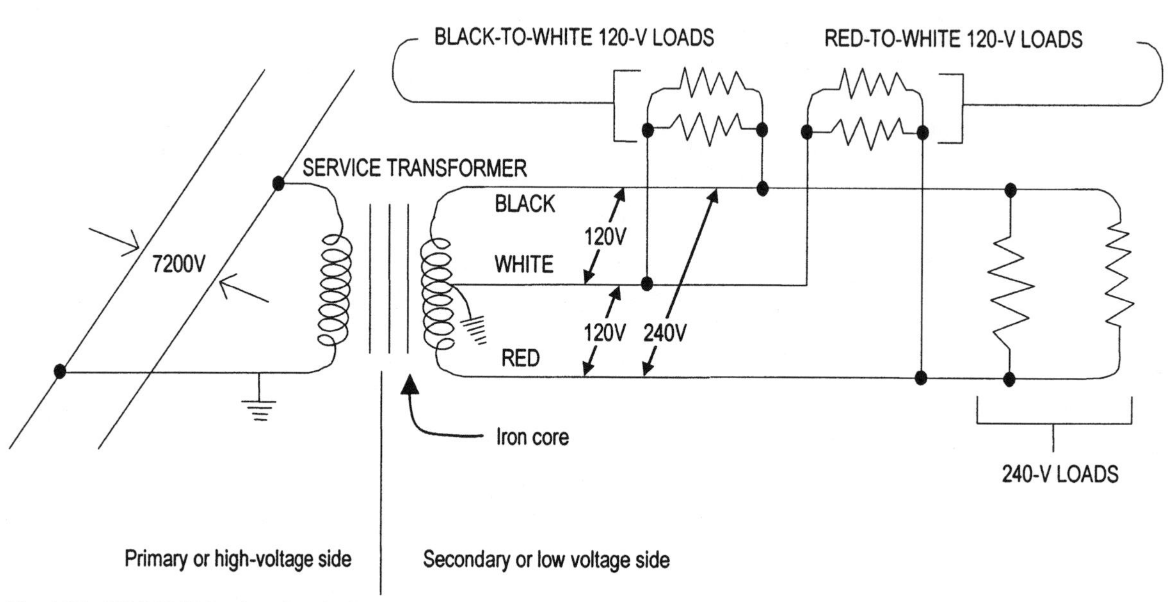

Fig. 1.25. 120/240-V, 3-wire electrical service.

- This center tap is connected to the earth through a ground connection.
- If the low-voltage wires are colored, they are black, white, and red.
- The white (center tap) is called a neutral conductor, since it is at ground or earth voltage.
- The black and red conductors are "hot" since there is 240 V between them and 120 V from either of them to the white (neutral).
- 240-V loads can be connected between black and red, and can operate with no connection to neutral.
- 120-V loads can be connected from black to white or from red to white.
- On the secondary or low-voltage side, none of the wires has a voltage to ground greater than 120 volts.

Three-Wire Circuit Analysis

Example 1.12 illustrates 3-wire circuit computation.

Example 1.12

Some 120-V and 240-V loads are connected to a 3-wire service as shown in Figure 1.26.

a. Compute currents in black, white, and red conductors.

Current in R_4 is computed: $I_4 = \dfrac{E_4}{R_4} = \dfrac{240}{80} = 3\text{ A}$

$I_5 = I_6 = I_4$, since these are the only wires leading to R_4.

Current in R_1 is computed: $I_1 = \dfrac{E_1}{R_1} = \dfrac{120}{40} = 3\text{ A}$

The black wire is the only source of current for R_1 and R_4.

Thus, $I_B = I_1 + I_5 = 3 + 3 = 6\text{ A}$

Current in the red wire can be found in the same way:

$$I_R = I_2 + I_3 + I_6 = \frac{120}{30} + \frac{120}{20} + 3$$
$$= 4 + 6 + 3 = 13\text{ A}$$

To find the current in the white, we need to assign current directions. Figure 1.27 shows the service transformer winding which is the source of all load currents. This winding carries AC current.

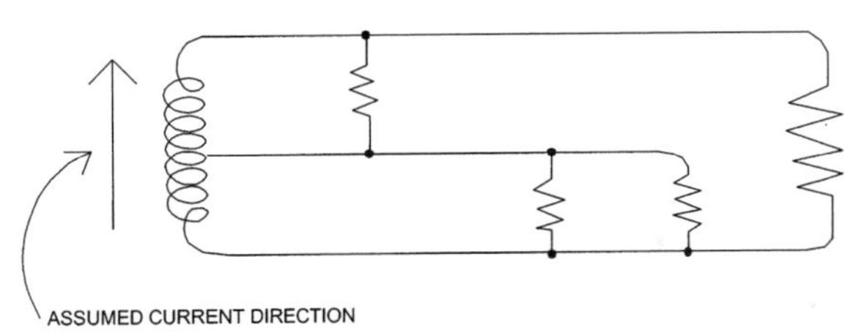

Fig. 1.27. Service transformer winding.

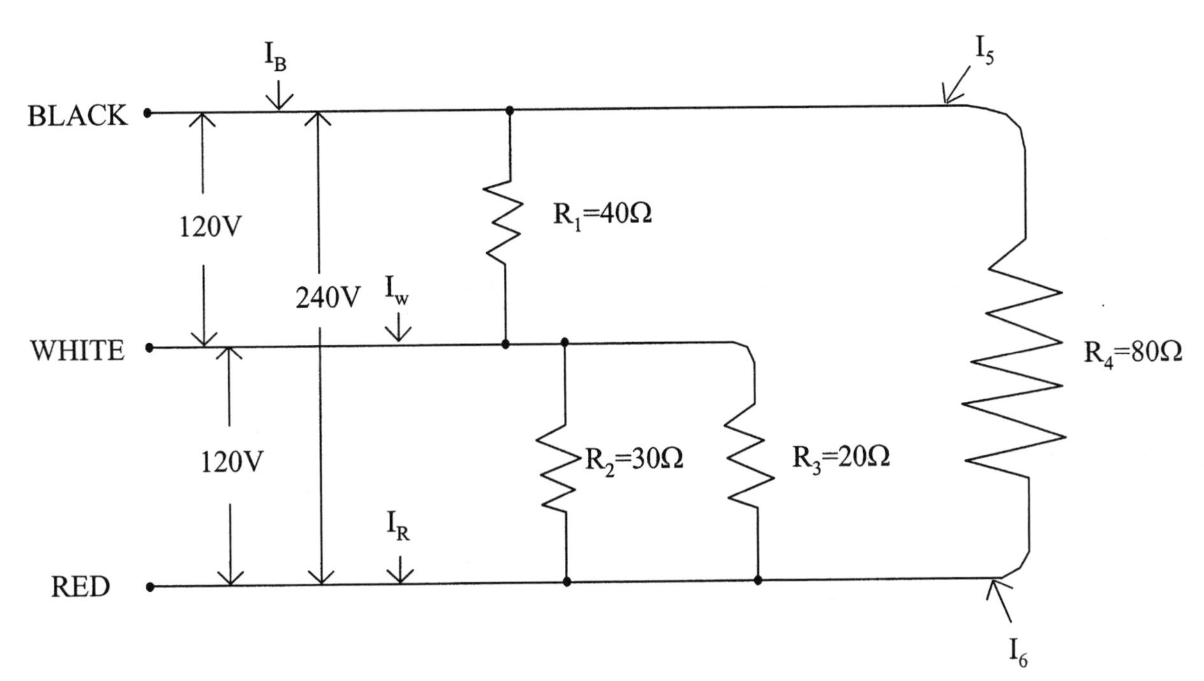

Fig. 1.26. Three-wire circuit.

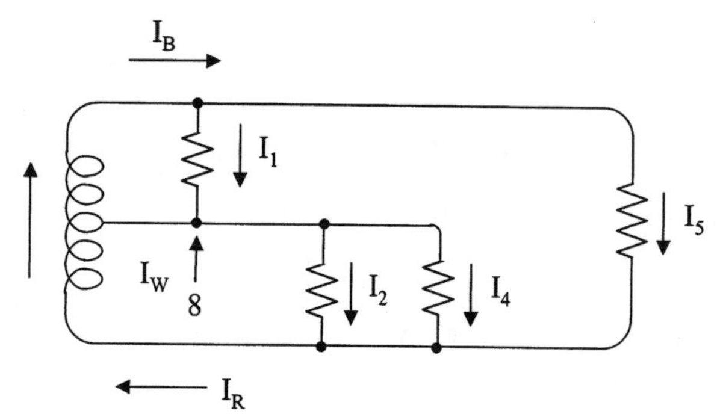

Fig. 1.28. Direction of currents, deduced from Figure 1.27.

We'll look at it at the instant when current is flowing up. With this current up, currents through all loads must be as shown in Figure 1.28.

Now, let's look at point 8. Since current can not be used up or created at point 8, the sum of the currents going into point 8 must equal the sum of currents coming out of point 8. We don't know which way current in the white (I_W) is flowing. Let's assume it's flowing out of point 8. Then,

Sum of currents out = Sum of currents in

$$I_W + I_2 + I_4 = I_1$$

Substituting in values,

$$I_W + 4 + 6 = 3$$

Solving for I_W:

$$I_W = -7\ A$$

This tells us that I_W is 7 A. Since the sign came out negative, we assumed its direction wrong. Thus,

$$I_W = 7A$$

Now we know all currents (Figure 1.29):

Note that:

- Current in white wire equals the difference between I_B and I_R.
- Current in white is not influenced by 240-V loads.

In equation form,

$$I_W = |I_R - I_B| \qquad \textbf{(1.29)}$$

In word form, Equation 1.29 says that I_W equals the absolute value of the difference between I_B and I_R (vertical lines mean absolute value). In a 3-wire 120/240-volt circuit, current in the neutral always equals the current difference between current in the two hot wires.

Note that:

- When current in hot wires is equal, current in white is zero.
- Current in hot wires is equal when only 240-V loads are connected or when 120-V loads between black and white equal 120-V loads between red and white.
- When I_W is zero, circuit is said to be balanced. This is desirable.

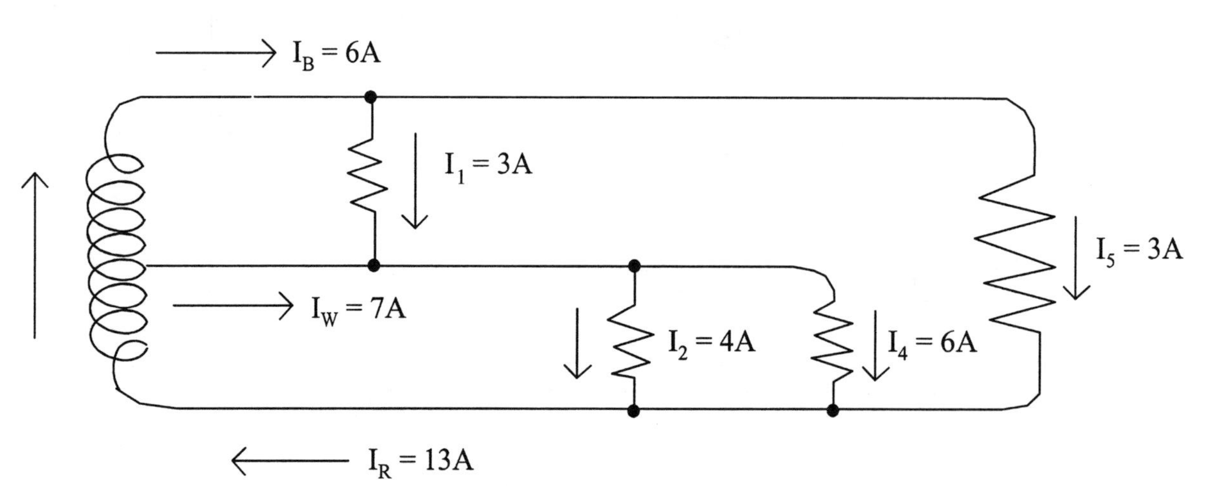

Fig. 1.29. Actual direction and value of all currents.

Problems

1.1. An aircraft electrical system operates at 400 Hz. How many cycles will the current go through in one minute?

1.2. The ammeter reads 0.5 A. What is the resistance of the load? Which way are electrons flowing? Which way is current flowing? How much power does the resistance use?

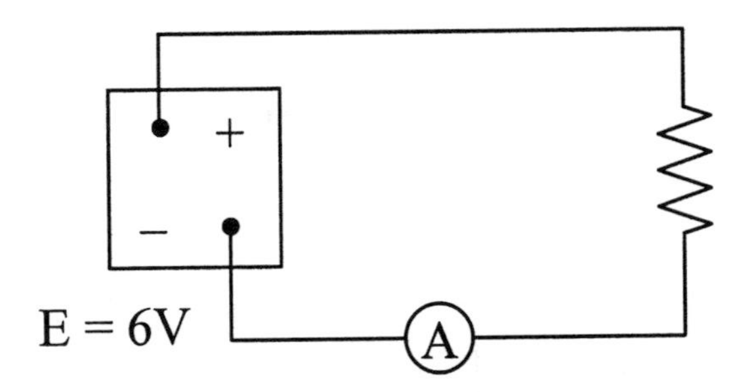

1.3. A 40-ohm resistor is connected across a voltage of 120 V.

(a) What current flows?

(b) How much power does this load use?

(c) How much energy would this load use if it is operated for 10 h per day for 30 days?

1.4. An instrument powered by a 9-V battery uses 1 W and, being DC, operates at a power factor of 1.

(a) What current does the instrument draw?

(b) What is its resistance?

1.5. Estimate the current, annual energy usage, and annual operating cost for a 1.5-hp 230-V motor operating at full load, which is used 17 h per week. Assume electrical energy costs 7¢/kWh. Refer to Appendix B.

1.6. Compute current and resistance values for each of these incandescent light bulb models. Assume operation is at rated voltage.

(a) Bulb ratings: 115 V 150 W

(b) Bulb ratings: 230 V 150 W

(c) Bulb ratings: 115 V 75 W

(d) Bulb ratings: 230 V 75 W

1.7. What is the current drawn by a light bulb rated at 150 W, 115 V when connected across 115 V? What is its resistance?

1.8. Using Ohm's Law and the power equation (with PF = 1),

(a) Derive an equation of $P = f(I,R)$

(b) Derive an equation of $P = f(E,R)$

1.9. An incandescent light bulb is rated at 100 W, 115 V. What power does this bulb use if operated at 90 V? Hint: Assume resistance stays constant.

1.10. How much electrical energy is used by:

(a) A 100-watt light bulb burning continuously for 10 hours?

(b) A 250-watt infrared heat lamp which is on 24 hours a day for eight weeks?

1.11. What current is drawn by a 10-kW, 230-V load operating at a power factor of 0.84?

1.12. An air conditioner unit contains a 3-hp electric motor to drive the compressor and a 0.1-hp electric motor to drive a fan. Assuming that both motors are fully loaded, estimate the current drawn by the unit and the hourly cost of operation. Assume 230-V rating, and 7¢/kWh cost for energy. Refer to Appendix B.

1.13. What is the month's electrical energy bill and the average cost per unit of electrical energy for an Ames customer who used 1720 kWh? (Assume rate structure listed in this chapter.)

1.14. Compute: Total resistance
Current
Voltage drop across R_1
Voltage drop across R_2

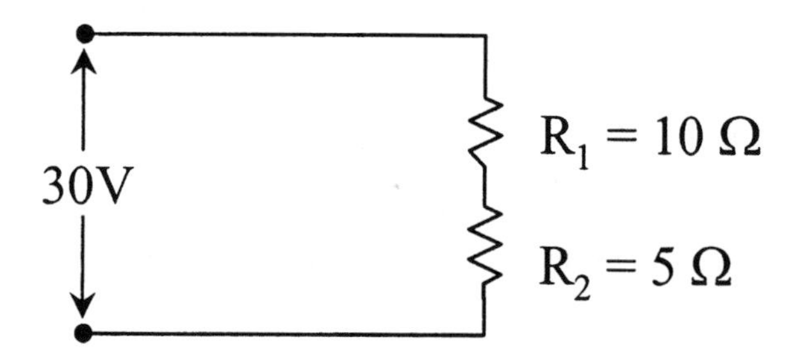

1.15. Compute R_2.

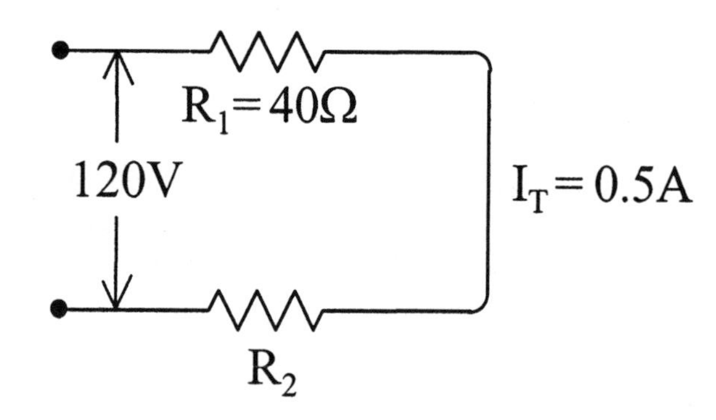

1.16. R_1 uses 400 W and R_2 uses 600 W. Compute R_1 and R_2.

1.17. A 115-V Christmas tree series light string is to have 10 identical lights, and use a total of 69 W.

(a) What current does the circuit draw?

(b) What should the resistance of each light bulb be?

(c) What is the voltage drop across each bulb?

(d) By mistake, a circuit is built containing only 9 bulbs, with each bulb having the resistance computed in (b). When this string is used, what is the current through, voltage across, and power used by each bulb?

(e) What is the total power used by the 9-bulb string?

1.18. For the circuit shown, find:

(a) Equivalent resistance of circuit.

(b) Current through each resistor.

(c) Power used by entire circuit.

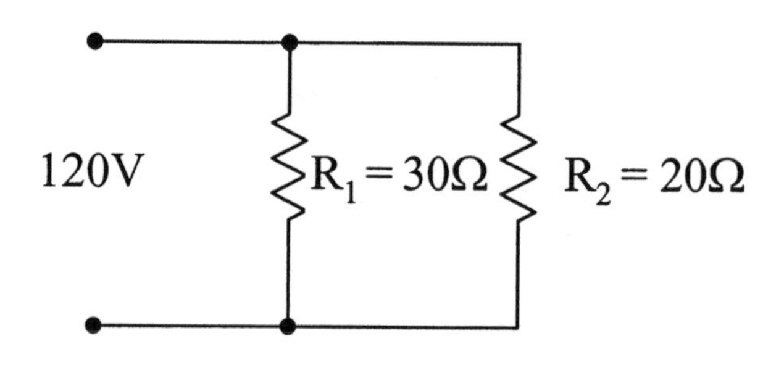

1.19. Assume ammeter has zero resistance. Calculate:

(a) R_4.

(b) Equivalent resistance of the circuit.

(c) Total current.

(d) Current through R_1, R_2, R_3, and R_4.

Compare total current with the sum of currents in (d).

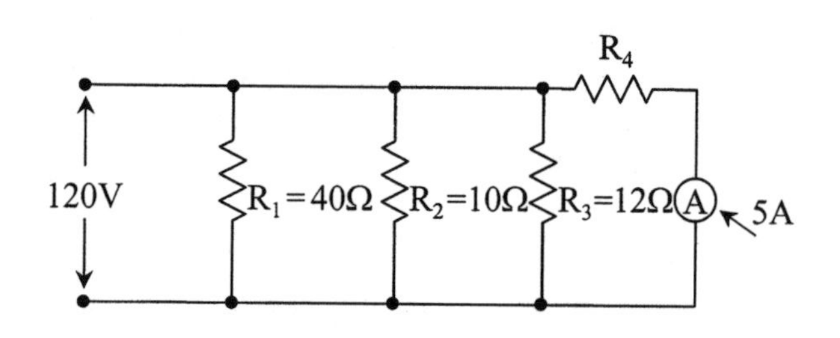

1.20. Derive an equation for R_T for a circuit composed only of parallel resistors R_1 and R_2.

That is, derive an equation of this form: $R_T = f(R_1, R_2)$.

1.21. Calculate:
Equivalent resistance of circuit.
Total current.
Voltage drop across R_1 and R_2.
Voltage drop across R_3.
I_1.
I_2.
I_3.

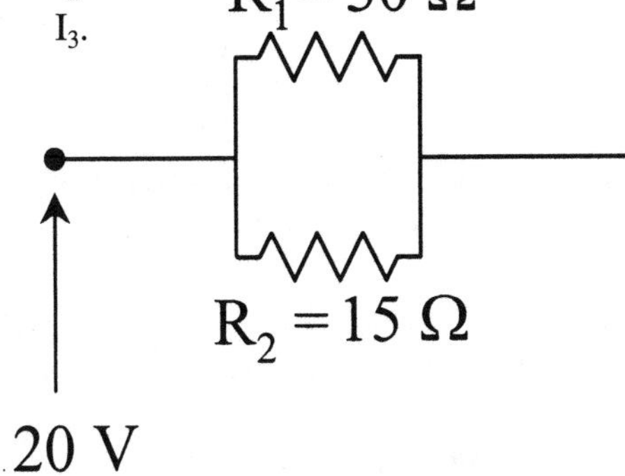

1.22. Compute: R_3, R_6, R_{89}, R_4, R_T, R_{10}, E_{123}, E_8, E_9, I_2, E_{10}.

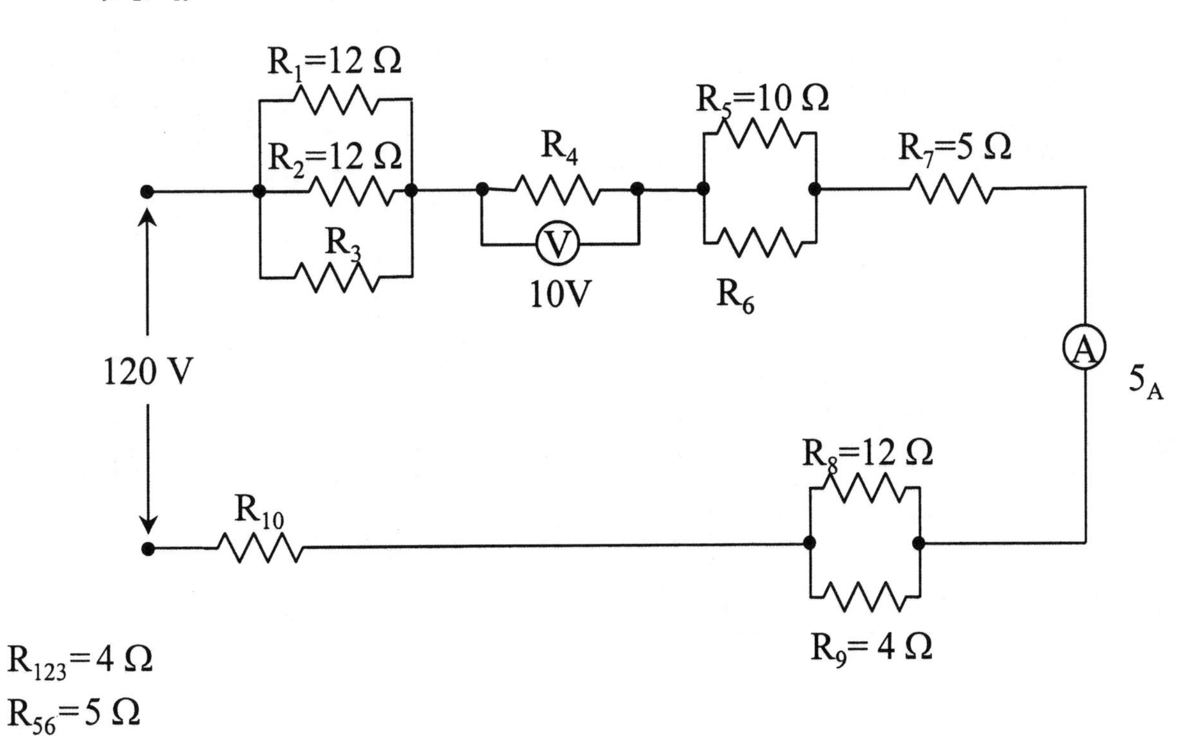

$R_{123} = 4\ \Omega$
$R_{56} = 5\ \Omega$

1.23. Compute circuit current and resistance of load using 320 W.*

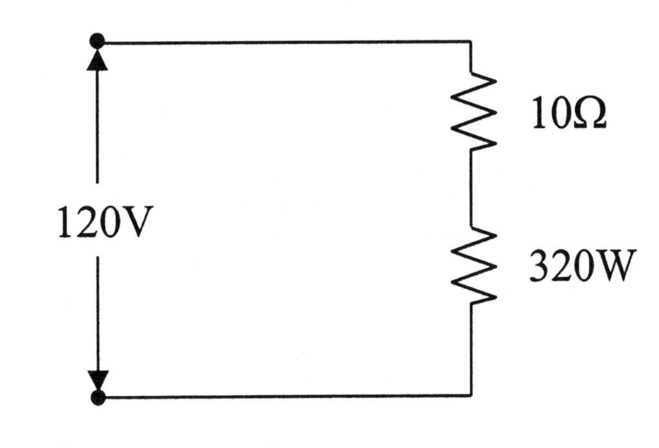

1.24. (a) What is the voltage across each bulb?

(b) What power is used by each bulb?

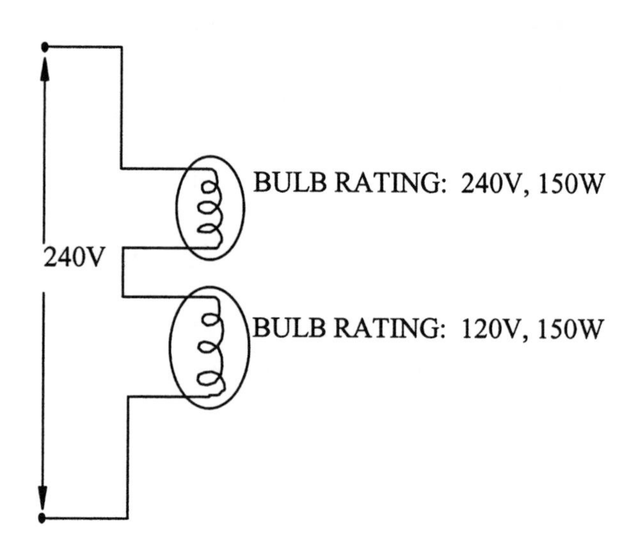

1.25. Compute values for R_1, R_2, and R_3.*

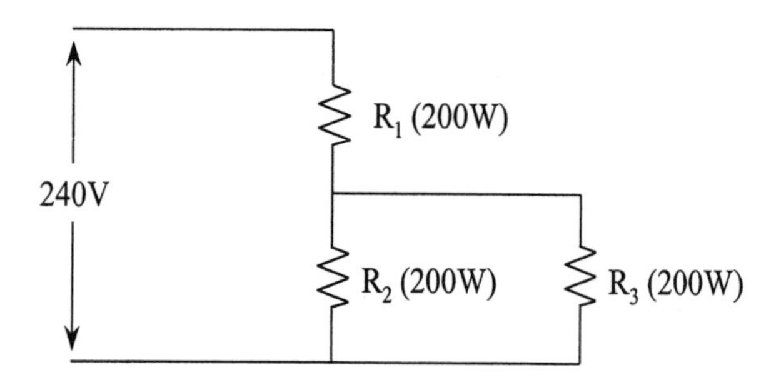

1.26. (a) What is the value of R?

(b) What power does the 20 Ω heater use?

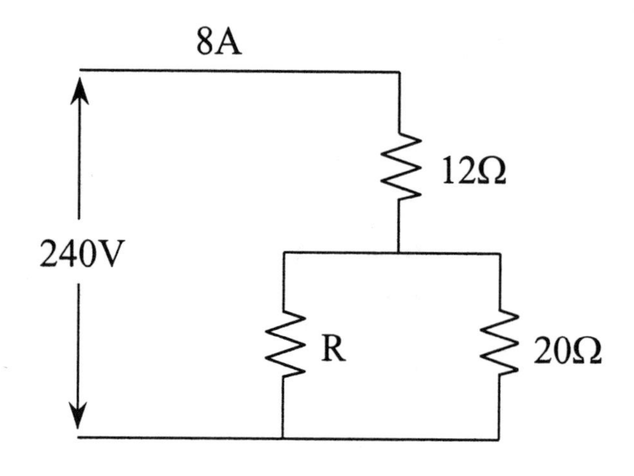

1.27. Complete circuit uses 1000 W.*

(a) Compute R.

(b) Compute power used by each load.

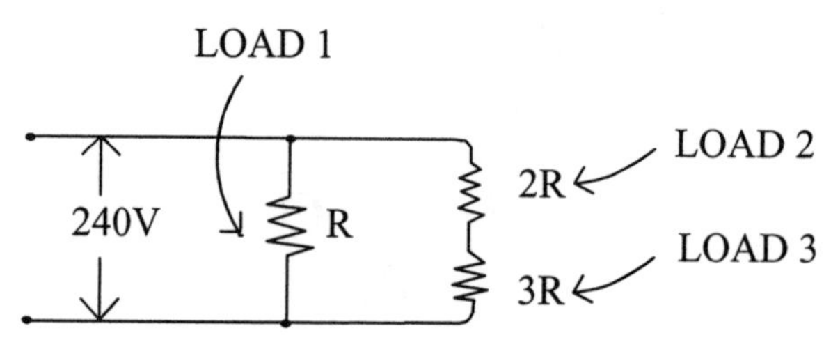

1.28. A service transformer is used to step down 7200 V to 240 V. The transformer is rated at 25 kVA.

(a) If there are 500 turns on the secondary, how many turns are on the primary?

(b) At rated load, what is the primary and secondary current?

1.29. Total power = 5760 W.
Compute R.
Compute I_1, I_2, and I_3.

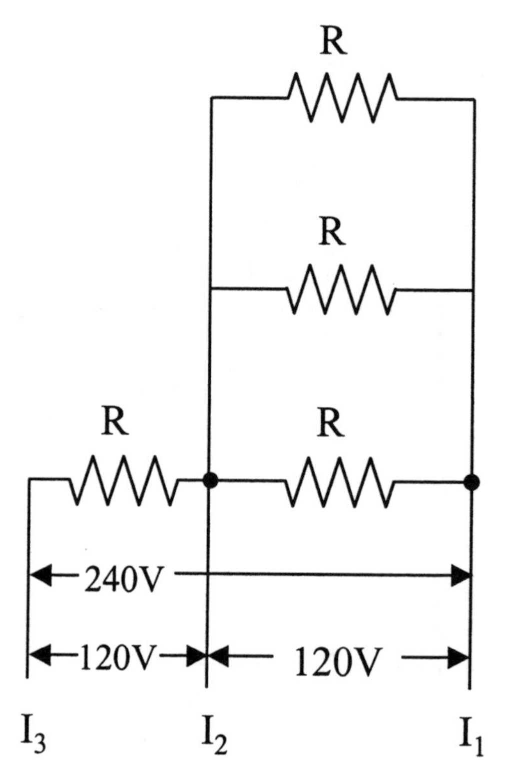

*Relatively difficult

1.30. Compute:

I_1	I_8
I_2	I_9
I_3	I_{10}
I_4	I_{11}
I_5	I_B
I_6	I_R
I_7	I_W

1.31. A 3-wire service has these resistive loads connected:
120-V: Black to white: 800 W, 400 W
120-V: Red to white: 900 W, 200 W, 600 W
240 V: 4 kW, 10 kW

Compute current in black, white, and red.

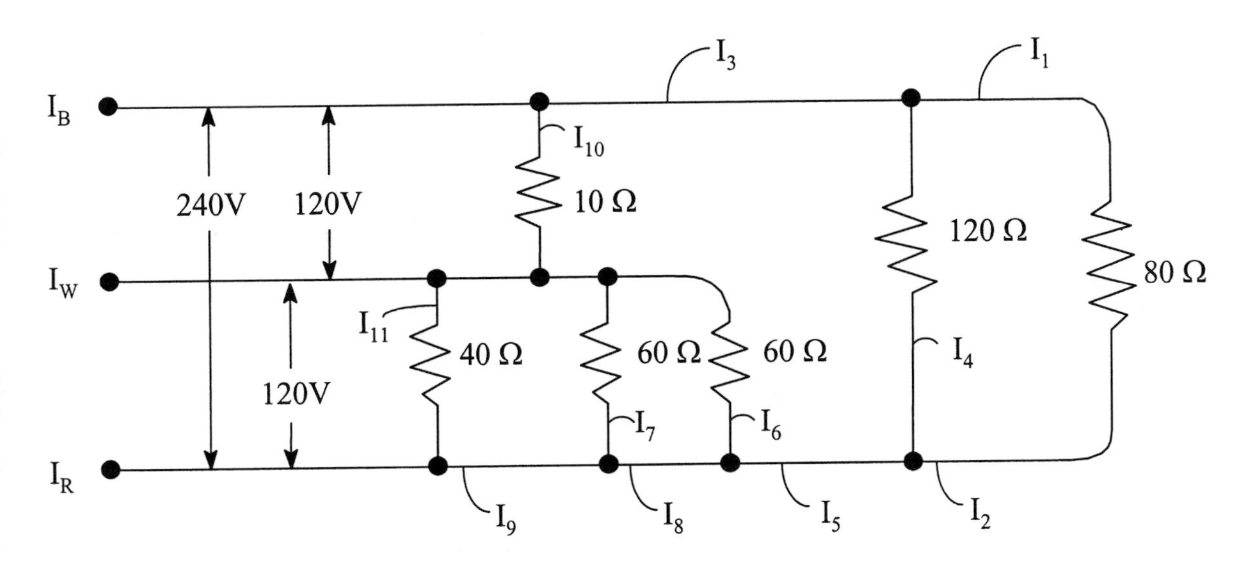

Chapter 2

REACTIVE AC CIRCUITS

Most of the electrical energy used in the United States is generated, distributed, and used as alternating current, or AC power. The main reason for this dominance is the ease with which AC voltages can be raised and lowered by use of transformers (see Figure 1.24). Since AC circuits exhibit a property called *reactance*, they must be analyzed using different techniques than those explained for DC circuits in chapter 1. Therefore, chapter 2 is devoted to learning about reactive AC circuits.

Effective Value of AC Current, Voltage, and Power

As was illustrated in Figure 1.4, current in an AC circuit varies with time in a sinusoidal fashion. The frequency of line power in the United States is 60 Hz (cycles per second). When an AC ammeter (a device to measure electrical current in an AC circuit) is placed in series with the load as in Figure 2.1, the dial reads a steady value even though the current through this meter is varying with time as illustrated in Figure 2.2. If the frequency is 60 Hz, the time for one cycle is 1/60 (or 0.0167) seconds.

The dial is obviously not indicating the instantaneous current, i. Nor is it indicating average current (since the average current during a cycle is zero).

The ammeter dial reads the root mean square (rms) or effective value of the current. Effective value is defined as follows: *An AC current has an effective value of* 1 *amp when this current will produce the same heat in a given resistance as* 1 *amp of DC current.*

In an AC circuit, what is the relationship between i_m, the maximum instantaneous current, and I, the rms (or effective) current?

Combining Equations 1.1 and 1.5, we obtain:

$$P = I^2R \quad \textbf{(2.1)}$$

For equal heating effect between a DC current and an AC current, it is necessary that:

$$I_{DC}^2R = I_{AC}^2R \quad \textbf{(2.2)}$$

This equation holds true when I_{AC} is the rms of the AC waveform. When describing AC electricity, capital letters I and E will always refer to rms or effective values of current and voltage in this text. Furthermore, this is true of most AC meters, and most electricity textbooks.

It can be shown mathematically that for a sinusoidal waveform, the rms or effective value is the maximum value divided by $\sqrt{2}$:

$$I = \frac{i_m}{\sqrt{2}} \quad \textbf{(2.3)}$$

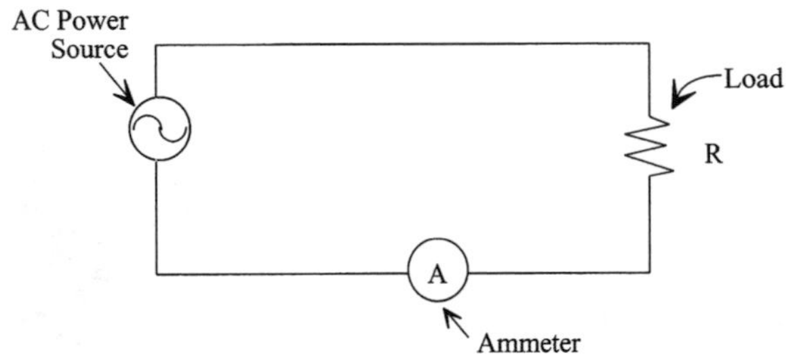

Fig. 2.1. AC circuit with ammeter.

and

$$E = \frac{e_m}{\sqrt{2}} \qquad (2.4)$$

For normal U.S. electrical service, E = 120 V and since E specifies an rms value,

$$e_m = 120\,(\sqrt{2}) = 170\text{ V}$$

The maximum instantaneous voltage to ground on a 120-V AC service is 170 V. A more intuitive notion of rms is explained in Figure 2.3.

As is illustrated in the figure, the rms value can be thought of as the (square) root of the mean of the squares of the instantaneous values.

Phasors

In the analysis of AC circuits, phase or time relationships among circuit parameters of current and voltage become important. Dealing with these relationships using wave form notation (such as Figure 1.10) is not convenient. The use of phasors, which are mathematically represented as vectors, makes analysis much easier.

Think of a phasor as an arrow with a defined length and direction. The length of the phasor is proportional to the rms value of the parameter (voltage or current) being described. The counterclockwise angle between horizontal and the phasor is the angle at which the waveform crosses zero on the way up. This is illustrated in Figure 2.4. The phasor is assumed to be spinning counterclockwise about the origin at 60 revolutions per second for 60 Hz line power.

Notice in Figure 2.4 that the *voltage* arrow has an *open* tip and the *current* arrow has a *solid* tip. This is a common convention. The arrow over the letter ($\vec{E}$, $\vec{I}$) means the letter defines a phasor.

In Figure 2.4a, the phasor $\vec{E}$ is 100 units long (signifying a magnitude of 100 volts) and is pointing horizontally to the right. Its angle with horizontal is therefore zero. This phasor is equivalent to a waveform of 100 volts rms, which crosses the zero volts line on the way up at the origin, or the zero degrees point (diagram on the right).

In Figure 2.4b, the phasor $\vec{I}$ is 100 units long (signifying a magnitude of 100 A) and is pointing down, 90° behind the reference position. This phasor is equivalent to a waveform of 100 amps rms (diagram on the right), which crosses the zero amps line on the way up 90 electrical degrees past 0° (the origin).

Also by convention, scales are chosen so that voltage phasors are longer than current phasors. Although analysis can be carried out graphically, analytical calculations with sketches having no precise scale is usually easier. This latter procedure will be employed in this chapter.

Phase Angle between Two Phasors

If the phase relationship between two parameters is being described, either phasor can be shown as being horizontal, and thus used as a reference. Figure 2.5 illustrates this.

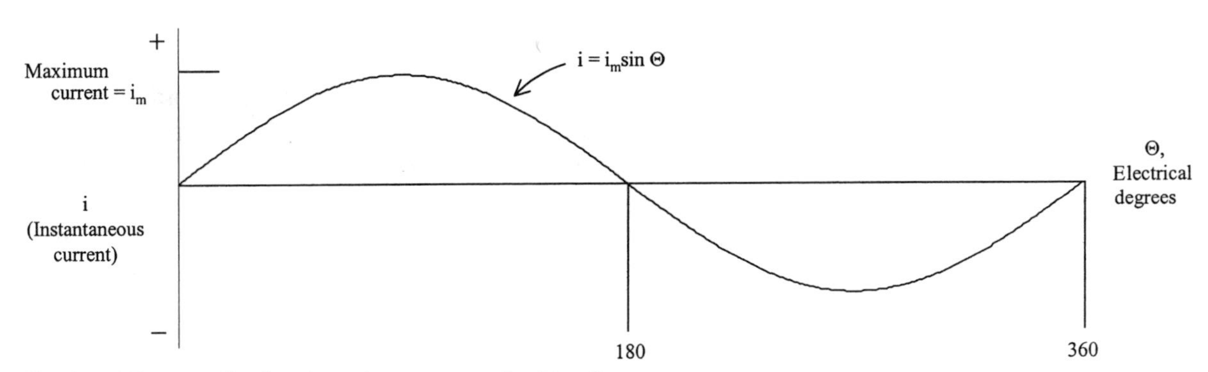

Fig. 2.2. AC current flowing through an ammeter (and load).

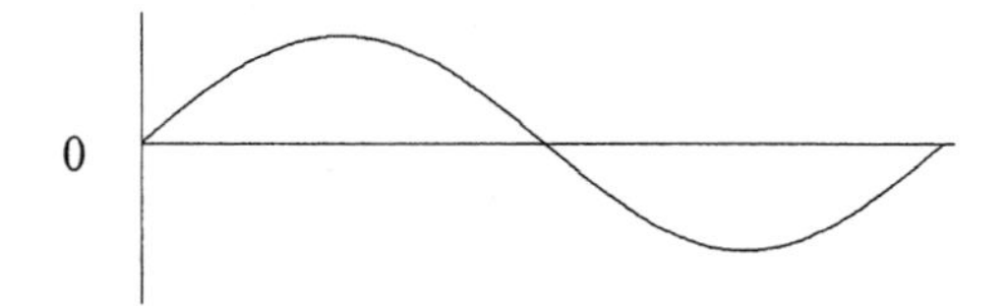

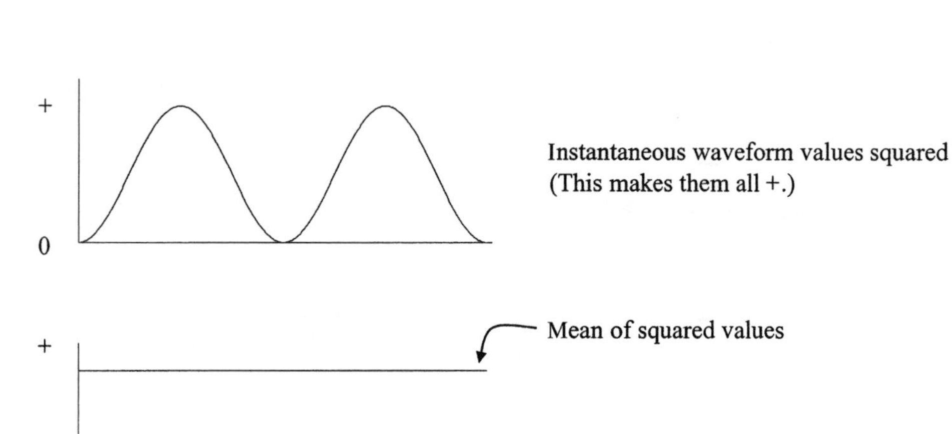

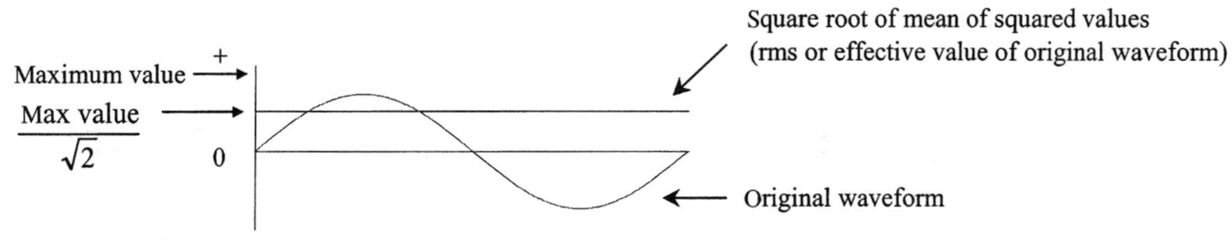

Fig. 2.3. Intuitive explanation of root mean square.

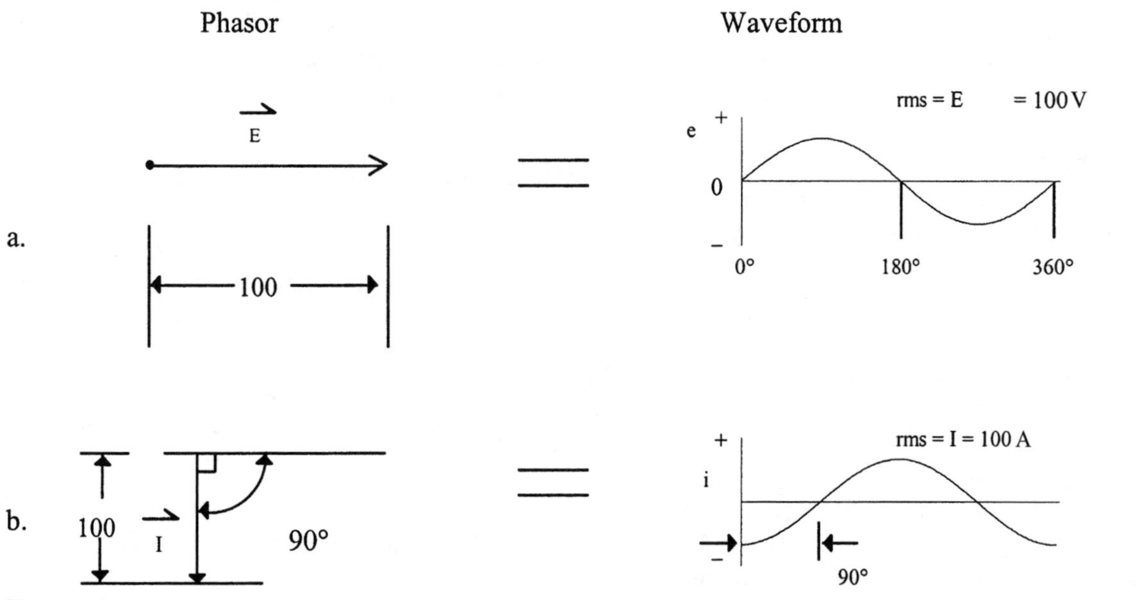

Fig. 2.4. Phasor notation and equivalent waveform.

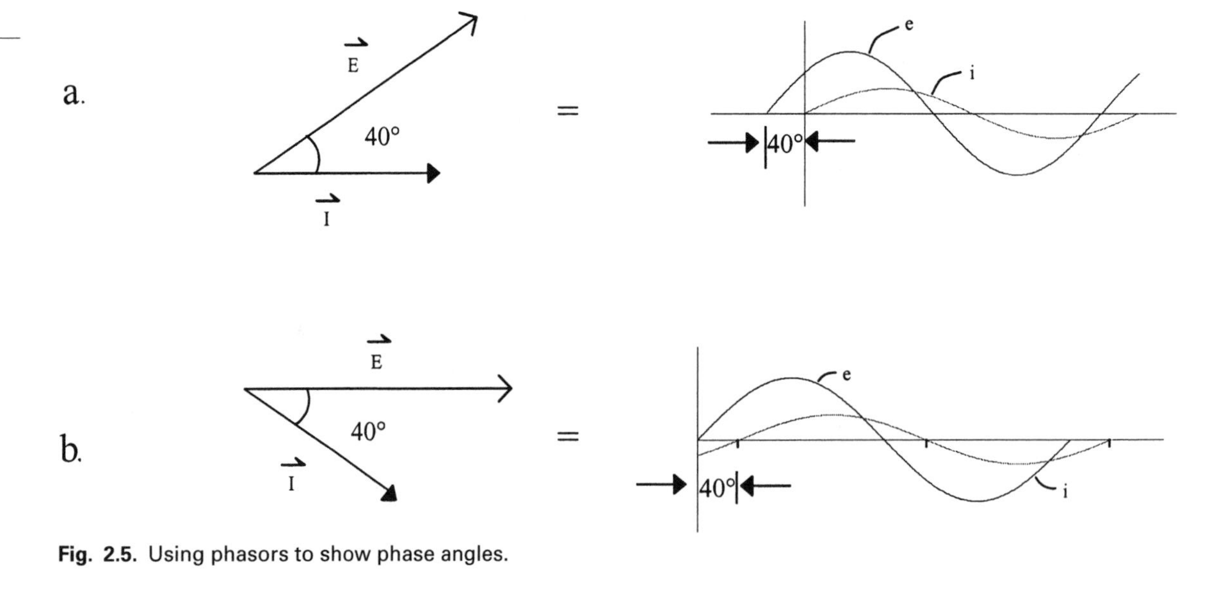

Fig. 2.5. Using phasors to show phase angles.

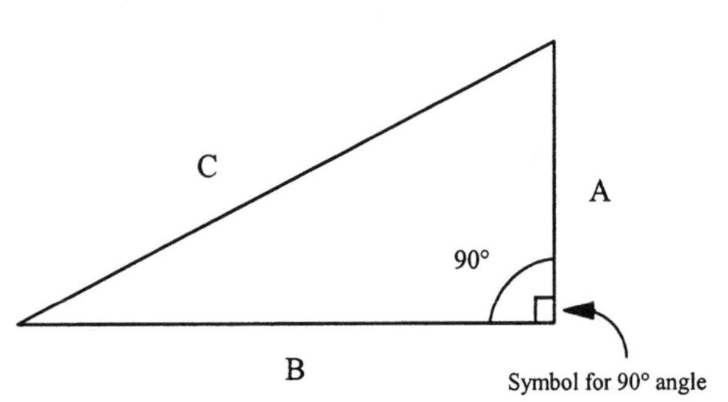

Fig. 2.6. Right triangle.

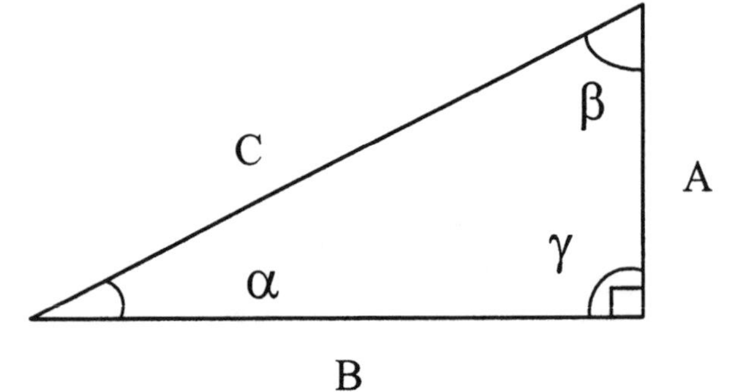

Fig. 2.7. Angle notation.

In Figure 2.5, a and b describe the same situation, that is, voltage leading current by an angle of 40°. In a the phasor diagram is referenced on $\vec{I}$; in b, the phasor diagram is referenced on $\vec{E}$. $\vec{E}$ is said to be leading $\vec{I}$ since as it spins counterclockwise, it passes through the reference position (horizontal to the right) 40 electrical degrees ahead of $\vec{I}$ every revolution.

Trigonometry

In order to do the computations related to phasor notation, it is necessary to use trigonometry, the study of triangles. A section illustrating use of trigonometry, as needed for use in phasor notation, is inserted here. The reader who is familiar with trigonometry can skip this section and problems 2.2 through 2.9 without loss of continuity.

Right Angle

A right angle triangle is one having a right (or 90°) interior angle (Figure 2.6). Letters A, B, and C represent lengths of the sides. The side of the triangle opposite to (not bordering) the right angle is called the hypotenuse. Angles are often designated by Greek letters. As previously noted, theta (Θ) is commonly used to symbolize the angle between current and voltage phasors.

Figure 2.7 is a right triangle with angles denoted as α (alpha), β (beta), and γ (gamma). The little box at angle γ signifies that it is a right (90°) angle.

Basic Trigonometric Functions

The basic trigonometric functions to be used with phasor notation are the sine (abbreviated sin), the cosine (abbreviated cos) and the tangent (abbreviated tan). They are defined for a right triangle as follows (referring to Figure 2.7).

$$\sin \alpha = \frac{A \leftarrow \text{length of side opposite the angle } \alpha}{C \leftarrow \text{length of hypotenuse}} \tag{2.5}$$

$$\cos \alpha = \frac{B \leftarrow \text{length of side adjacent to angle } \alpha}{C} \tag{2.6}$$

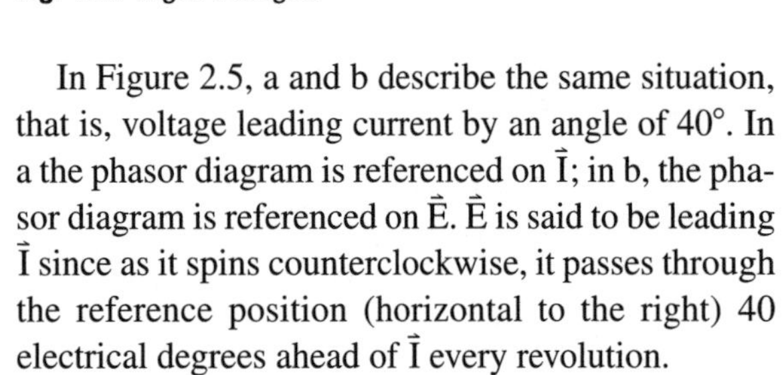

$$\tan \alpha = \frac{A}{B}$$ If the triangle is laid out so that the adjacent side is horizontal, the tangent is always the rise (vertical distance A) over the run (horizontal distance B) **(2.7)**

In the language of land surveyors, the tangent is defined as the rise over the run. The trigonometry needed with phasors consists of knowing how to solve any of these equations for an unknown quantity when the other two are known. Use of a hand calculator with trigonometric and inverse trigonometric functions is assumed.

Examples follow.

Example 2.1

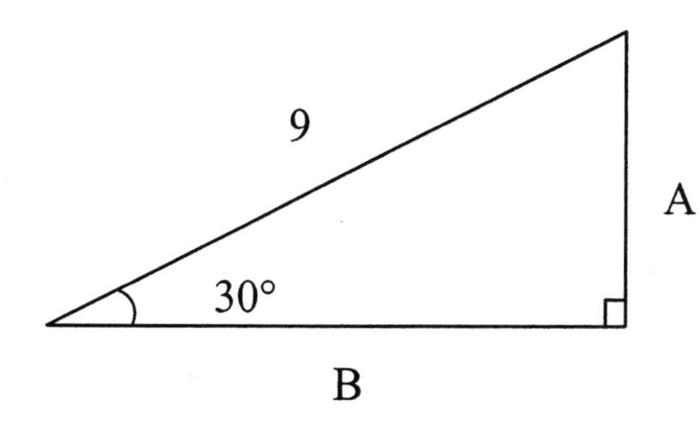

What is the length of side A?

Considering what we know, the sin equation is selected:

$$\sin 30° = \frac{A}{9}$$

$$A = (9)(\sin 30°) = (9)(0.50) = 4.5$$

Example 2.2

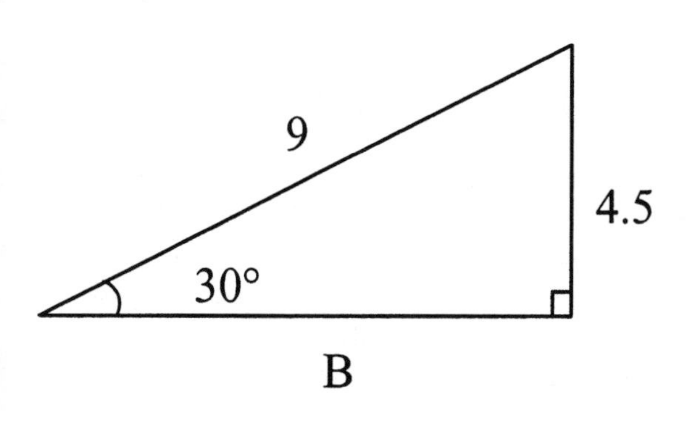

What is the length of side B?

$$\cos 30° = \frac{B}{9}$$

$$B = (9)(\cos 30°) = (9)(0.866) = 7.79$$

Example 2.3

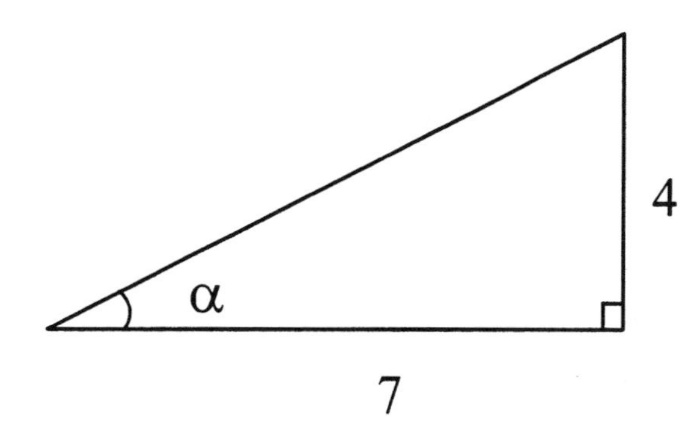

What is angle α?

$$\tan \alpha = \frac{4}{7} = 0.5714$$

$$\alpha = \tan^{-1}(0.5714)$$

(Read this as α equals the inverse tangent of 0.5714. It means α equals the angle whose tangent is 0.5714.)

$$\alpha = 29.74°$$

Your calculator should have inverse trig functions available.

Example 2.4

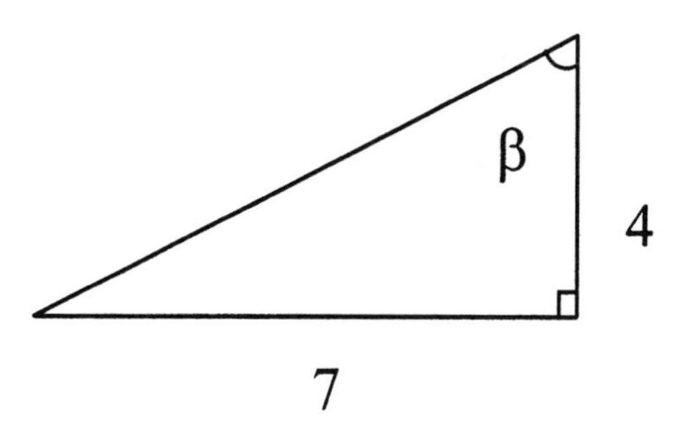

What is angle β?

Sometimes the angle of interest is in the opposite position. The triangle can be turned around for convenience:

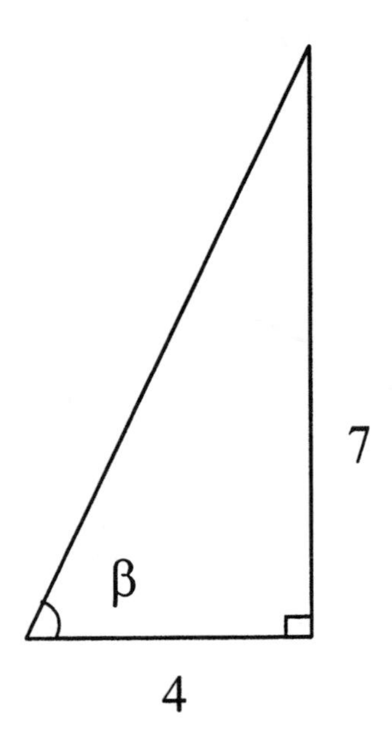

$$\tan \beta = \frac{7}{4} = 1.75$$

$$\beta = \tan^{-1}(1.75) = 60.26°$$

(Note that 29.74 + 60.26 + 90 = 180°. The interior angles always sum to 180°.)

Example 2.5

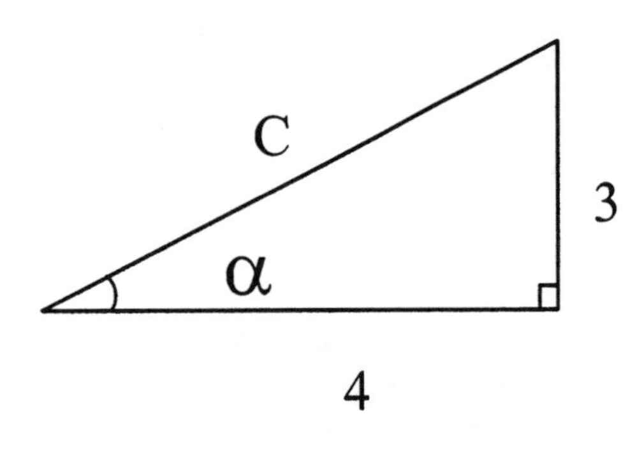

What is distance C?

$$\tan \alpha = \frac{3}{4} = 0.75$$

$$\alpha = 36.87°$$

Note: This could also be solved using the Pythagorean Theorem:

$$C^2 = 4^2 + 3^2$$

$$C = \sqrt{4^2 + 3^2}$$

$$C = \sqrt{25} = 5$$

$$\cos(36.87°) = \frac{4}{C}$$

$$C = \frac{4}{\cos(36.87)} = \frac{4}{0.80} = 5.0$$

Example 2.6

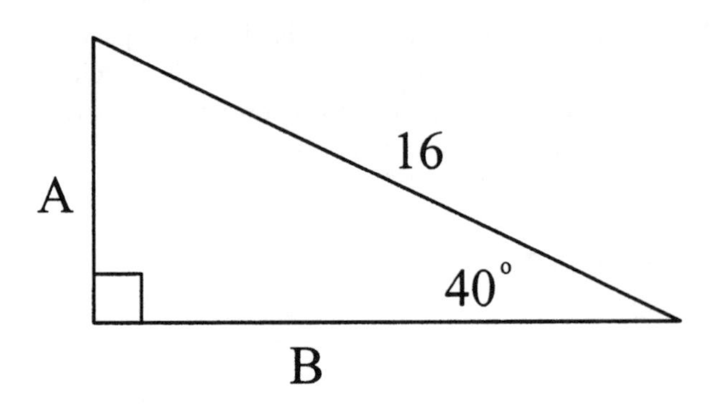

Find distances A and B.

$$\cos 40° = \frac{B}{16}$$

$$B = 16(\cos 40°) = 12.26$$

$$\sin 40° = \frac{A}{16}$$

$$A = 16(\sin 40°) = 10.28$$

Phasor Notation

Phasors can be specified in polar or rectangular notation.

Rectangular Notation

A phasor specified in rectangular notation takes this form:

$$a + jb \qquad \textbf{(2.8)}$$

where:

a = magnitude of horizontal component
+ a means →
− a means ←

b = magnitude of vertical component
+ b means ↑
− b means ↓

j = operator (mathematically, an "imaginary number" equal to $\sqrt{-1}$)

Addition and subtraction operations are carried out with phasors in rectangular form. Horizontal components of phasors can be added (or subtracted) algebraically, as can vertical components. Example 2.7 illustrates this.

Polar Notation

In polar notation, a phasor takes this form:

$$A\angle\Theta \quad \textbf{(2.9)}$$

where A is the magnitude of the phasor. Θ is the counterclockwise angle from horizontal. Multiplication and division operations are carried out in polar notation:

$$A\angle\Theta_1 \times B\angle\Theta_2 = A \times B\angle(\Theta_1 + \Theta_2) \quad \textbf{(2.10)}$$

$$\frac{A\angle\Theta_1}{B\angle\Theta_2} = \frac{A}{B}\angle(\Theta_1 - \Theta_2) \quad \textbf{(2.11)}$$

Phasor Computations

An example will illustrate manipulation of phasor quantities.

Example 2.7

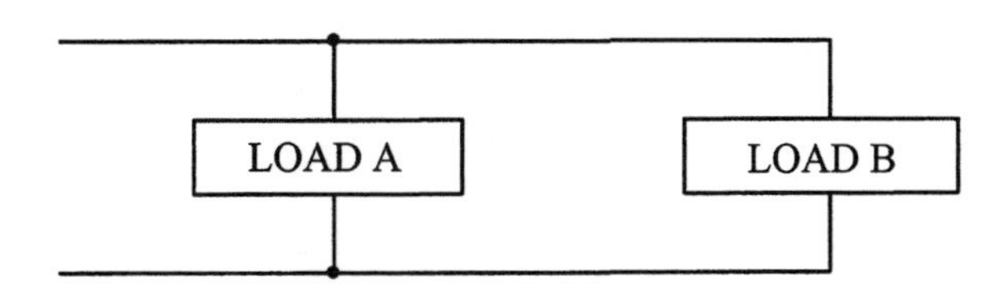

A conductor is carrying current to loads A and B. The current to load A is 10 A and is at a phase angle of 0°. The current to load B is 20 A and leads current A by 40°. Find the sum or resultant of these currents and express in polar and rectangular form. First, the phasor diagram is sketched:

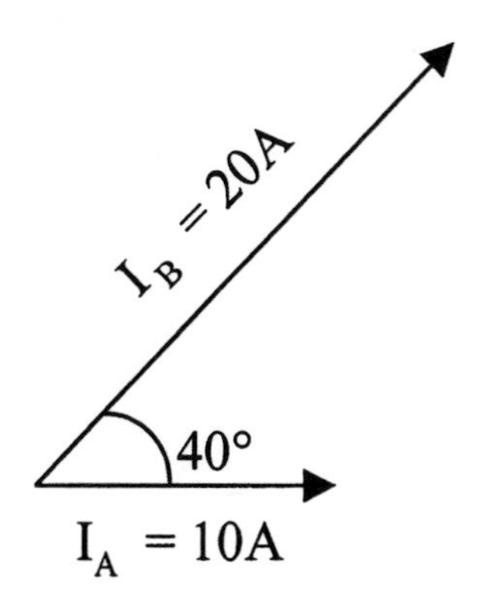

Converting to rectangular notation:

$$\vec{I}_A = 10 + j0$$

$I_B \sin 40 = 12.9A$

$I_B = 20$

40°

$I_B \cos 40 = 15.3A$

$$\vec{I}_B = 15.3 + j12.9$$

The resultant current in rectangular notation is:

$$\vec{I}_R = 25.3 + j12.9$$

Graphically, I_R is the bisector of the parallelogram formed from $\vec{I}_A$ and $\vec{I}_B$:

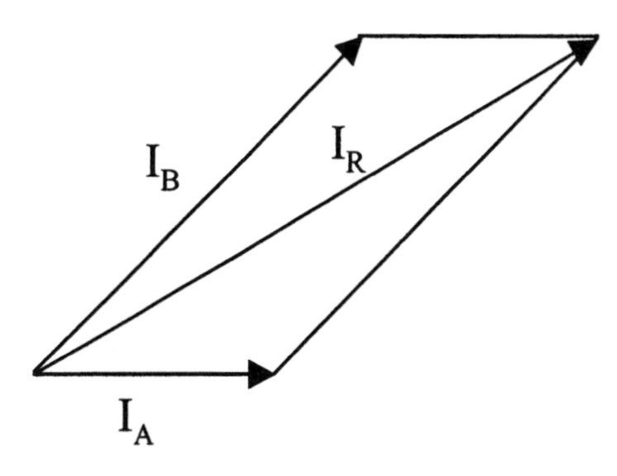

Now we can convert I_R to polar notation:

$$\tan\theta = \frac{12.9}{25.3} = 0.51$$

$$\tan^{-1}(0.51) = 27.0°$$

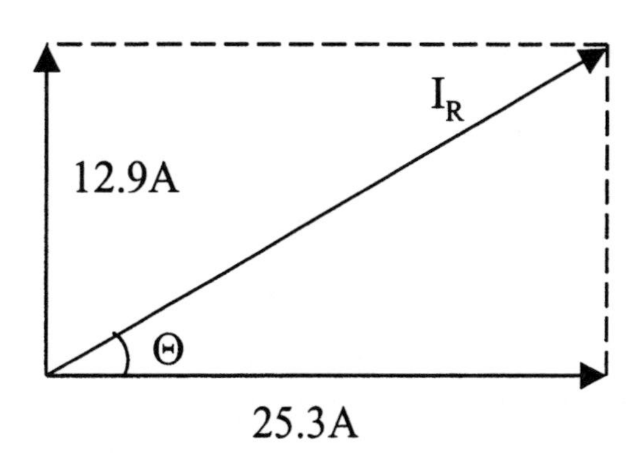

Now we can find the value of I_R:

$$I_R \cos 27° = 25.3$$

$$I_R = \frac{25.3}{\cos 27°} = 28.4A$$

Thus in polar notation, $I_R = 28.4 \angle 27.0°$

By noting what ammeters would read on this circuit we can get a better understanding of adding out-of-phase currents:

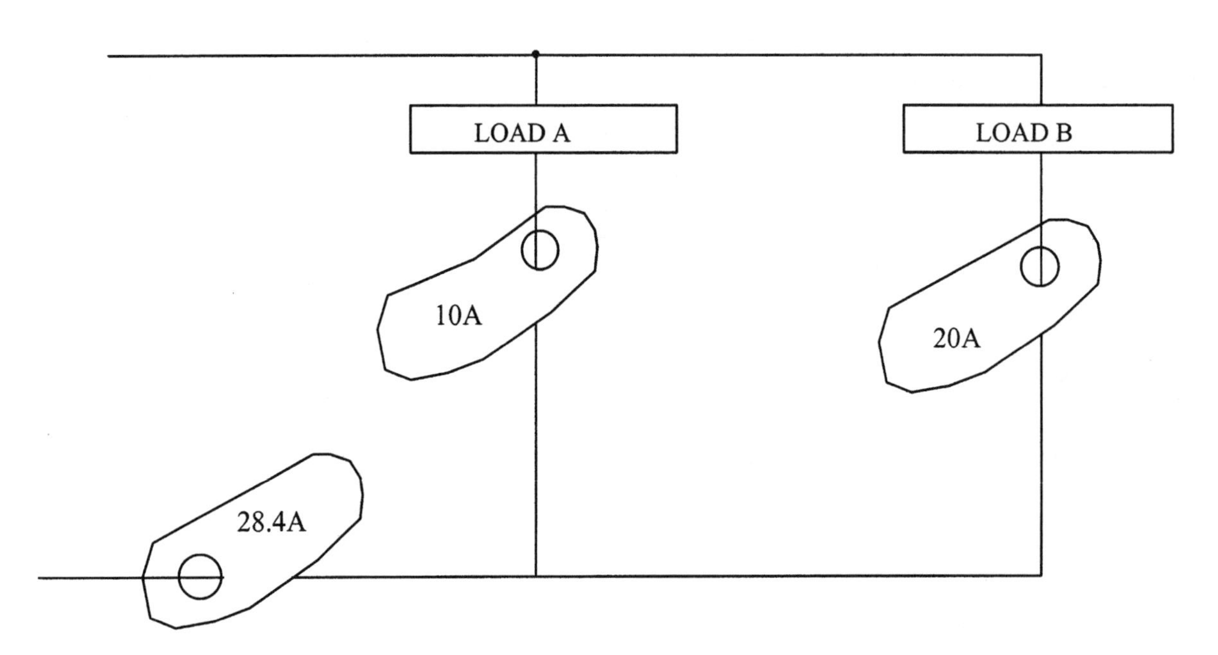

Note that since these currents are out of phase: $I_A + I_B \neq I_R$ or 10 A + 20 A ≠ 28.4 A. However, they can be added using phasor notation. That is, $\vec{I}_A + \vec{I}_B = \vec{I}_R$, as was done at the beginning of this example.

Example 2.8

A voltage phasor is written as: $\vec{E} = -6 + j8$. The components are $-6 \leftarrow$ and $+ 8 \uparrow$. Graphically, it is

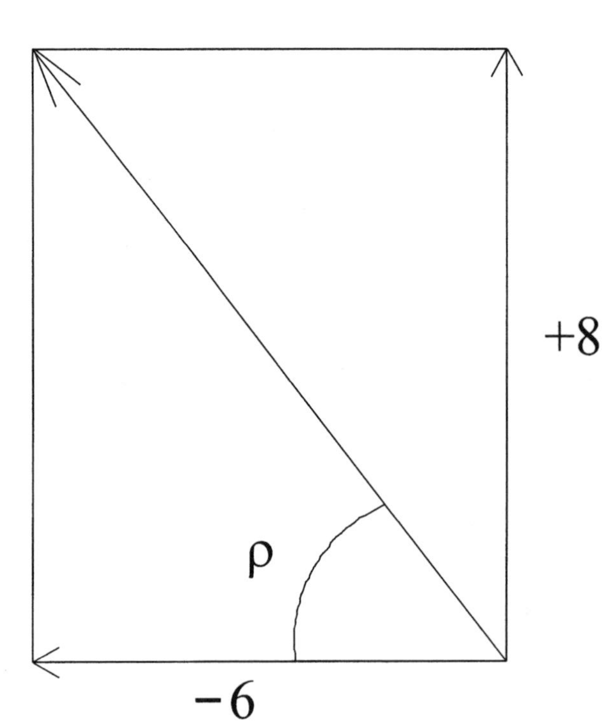

$$\tan \rho = \frac{8}{-6} = -1.33$$

$$\rho = -53.1°$$

Knowing the angle, we can compute the phasor length:

$$E \cos -53.1 = -6 \qquad E \sin -53.1 = 8$$

or

$$E = \frac{-6}{\cos -53.1} = 10.0 \qquad E = \frac{8}{\sin -53.1} = 10.0$$

In polar notation, this voltage can be expressed as:

$$\vec{E} = 10\angle -233.1° \text{ or } \vec{E} = 10\angle 126.9°$$

Impedance

In a DC circuit, the characteristic of the load which influences current flow is the resistance. In an AC circuit, resistance along with capacitive reactance and inductive reactance influence current flow. Circuit impedance is a vector quantity analogous to resistance in a DC circuit. Impedance is defined as follows:

$$Z = R + j(X_L - X_c) \qquad \textbf{(2.12)}$$

where:

- Z = impedance, ohms
- R = resistance, ohms
- X_L = inductive reactance, ohms
- X_c = capacitive reactance, ohms

The form of this equation is seen to be that of phasor rectangular notation. X_L and X_c will be defined later. Ohm's Law now becomes:

$$\vec{I} = \frac{\vec{E}}{\vec{Z}} \qquad \textbf{(2.13)}$$

Resistance

An example illustrates use of this equation for a simple resistive circuit.

Example 2.9

Compute current and express current, voltage, and impedance for the circuit shown in both polar and rectangular notation:

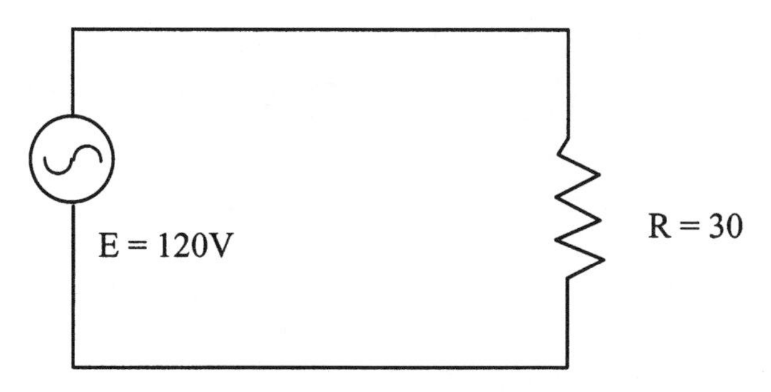

$$Z = R + j(X_L - X_c)$$

$X_L = 0$
$X_c = 0$

$$Z = 30 + j\,0 = 30 \angle 0°$$

Being a series circuit, it is desirable to reference the phasor diagram on I, since I is the same through all components in a series circuit. Thus:

$$\vec{I} \angle 0° = \frac{120 \angle \Theta_E}{30 \angle 0°}$$

following Equation 2.11, $0° = \Theta_E - 0°$

and $\Theta_E = 0°$

thus, $\vec{I} \angle 0° = \dfrac{120 \angle 0°}{30 \angle 0°} = 4 \angle 0°$

The phasor diagram:

The phasor solution shows us, as expected, that in this resistive circuit, current and voltage are in phase. Expressing all quantities in polar and rectangular:

$$\vec{Z} = 30 + j\,0 = 30 \angle 0°$$
$$\vec{E} = 120 + j\,0 = 120 \angle 0°$$
$$\vec{I} = 4 + j\,0 = 4 \angle 0°$$

Inductance

Inductance is the characteristic of an electrical circuit which makes it oppose a change in current. Inductance is, thus, somewhat analogous to inertia in a mechanical system. A mechanical system is capable of storing energy as kinetic energy. An inductive circuit is capable of storing energy in the form of a magnetic field. Any circuit containing coils (motors or fluorescent lamps, for example) is inductive.

The unit of inductance (symbol L) is the henry (H). Inductive reactance (X_L in Equation 2.12) is defined as:

$$X_L = 2\,\pi\,f\,L \qquad \textbf{(2.14)}$$

where:

- X_L = Inductive reactance, ohms
- f = Frequency of power, hertz
- L = Inductance, henrys

An example illustrates the effect of inductance in an AC circuit.

Example 2.10

Compute current and sketch a phasor diagram for the circuit shown. Express I, E, and Z in both polar and rectangular notation.

$$X_L = 2\,\pi\,f\,L$$
$$X_L = 2\,\pi\,(60)(0.2)$$
$$X_L = 75.4\ \Omega$$

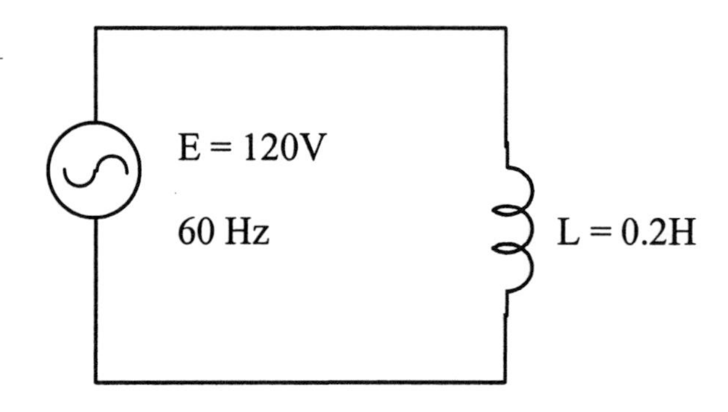

$$Z = R + j(X_L - X_c) \quad \textbf{(2.15)}$$
$$Z = 0 + j\,75.4 = 75.4\ \angle 90°$$

Referencing on I since this is again a series circuit,

$$\vec{I} = I\ \angle 0° = \frac{120\ \angle \Theta_E}{75.4\ \angle 90°} \qquad 0° = \Theta_E - 90° \qquad \Theta_E = 90°$$
$$I\ \angle 0° = 1.59\ \angle 0° = 1.59 + j\,0$$
$$\vec{E} = 120\ \angle 90° \quad = 0 + j\,120$$

Expressing as a phasor diagram and a waveform:

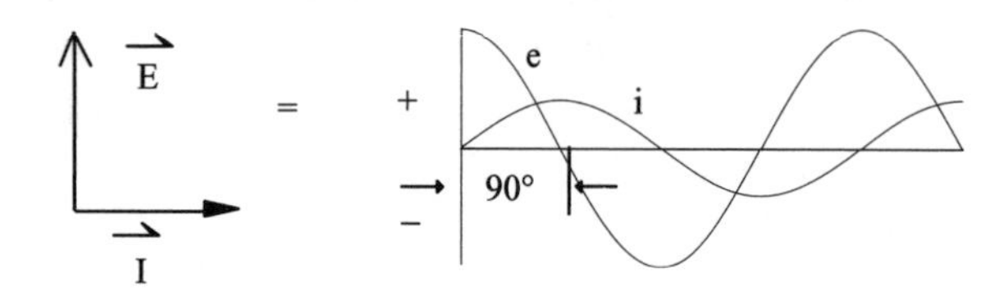

The current is seen to lag voltage by 90° in a purely inductive circuit.

Capacitance

Two electrical conductors separated by a nonconductor exhibit a property called capacitance. A capacitor is a device which is capable of storing an electrical charge, that is, it can store energy in an electrostatic field. A capacitive load tends to oppose a change in voltage. The unit of capacitance (symbol C) is the farad (F). Capacitive reactance (X_c in Equation 2.12) is defined as:

$$X_c = \frac{1}{2\pi\, fC}$$

where:

X_c = capacitive reactance, ohms
f = frequency of power, hertz
C = capacitance, farads

Example 2.11 illustrates the effect of capacitance in an AC circuit.

Example 2.11

Calculate the current and sketch the phasor diagram for the circuit shown.

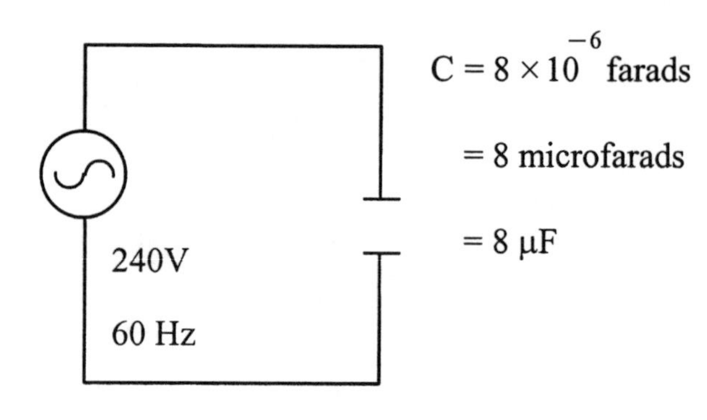

$$X_c = \frac{1}{2\pi(60)(8 \times 10^{-6})}$$
$$X_C = 331.6\ \Omega \quad Z = 0 - j\,331.6 = 331.6\ \angle -90°$$

Referencing on I:

$$\vec{I} = I\ \angle 0° = \frac{240\ \angle \Theta_E}{331.6\ \angle -90°} \qquad 0° = \Theta_E - -90° \qquad \Theta_E = -90°$$
$$\vec{I} = 0.72\ \angle 0° = 0.72 + j\,0$$
$$\vec{E} = 240\ \angle -90° = 0 - j\,240$$

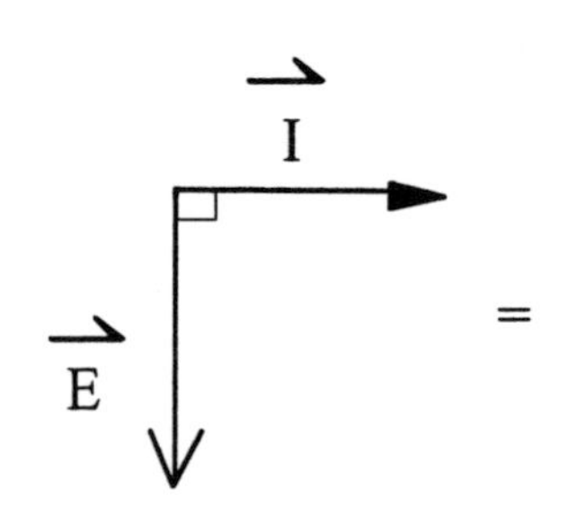

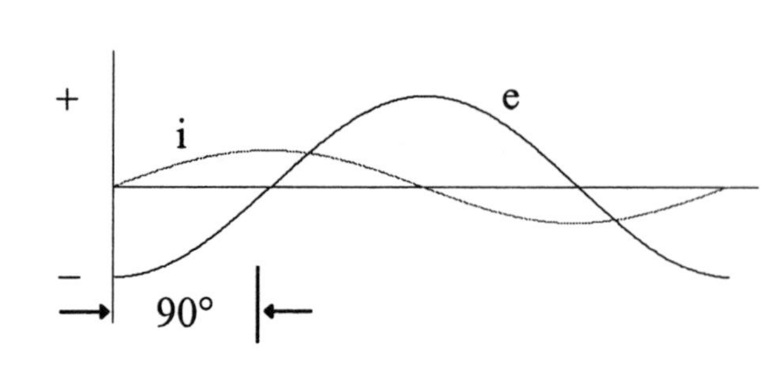

In an AC circuit, a pure capacitance is seen to cause voltage to lag current by 90°. Notice that the symbol for a capacitor shows a break or an open place in the circuit. This is in fact true. In an AC circuit, no electrons actually flow through the capacitor.

Series RL Circuits

We will now look at circuits involving combinations of R, L, and C in series.

Example 2.12

Sketch a phasor diagram for this circuit. Then compute I and determine the phase angle between E and I.

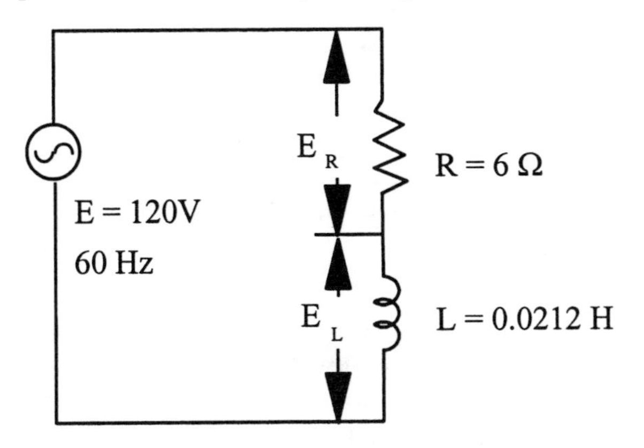

$$X_L = 2\ \pi\ (60)\,(0.0212) = 8.0\ \Omega$$

Step 1—The current in all parts of a series circuit is the same and for this reason we will again reference on I. We can then sketch the current phasor in the reference position. At this point, we know its direction but not its length:

$\vec{I}$

Step 2—We now substitute into the impedance Equation 2.12:

$$\vec{Z} = 6 + j(8 - 0) = 6 + j8$$

Converting this from rectangular to polar notation using the procedure of Example 2.7, we get:

$$\vec{Z} = 10\angle 53.1°$$

Step 3—Applying Ohm's Law for phasors (Equation 2.13), and noting that we know the angle, but not the magnitude, of $\vec{I}$:

$$I\angle 0° = \frac{120\angle\theta_E}{10\angle 53.1°}$$

Step 4—Compute the unknown angle:

$$0° = \theta_E - 53.1°$$
$$\theta_E = 53.1°$$

Step 5—Sketch the $\vec{E}$ phasor:

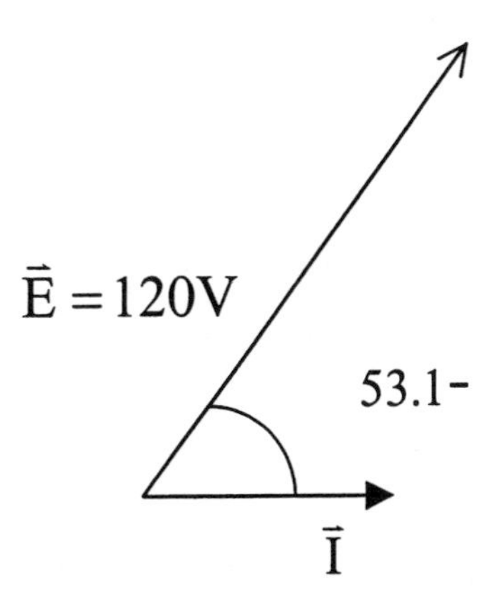

Step 6—Compute the magnitude of $\vec{I}$:

$$I\angle 0° = \frac{120\angle 53.1°}{10\angle 53.1°} = 12\angle 0°$$

Step 7—Noting that E_R is in phase with I (Example 2.9) and that E_L leads I by 90° (Example 2.10), we can add E_R and E_L to the phasor diagram and compute their magnitudes:

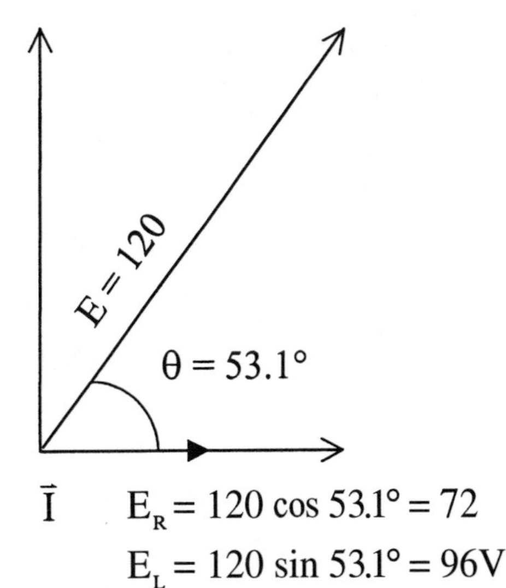

Step 8—We can show voltmeters connected to the original circuit illustrating where computed voltages exist:

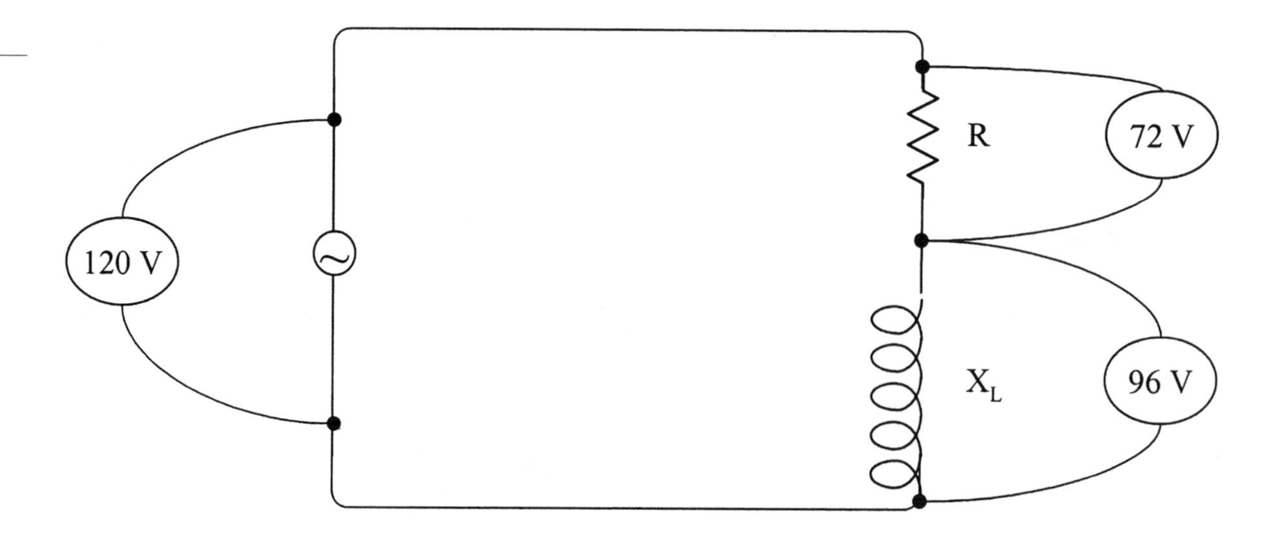

Power

Equation 1.9 showed that power in an AC circuit is a function of current, voltage, and power factor. With the background of chapter 2 to this point, it is appropriate to deal further with the concept of power factor. We can now define power factor:

Power factor = cos Θ **(2.16)**

Theta is the phase angle between the voltage applied to the load and the current through the load. Power factor, like cos, must be between 0 and 1. If a load is predominantly inductive (as are most loads) the power factor is said to be *lagging*, inferring that current is lagging voltage (or behind voltage in time). If the load is capacitive, it has a leading power factor. Here are the AC power equations:

Total power (power carried by lines)
VA = IE, volt amps **(2.17)**

Actual power (power used by load)
P = IE cos Θ, watts **(2.18)**

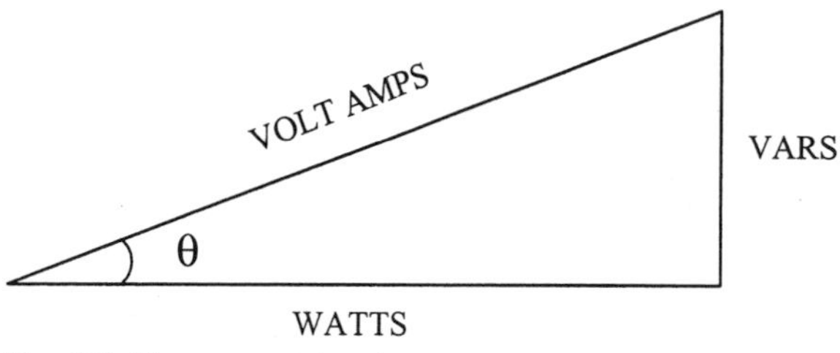

Fig. 2.8. The power triangle.

Reactive power (power carried, not used)
P_R = IE sin Θ, vars **(2.19)**

Electrical machines are often rated in volt amps (rather than watts) because the volt amp rating gives a truer idea of what power must be supplied to the machine. Notice that IE cos Θ = IE_R. Thus, the power used by the load is used only by the resistive component. The unit var is an acronym of volt amps reactive. If these power magnitudes are represented by line lengths, volt amps, watts, and vars define a right triangle (Figure 2.8).

Example 2.13 illustrates use of the power equations.

Example 2.13

Compute total power, actual power, and reactive power for the load of Example 2.12.

VA = (12)(120) = 1440 volt amps
P = (12)(120)(cos 53.1°) = 865 watts
Power factor = cos 53.1° = 0.60 (lagging)
P_R = 12(120)(sin 53.1°) = 1152 vars

Modeling Motors

An induction motor appears to the generator like a combination R-L circuit. We can use the tools previously learned to define this circuit.

Example 2.14

Model the motor in the test circuit shown by a series R-L circuit:

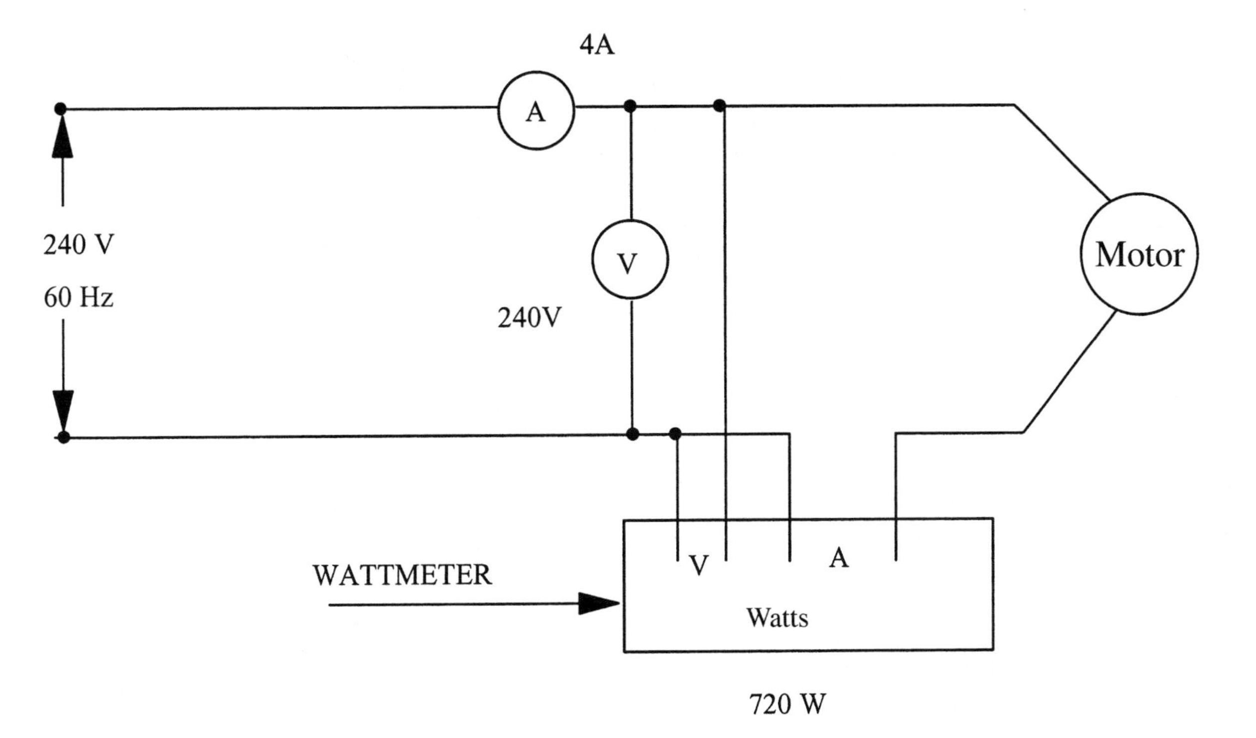

P = IE cos Θ
720 = (4)(240) cos Θ
cos Θ = 0.75 (assumed to be *lagging* because load is a motor)
Θ = 41.4°

At this point it may be helpful to take another look at a waveform graph:

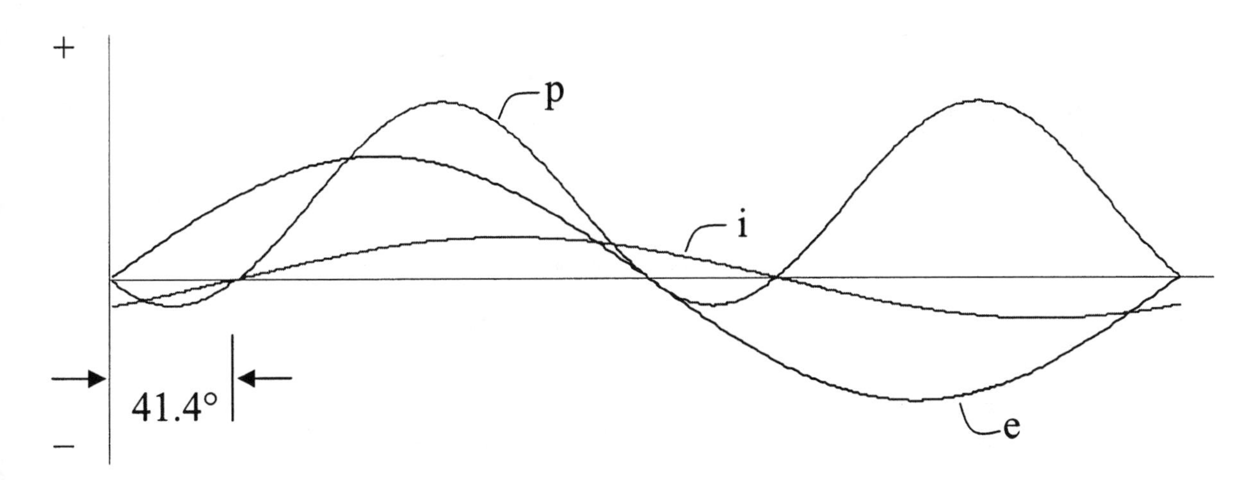

The current wave is seen to cross the zero line (going up) 41.4° after the voltage wave. Thus current lags voltage by 41.4°. The instantaneous power curve is also shown. At every point, instantaneous power (p) equals instantaneous voltage (v) times instantaneous current (i). In regions where current and voltage are both positive or both negative, their product is positive (negative × negative = positive). In regions where one is positive and one is negative, the power curve is seen to go negative. Power is negative when the flow is from load to generator. During the portion of the cycle when the power curve is positive, energy is flowing from the generator to the load. Some of this energy is used by the load and some is stored in the form of a magnetic field. During the portion of the cycle when power is negative, the energy which has been stored is released as the magnetic field collapses and is passed back to the generator through the lines and transformers. Because the power supplier must build generators, lines, transformers, switches, and other equipment to handle the total power (volt amps), but is paid for only the actual power (watts), the power supplier is concerned about the power factor of the power user's loads. Power factor may be changed to a value nearer unity (improved) by installing capacitors at the load. We will explore this more later.

From what was learned in Example 2.12, we can sketch the phasor diagram for this motor. The phasor diagram is sketched, at this point, without regard to precise lengths. Exact values will be calculated later.

- Since our model is a series circuit, we reference on I, and draw it horizontally.
- Since the voltage across the resistor (E_R) is in phase with current, we draw it in phase with I.
- Since the voltage across the inductance (E_L) leads current by 90 degrees, we draw it vertically.
- Voltage (E) is drawn so that it equals the phasor sum of E_L and E_R.

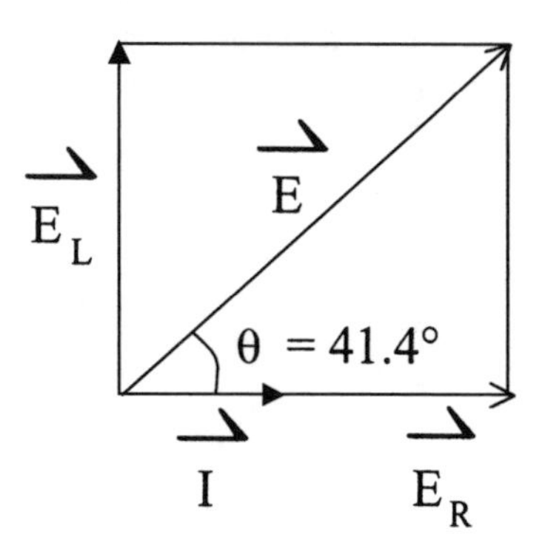

$$E_R = E \cos \Theta = 180\text{V}$$

$$Z_R = \frac{\hat{E}_R}{\vec{I}} = \frac{E_R}{4\angle 0^\circ} = \frac{180\angle 0^\circ}{4\angle 0^\circ} = 45\angle 0^\circ$$

thus,

$$R = 45\ \Omega$$

likewise,

$$E_L = E \sin \Theta = 158.7\ \text{V}$$

$$Z_L = \frac{E_L}{I} = \frac{158.7\angle 90^\circ}{4\angle 0^\circ} = 39.7\angle 90^\circ$$

thus,

$$X_L = 39.7\ \Omega$$

$$L = \frac{X_L}{2\pi f} = \frac{39.7}{2\pi(60)} = 0.105\ \text{H}$$

Here is the circuit, which looks exactly the same to the generator as the motor under test:

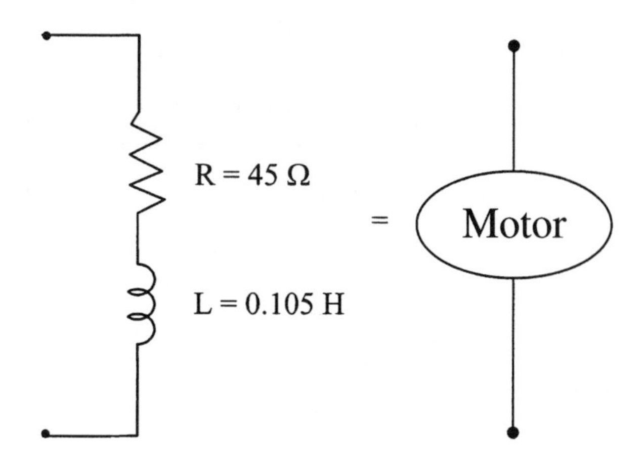

Series RC Circuits

The behavior of an RC series circuit differs from that of an RL series circuit in the direction of the phase shift. An example illustrates this.

Example 2.15

Sketch the phase diagram and compute the phase angle, power factor, actual power, and reactive power for the circuit shown.

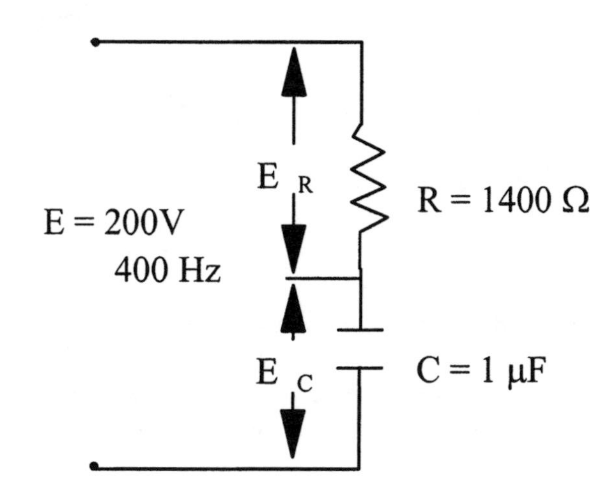

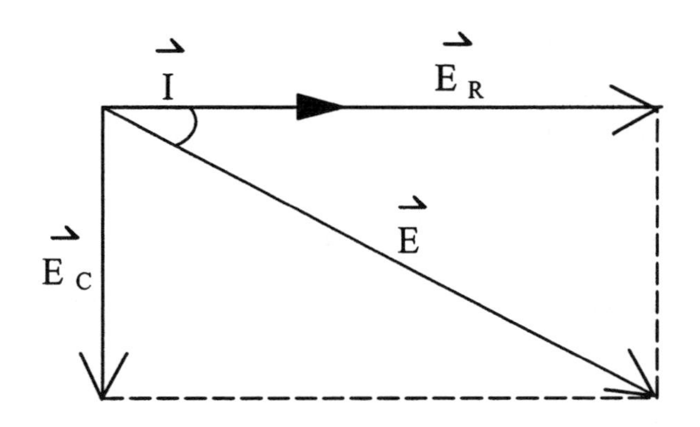

$$X_c = \frac{1}{2\pi fC} = \frac{1}{2\pi(400)(1.0 - 10^{-6})} = 398\ \Omega$$

$$Z = 1400 - j\,398$$

$$\tan\Theta = \frac{-398}{1400} = -0.28$$

$$\Theta = -15.9°$$

$$Z\cos - 15.9° = 1400$$

$$Z = 1456 \qquad Z = 1456\ \angle -15.9°$$

$$\vec{I} = I\angle 0° = \frac{200\angle\Theta_E}{1456\angle -15.9} = 0.137\angle 0°$$

$$0° = \Theta_E - (-15.9°)$$

$$\Theta_E = -15.9° \qquad \vec{E} = 200\ \angle -15.9°$$

$$P = IE\cos\Theta = (0.137)(200)(0.96) = 26.3 \text{ watts}$$

$$P_R = IE\sin\Theta = (0.137)(200)(-0.27) = -7.5 = 7.5 \text{ vars}$$

Parallel Circuits

Parallel circuits involving reactance are analyzed in a manner similar to that used for parallel resistive circuits. It is important to remember that:

- All parallel loads are subjected to the same voltage. Because of this, it is convenient to reference phasor diagrams of parallel circuits on voltage.
- The phasor sum of the currents through all the parallel branches equals the current from the source.

Some examples will illustrate analysis of parallel circuits.

Example 2.16

An inductance of 0.05 H is connected in parallel with a resistance of 25 ohms. A voltage of 1000 V at 60 Hz is applied. Sketch the phasor diagram and compute the actual power, total power, and reactive power. Compute the load impedance.

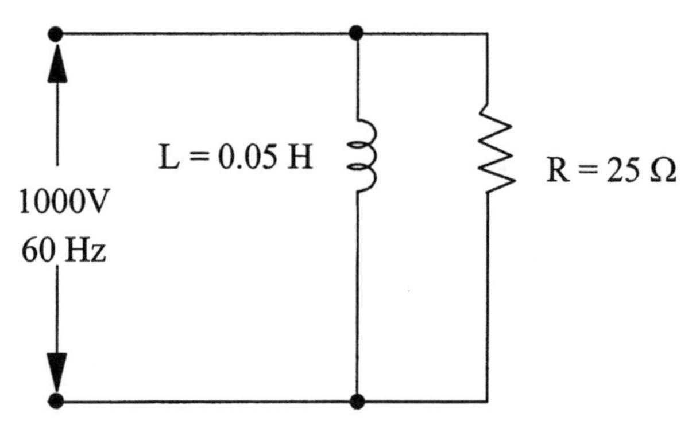

Referencing on voltage (which is constant) we begin the phasor diagram:

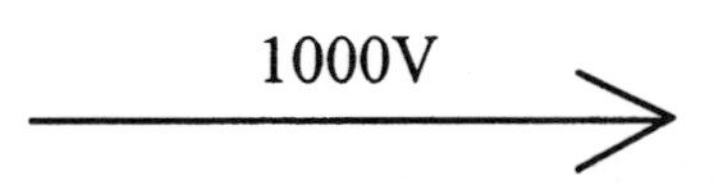

Current through the resistance will be in phase with voltage, and current through the inductance will lag voltage by 90°. The resultant current is the phasor sum of these two currents:

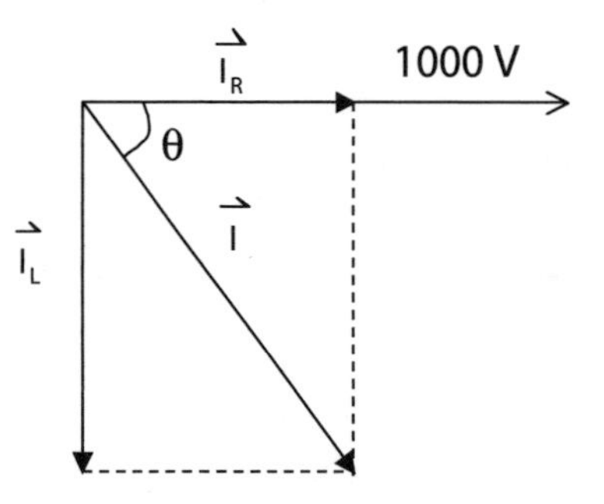

$$I_R = \frac{E}{R} = \frac{1000}{25} = 40\text{ A}$$

$$X_L = 2\pi fL = 2\pi(60)(0.05) = 18.8\ \Omega$$

$$I_L = \frac{1000}{X_L} = \frac{1000}{18.8} = 53.2\text{ A}$$

$$\tan\Theta = \frac{-53.2}{40} = -1.33$$

$$\vec{I} = 40 - j\,53.2$$

$$\Theta = -53°$$

$$I\cos\Theta = 40$$

$$I = 66.5\text{ A}$$

$$\vec{I} = 66.5\ \angle -53°$$

Power factor $= \cos\Theta$
$= 0.6$ (lagging)

$VA = (1000)(66.5) = 66{,}500\ VA$

$P = (1000)(66.5)(\cos -53°) = 40{,}021\ W$

$P_R = (1000)(66.5)(\sin -53°) = -53{,}109\ vars$

$$\vec{Z} = \frac{\vec{E}}{\vec{I}} = \frac{1000\ \angle 0°}{66.5\ \angle -53°} = 15.04\ \angle 53°$$

$$15.04 \cos 53° = 9.05$$

$$15.04 \sin 53° = 12.01$$

$$Z = 9.05 + j\ 12.01 \quad L = \frac{X_L}{2\pi f} = \frac{12.01}{2\pi 60} = 0.0319\ H$$

Computing the impedance and expressing it in rectangular notation shows that the 0.05 H inductance in parallel with a 25 Ω resistance appears to the source the same as a 9.05 Ω resistance in series with a 0.032 H inductance.

Example 2.17

We will go through the analysis of Example 2.16 with the inductance replaced by a capacitance of equal reactance:

$$C = \frac{1}{2\pi f X_c} = \frac{1}{2\pi (60)(18.8)} = 141\ \mu F$$

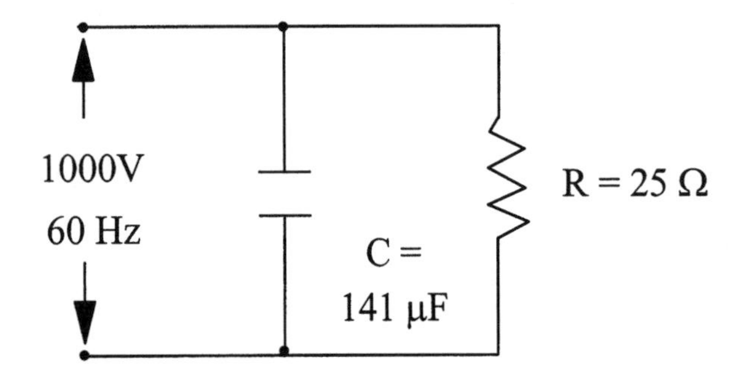

Drawing the phasor diagram:

$$I_R = \frac{E}{R} = \frac{1000}{25} = 40\ A$$

$$X_C = \frac{1}{2\pi fC} = \frac{1}{2\pi(60)(141 \times 10^{-6})} = 18.8\ \Omega$$

$$I_C = \frac{1000}{X_c} = \frac{1000}{18.8} = 53.2\ A$$

$$\vec{I} = 40 + j\ 53.2 \qquad \tan \Theta = \frac{+53.2}{40} = +1.33$$

$$\Theta = +53°$$

$$I \cos \Theta = 40$$

$$I = 66.5\ A$$

$$\vec{I} = 66.5\ \angle 53° \quad \text{Power factor} = \cos \Theta = 0.6 \text{ leading}$$

$VA = (1000)(66.5) = 66{,}500\ VA$

$P = (1000)(66.5)(\cos 53°) = 40{,}021\ W$

$P_R = (1000)(66.5)(\sin 53°) = -53{,}100\ vars$

$$\vec{Z} = \frac{E}{I} = \frac{1000\ \angle 0°}{66.5\ \angle 53°} = 15.04\ \angle 53°$$

$$15.04 \cos 53° = 9.05\ \Omega$$

$$15.04 \sin 53° = 12.01\ \Omega$$

$$\vec{Z} = 9.05 + j\ 12.01 \quad C = \frac{1}{2\pi f X_c} = \frac{1}{2\pi(60)(12.01)}$$

$$C = 221\ \mu F$$

This example shows that the effect of replacing the inductance with a capacitance of the same reactance is to cause the same phase shift, but in the opposite direction. This effect suggests that inductances and capacitances tend to cancel the phase shifting effects of each other. The next example will illustrate this.

The last part of Example 2.17 shows that a 221 μF capacitor in series with a 9.05 Ω resistance appears the same to the power source as the original parallel load.

Example 2.18

Draw the phasor diagram and compute the current and power factor for the load shown:

$$X_c = \frac{1}{2\pi(60)(100 \times 10^{-6})} = 26.5\ \Omega$$

$$X_L = 2\pi(60)(.025) = 9.42\ \Omega$$

$$I_R = \frac{240}{35} = 6.86\ A = 6.86 + j0$$

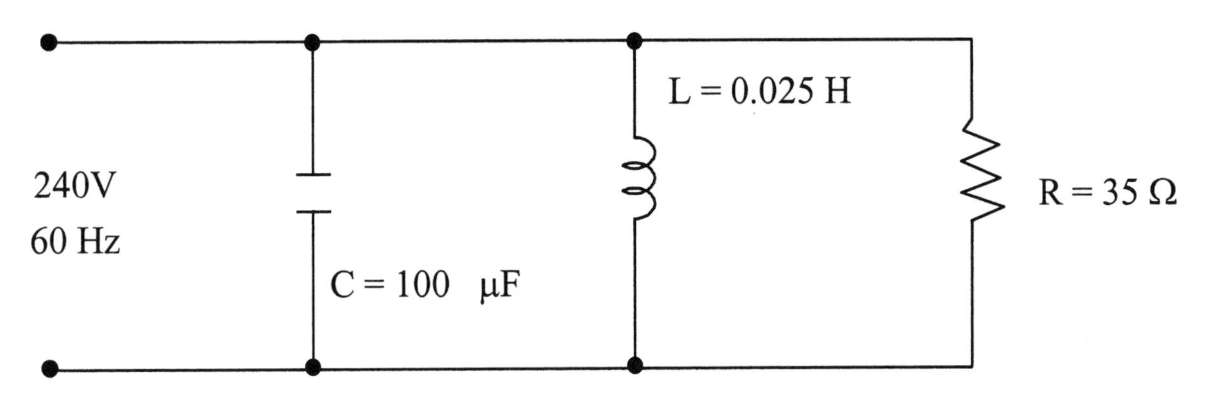

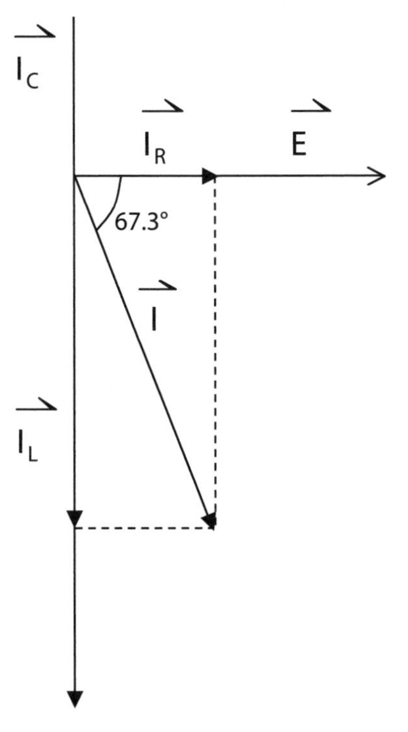

$$I_c = \frac{240}{26.5} = 9.06 = 0 + j\,9.06$$

$$I_L = \frac{240}{9.42} = 25.48 = 0 - j\,25.48$$

$$\vec{I} = 6.86 - j\,16.42 = 17.79\angle{-67.3^\circ}$$

Power factor $= \cos -(67.3^\circ) = 0.39$ (lagging)

Notice that the effect of the capacitor is to increase (improve) the power factor of the load.

Power Factor Improvement

Loads such as motors, fluorescent lamps, and transformer welders can have low (bad) power factors. The previous example illustrated the effect of adding a capacitor in parallel with a load with a lagging power factor. By choosing a capacitor with a suitable reactance, the power factor can be corrected to 1, in which case the load appears as totally resistive. An example illustrates this:

Example 2.19

Specify a capacitor of suitable value to correct the power factor of the motor of Example 2.14 to unity, when placed in parallel with the motor. Here is the phasor diagram of the motor:

E = 240 V
I = 4 A

$\vec{E}$

41.4°

$\vec{I}$

In order to reference on E, we rotate the diagram through 41.4°.

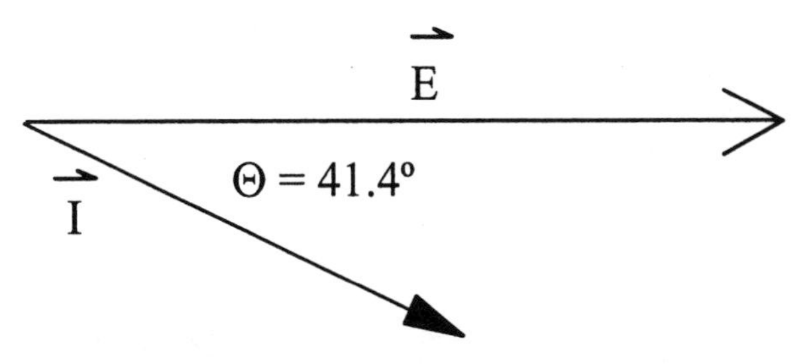

Now $\vec{I} = 4\angle{-41.4^\circ}$
$E = 240\angle{0^\circ}$

We need to divide I into its components:

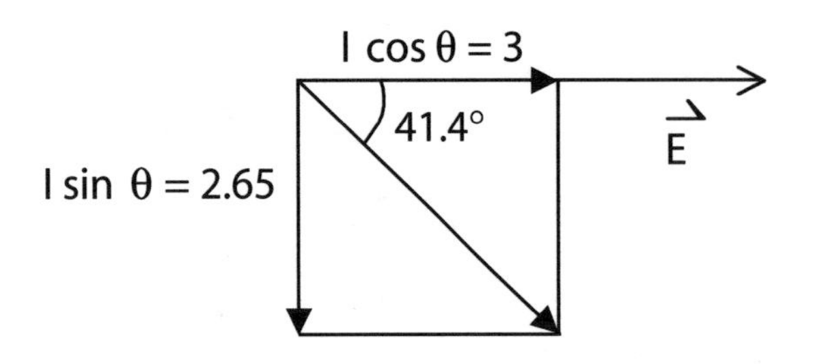

A capacitor in parallel will cause a current component opposite in direction to I sin Q:

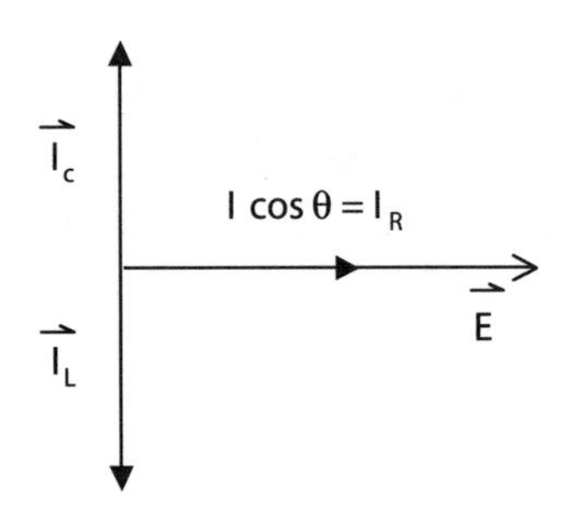

If I_c is numerically equal to I sin Θ (that is, $|I_c| = |I \sin \Theta|$), then the two will cancel each other.

$$I \sin \Theta = (4)(\sin 41.4°) = 2.65$$

$$X_c = \frac{E}{I \sin \Theta} = \frac{240}{2.65} = 90.7\ \Omega$$

$$C = \frac{1}{2\pi f X_c} = \frac{1}{2\pi (60)(90.7)} = 29.3\ \mu F$$

Here is what the motor instrumentation would read with the capacitor in place:

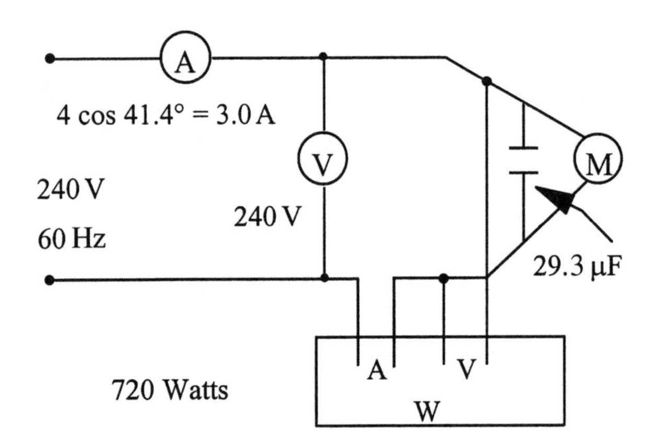

With the capacitor in place, the motor–capacitor combination looks like a resistance to the power source. Although the watts used by the motor remain the same, the current in the line is reduced from 4.0 to 3.0 amps.

When the capacitor is added, the capacitor and the inductive reactance of the motor exchange reactive power with each other instead of the motor exchanging reactive power with the power source as was the case in Example 2.14.

Because the power factor of a motor normally increases with load on the motor, correction of the power factor to 1.0 would not be maintained if motor load was changed. Notice that if a capacitor larger than 29.3 μF was added, the effect would be to cause the power factor to worsen and, in this case, to become leading.

Decisions about power factor improvement are often based on costs. West Central Coop's Ralston, Iowa elevator was operating at a power factor of 69% or 70% (power factors are often specified as percents rather than decimals). The turnkey cost of a power factor improvement system was $80,000. This was weighted against the cost of paying the power company a penalty for a power factor below 85% and receiving a credit if it is at or above 95%. The power factor correction system was installed and savings totaled more than the purchase price after 11 months (West Central 1998).

References

West Central. 1998. West Central improves efficiency of operations through electrical power factor. West Central News, April. West Central Coop, Ralston, Iowa.

Problems

2.1. Why are most AC meters built to display the rms of the parameter being measured?

2.2. Use a spreadsheet to show that the rms value of a sinusoidal waveform is the maximum value divided by $\sqrt{2}$.

- Assume the maximum value is 100 and form a column of 100 sinΘ from 0 to 360. This is column A.
- Form column B as (col A)^2.
- Compute the mean of col B.
- Compute the square root of this mean, and verify that it is $100/\sqrt{2} = 70.7$.

2.3. Find length of A:

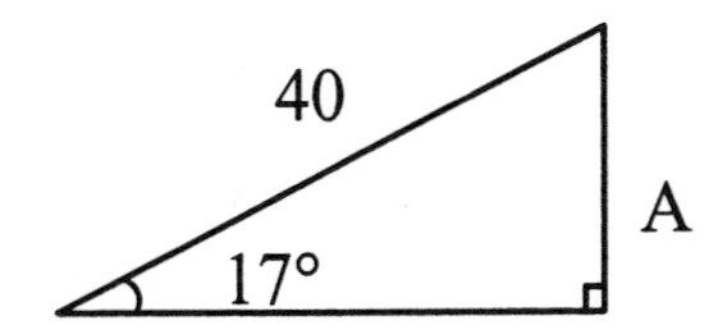

2.4. Find length of B:

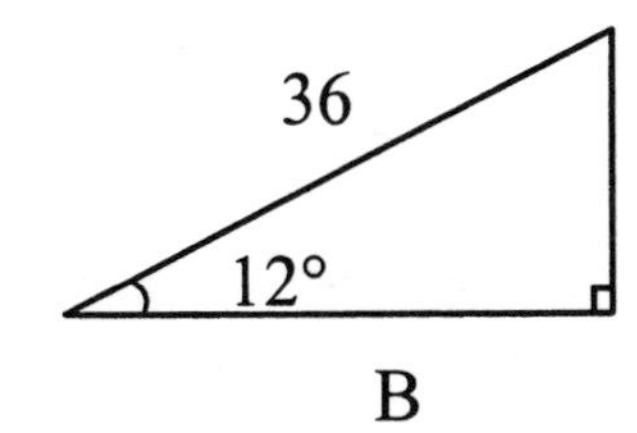

2.5. What is angle α?

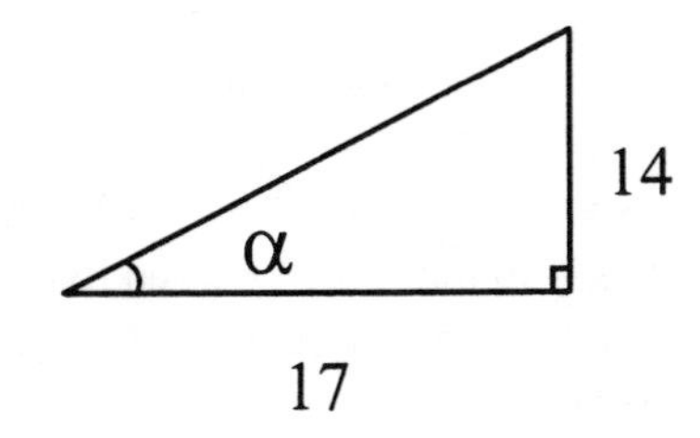

2.6. What is angle β?

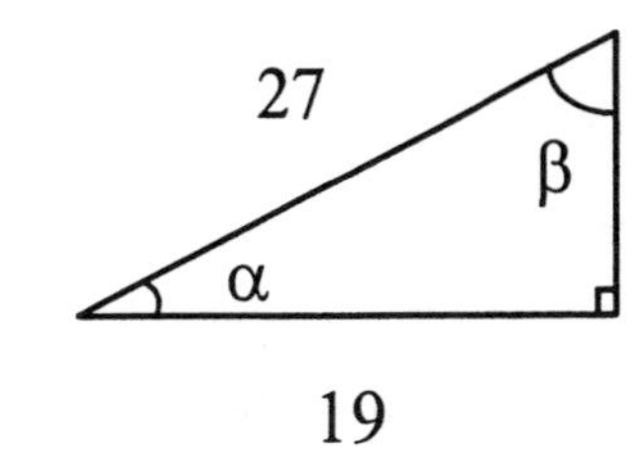

2.7. What is length C?

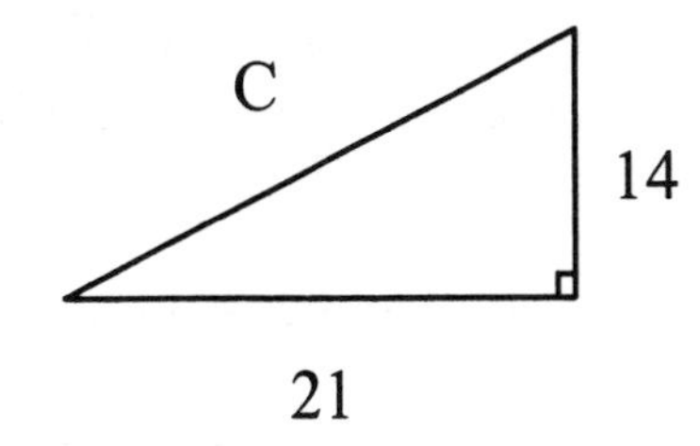

2.8. What is length B?

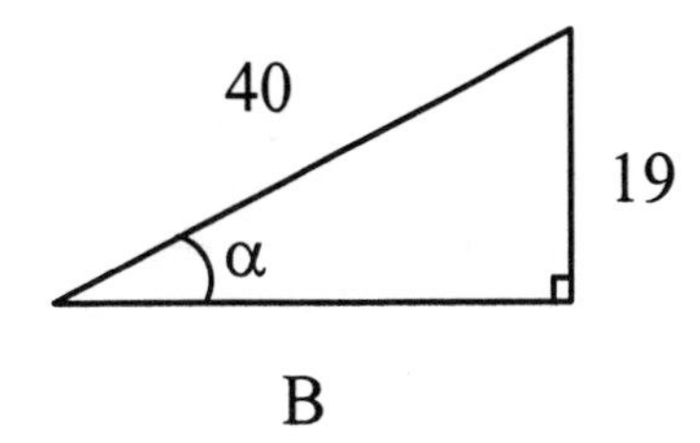

2.9. What is angle β?

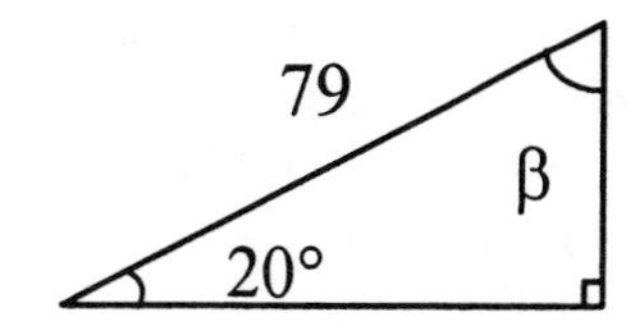

2.10. What is angle α?

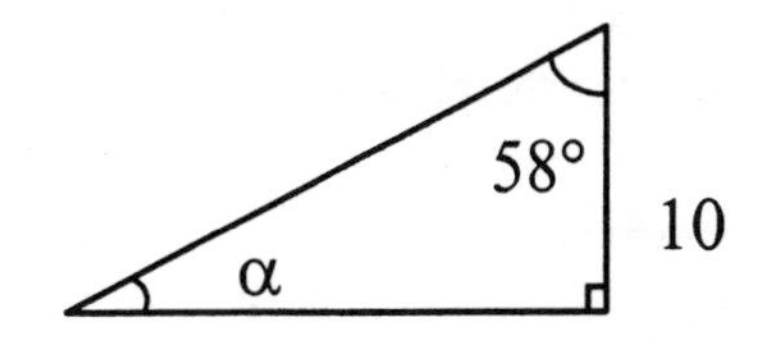

2.11. Convert this phasor to polar notation: $\vec{I} = 10 + j\,40$

2.12. Convert this phasor to rectangular notation: $\vec{E} = 120\angle -30°$

2.13. Voltage E_R is 120 volts and leads voltage E_B (also 120 volts) by 180°. This describes 120/240-V single-phase service.

a. Sketch the phasor diagram, referencing on E_B.

b. Sketch the wave diagram of E_B and E_R referencing on E_B and assuming the horizontal axis is at 0 volts.

c. Compute the phasor sum, $\vec{E}_B + E_R$.

d. Compute the phasor difference, $\vec{E}_B - E_R$.

e. Sketch the wave diagram of E_B assuming the horizontal axis is at E_R volts.

2.14. Current A, which is 12 amps, lags current B, which is 4 amps, by 45°. Sketch two phasor diagrams of these currents. Reference the first on current A and the second on current B. For each of these two cases, find the phasor sum, or resultant, of these two currents and express it in both polar and rectangular notations.

2.15. A 10-kW resistance heater is designed for connection across a 240-V, 60-Hz source. Compute the current drawn and the impedance of this load. Sketch the phasor diagram. Express Z, I, and E in polar and rectangular notations.

2.16. What is the inductive reactance and impedance (expressed in polar and rectangular notations) of a load consisting of a pure inductance that allows a 4-A current to flow when it is connected to a 117-V, 60-Hz source? Sketch the phasor diagram, referencing on I.

2.17. At what frequency will an 8-μF capacitor have a reactance of 49.73 ohms?

2.18. A solenoid having an inductance of 0.5 H and a resistance of 100 is connected to a 120-V, 60-Hz source. To the source, then, it appears as a series R-L circuit. Compute the current and sketch the phasor diagram referencing on current. Express I, E, and Z in polar and rectangular notations. Compute the total power, actual power, and reactive power.

2.19. A resistance of 100 ohms in series with a 60-μF capacitor is connected to a 120-V, 60-Hz source. Sketch the phasor diagram and find the load impedance, current, power factor, and actual power. Express impedance and current in polar and rectangular notations.

2.20. Repeat problem 19 with a 0.2-Hz inductance in series with the other two components.

2.21.

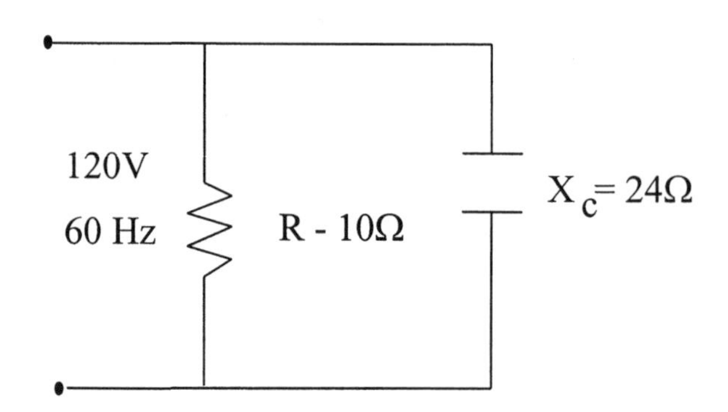

(a) Sketch the phasor diagram, referencing on voltage.

(b) Compute total circuit current and express in polar notation.

(c) Compute actual power use.

(d) Compute circuit impedance and express in polar notation.

2.22. A 0.5-hp motor is observed to draw 1000 W and 10 A when connected across a 120-V, 60-Hz source.

(a) This motor is to be modeled as a parallel R-L circuit. Sketch the phasor diagram and find the values of R and L.

(b) What size of capacitor (μF) placed in parallel with this motor will correct the power factor to 1.0? With this in place, what current and wattage will the motor draw?

(c) The capacitor in (b) is replaced by a 400-μF unit. Sketch the phasor diagram and compute the power factor, current, and wattage under this condition.

(d) Find values of the components (ohms and microfarads) which will model the load of (c) as a series circuit.

2.23.

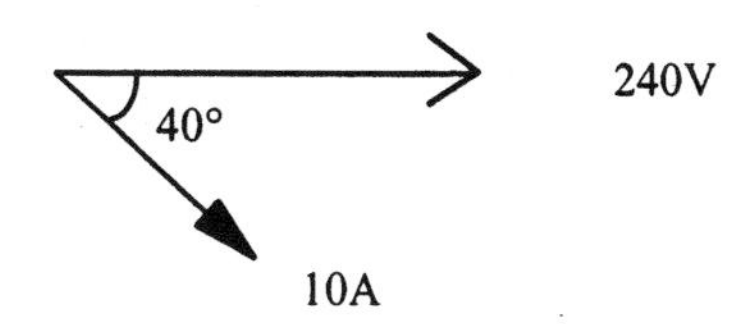

(a) Define the series and parallel R-L circuits which this phasor diagram can represent.

(b) Compute actual power used by each circuit.

(c) Compute actual power used by the resistor in each circuit.

2.24. Instruments wired to a test motor give these results:

(a) Compute the phase angle between current and voltage and sketch a phasor diagram for this motor.

(b) Compute the impedance of this motor and express in polar form.

(c) Compute the resistance and reactance values (in ohms) of a series circuit which models this motor.

(d) Compute the resistance and reactance values (in ohms) of a parallel circuit which models this motor.

(e) Compute the reactance of a capacitor which will change the power factor of this motor to unity if placed in parallel with the motor.

(f) If the capacitor from (e) is installed, what is the current reading on the ammeter?

2.25. An R-L-C parallel circuit contains a capacitor with $X_c = 10\ \Omega$. The circuit draws 20 amps and 2.4 kW when connected across 120 V, 60 Hz. Find R and X_L.

2.26. A fully loaded 10-hp motor is drawing 50 A on a 230-V line. The motor operates at a power factor of 0.88. When a bank of capacitors is added in parallel with the motor, the current

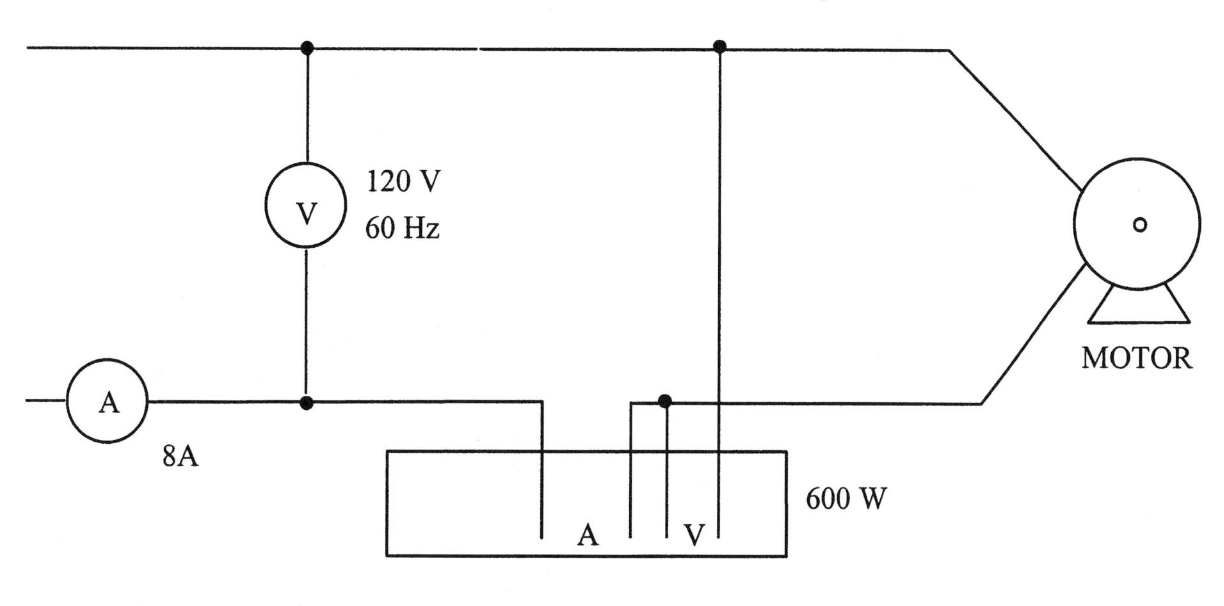

decreases to 45 A. Two different capacitance values can give this current. Compute both.

2.27. A 240-V, 42-A single-phase motor (PF = 0.6) and a 12-kW, 120-V resistive heater are connected on a 3-wire, single-phase 120/240-V circuit. The resistor is connected to the red conductor. Compute the magnitude of the current carried by the black, white, and red conductors.

Chapter 3

BIOLOGICAL EFFECTS OF ELECTRIC CURRENT

Introduction

An important component of a course on electric power applications is a discussion of the biological effects of electrical current. Because energized electrical components are usually quiet, cool, non-luminescent, and stationary, it is easy to develop a casual attitude about them and to ignore their possible danger to people and animals. This chapter discusses what happens when a person or animal becomes part of an energized electrical circuit.

The Body as a Conductor

The body of a human or an animal is an electric conductor and, therefore, when a voltage is present between two body contact points, electrons will flow through the body between the contact points. At frequencies of 50 to 60 Hz and voltages of 120 to 240 volts, the human body does not exhibit reactance (Dalziel 1972) and can be analyzed as a resistive load (Figure 3.1). The body resistance, R_B, which limits current flow, is then the sum of three series resistances (Equation 3.1).

$$R_B = R_S + R_F + R_S \tag{3.1}$$

Current through the body between the contact points can then be predicted by Ohm's Law (Equation 3.2):

$$I_B = \frac{E_B\,(1000)}{R_B} \tag{3.2}$$

where: I_B = current through the body, mA
E_B = voltage difference between contact points, V

Table 3.1 shows some resistance values. Skin resistance at the contact points is the major component of body resistance. The outer skin, or epidermis, which is from 0.05 to 0.2 mm thick, contributes most of the skin resistance. The inner skin (derma) has low resistance because of the salinity of blood and other body fluids. Except for bone and fat, which are poor conductors, resistance within the body is relatively low for the same reason.

Skin resistance is not always the same. Table 3.1 values show that wet skin may have 1% of the resistance of dry skin. Skin resistance decreases rapidly as current continues to flow (Kouwenhoven 1949). The skin resistance of females is characteristically lower than the skin of males. Skin resistance approximately doubles

TABLE 3.1 Human and animal resistances to current flow (Dalziel 1972)

Conditions	Resistance
Minimum resistance between major body extremities	500 Ω
Resistance between normal perspiring hands of a worker	1,500 Ω
Dry skin	100,000 to 300,000 Ω/cm²
Wet skin	1,000 to 3,000 Ω/cm²

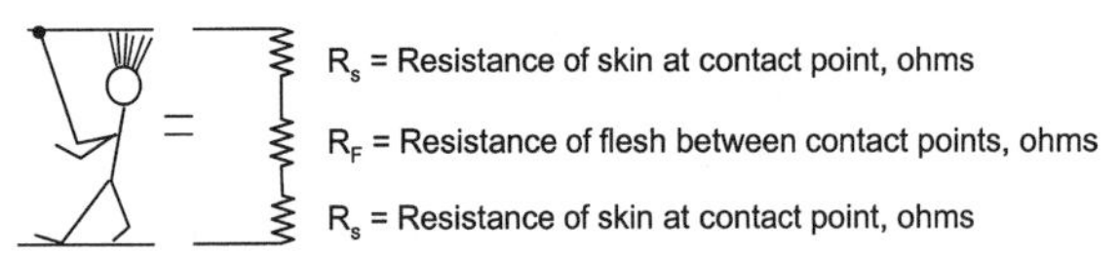

Fig. 3.1. Body resistance model.

while a person sleeps. On the body, skin resistances of the palms of the hands, the soles of the feet, a band around the waist, the center of the face, and the inner surfaces of the elbows and knees are lower than at other locations.

Voltage affects body resistance. Kouwenhoven (1969) found in tests on cadavers that hand-to-foot resistance decreased from 10,000 Ω to 1200 Ω as voltage was increased from 50 to 500 V. He attributed this to a breakdown of skin resistance by the higher voltage.

Effects of Current

Effects of various current levels through the body are summarized in Table 3.2.

Current becomes perceptible at about 1 mA; at about 5 mA it becomes disturbing, and is, by definition, an electrical shock. At currents above the average let-go values, muscular control is lost and you can no longer voluntarily let go of the electrode. Currents above the let-go values are dangerous due mainly to effects on the heart (the heart is an electrically controlled pump). Shock current flowing through the heart can disturb its rhythm and coordination and cause ventricular fibrillation—a state in which blood circulation ceases, and pulse disappears. Normal beating seldom resumes spontaneously; death occurs within a few minutes, if no assistance is given. At about 4,000 mA, the heart is paralyzed during the shock but may restart if the shock time is short. Burns occur for current over 5,000 mA. Such shocks may not be fatal unless vital organs are damaged. These currents are far below fuse or circuit breaker ratings, which are seldom lower than 15 A.

TABLE 3.2 Effects of 60-Hz current in the human body (Kouwenhoven 1969, Dalziel and Lee 1968, Cadick 1994)

Current, mA	Effect
1	Threshold of sensation
5	Disturbing shock, muscular control maintained
6–25	Muscular control lost in women, maximum let-go range for women
9–30	Muscular control lost in men, maximum let-go range for men
30	Respiratory paralysis
75	Fibrillation threshold 0.5%
250	Fibrillation threshold 99.5%
4,000	Heart paralysis
>5,000	Tissue burning

Defibrillation may be able to restore proper rhythm and coordination if the heart is in fibrillation. In this process, a current of about 10,000 mA is applied between electrodes placed against the person's chest. This current causes all heart muscles to violently contract simultaneously. Following this shock, the heart may restart normal pumping.

Factors Determining Shock Severity

Factors determining shock severity include:

1. Magnitude of current
2. Duration of shock
3. Frequency of power source
4. Pathway of current through body
5. Phase of the heart cycle
6. Body mass

Current Magnitude and Duration of Shock

The effect of current magnitude and shock duration are interrelated. The higher the current, the lower is the maximum safe duration. Dalziel and Lee (1968) express the current–time relationship in an equation. Based on research with animals and for current pathways between major extremities (hand-to-hand, hand-to-foot), they state that ventricular fibrillation is unlikely if:

$$I < \frac{116}{\sqrt{T}} \tag{3.3}$$

where: I = shock current, mA
T = shock duration, s

Figure 3.2 is a plot of Equation 3.3 on logarithmic scales. According to Equation 3.3, a shock of 116 mA is not dangerous if the duration is less than 1 s (plug in T = 1 and solve). However, if the duration is 5 s (plug in T = 5 and solve), a shock of over 51 mA may be dangerous. The equation has been found valid in the range of .008 to 5 s.

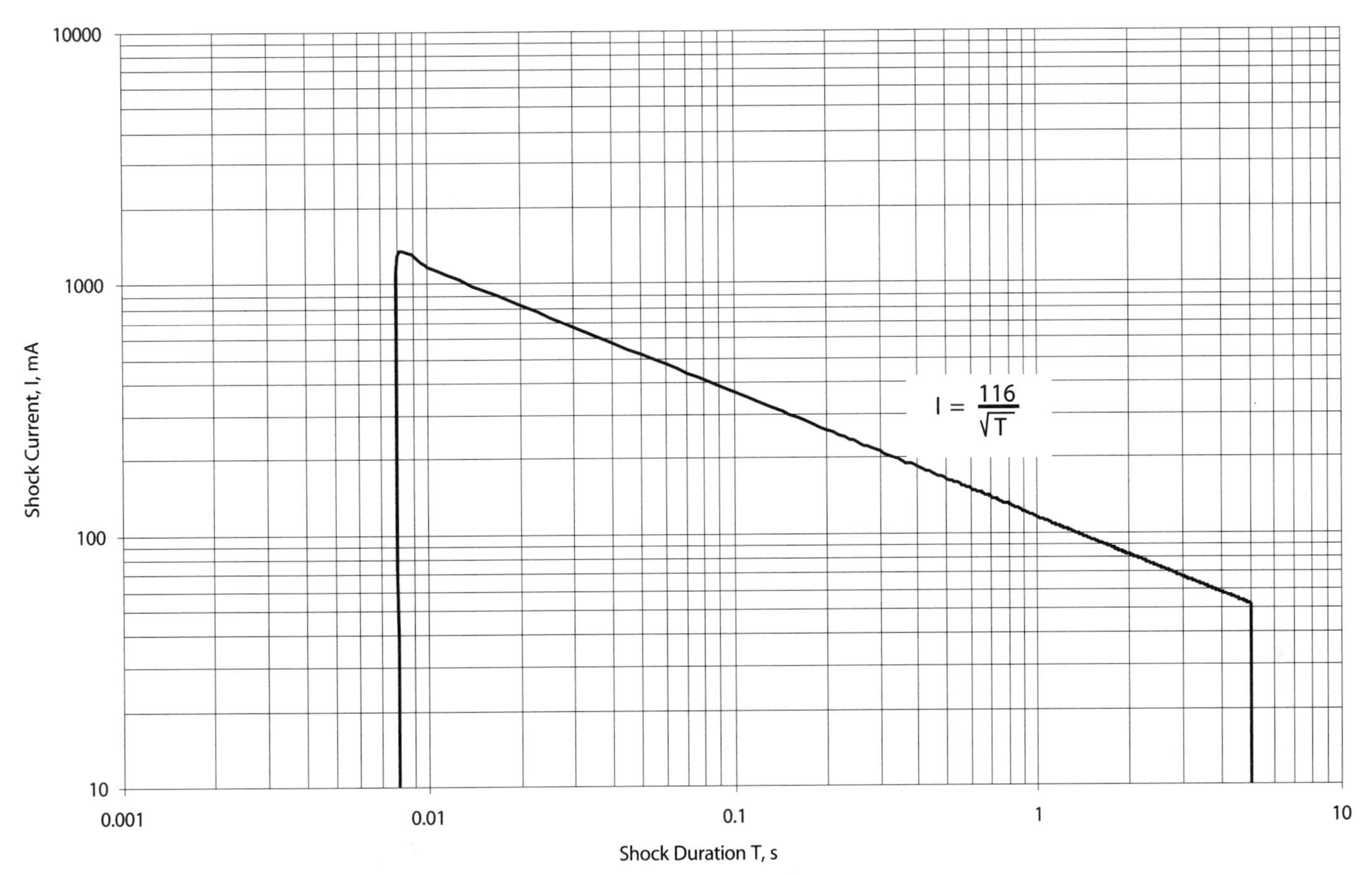

Fig. 3.2. Dalziel fibrillation line (Dalziel 1972).

Frequency of Power Source

According to Kouwenhoven (1949), reactions to shocks of a certain magnitude are less severe at frequencies much lower and much higher than 60 Hz. Response is practically uniform for frequencies between 10 and 300 Hz. DC circuits (0 Hz) of low voltage tend to be less dangerous than AC circuits of the same voltage. Kouwenhoven (1949) states that DC circuits do not produce the strong muscular contraction experienced with AC-circuit shocks.

As frequency is increased above 300 Hz, the effect of current is progressively less. Dalziel and Lee (1968) found that the average let-go current for men increased from about 16 mA at 60 Hz to about 75 mA at 10,000 Hz. It is likely that higher currents are required to produce fibrillation at higher power frequencies. Studies show that the 3,000-Hz current required to produce fibrillation in dogs is 22 to 28 times that at 60 Hz (Dalziel 1972). Subsequent discussion in this chapter assumes 60-Hz frequency.

Pathway of Current through Body

Kouwenhoven (1949) states that effects of a shocking current which does not involve current flow through the heart or lungs are usually limited to burns. He found on tests with cadavers that about 10% of hand-to-foot current traversed the heart. Hand-to-hand and hand-to-foot shocks are thus especially dangerous. In order to avoid either of these pathways through the body, Kline and Friauf (1954) recommend that persons working around live conductors stand on an insulated mat and work with only one hand. If the working hand contacts an energized part, these precautions can prevent a second point of contact which would cause a hand-to-foot or hand-to-hand shock.

Phase of the Heart Cycle

The "T" phase of the heart cycle is particularly sensitive to electrical shock (Kouwenhoven 1969). This phase, which occupies about 25% of the heart cycle time, and lasts about 0.25 s, is the recharging period of the heart, and follows the contraction or pumping phase. Data show that random 0.01-s shocks of 500 to 3,000 mA do not result in fibrillation if they occur outside the "T" phase. Shocks of more than 1-s duration do not exhibit a vulnerable period since they encompass an entire heart cycle.

Body Mass

In tests with dogs, sheep, calves, and pigs, Dalziel and Lee (1968) concluded that the minimum current to produce ventricular fibrillation is approximately proportional to body mass. They state that this is probably also true for humans.

Shock-Related Injury

Burns are the most common shock-related injury (OSHA 1997). *Electrical burns* result from current flowing through tissues or bone. Tissue damage results from I^2R heating as current flows through the body. *Arc burns* can result when current arcs to a part of the body, or when some body part comes close to an established arc. *Thermal contact burns* occur when flesh comes in contact with hot surfaces produced by current flow. All three can occur simultaneously.

Strong involuntary reactions of muscles to electric currents can cause injury when they cause entanglements, falls, or blows to the body. Such reactions may also prevent injury by throwing the person clear of the circuit.

Help for Victims

If you come upon an unconscious victim of electrical shock, you may be able to save the person's life if you know what to do:

1. Begin shouting for medical help
   and at the same time:
2. Make sure current flow is stopped
   then
3. Keep person alive until Emergency Medical Service help arrives

Stopping Current Flow

Be careful not to become a shock victim in your attempt to stop current flow! Bete (1971) suggests this procedure:

If the victim is in contact with live low-voltage equipment in a home or building, shut off the power by pulling the plug on the supply cord or by turning off a switch. If this is not possible, use a dry rope or a dry board or stick to free the victim. Make sure your hands are dry and you are standing on a dry surface.

If the victim is in contact with a live outdoor fallen wire, the only safe procedure is to call the power company to turn off the power. (The outdoor conductor may also be low voltage if it is on the secondary side of the service transformer. However, to avoid a dangerous error in judgment here, it is recommended that all outdoor conductors be regarded as high voltage.)

Cardiopulmonary Resuscitation

If the victim is unconscious, is not breathing, and has no pulse, proper application of cardiopulmonary resuscitation (CPR) may keep oxygenated blood flowing to vital organs until medical help arrives. CPR is a combination of chest compressions and rescue breathing. If the person is in a state of ventricular fibrillation, as was mentioned earlier, spontaneous recovery is unlikely. Recovery depends on sustaining life until medical professionals can carry out procedures for defibrillation. Irreversible brain damage occurs after about four minutes (Thygerson 2001).

The following is a list of procedures for rescue breathing and CPR for a person over 8 years of age who is unconscious, not breathing, and has no pulse (Thygerson 2001). This is not intended to replace formal instruction in CPR, which is available from other sources. Formal instruction is recommended to the reader.

1. Call 911 to summon professional help.
2. If an AED* is available, get it.
3. Tilt the head back, lift the chin, remove airway obstructions, and give 2 slow rescue breaths through the mouth, while pinching the nose.
4. If there are no signs of circulation or breath, perform CPR.

   a. Place heel of one hand on the lower half of the sternum between nipples of the victim.
   b. Using two hands, depress chest downward $1^1/_2$ to 2 inches.
   c. Give 15 chest compressions at the rate of about 100 per minute.
   d. Give 2 slow breaths.
   e. Continue cycles of 15 chest compressions and 2 slow breaths.

5. After four cycles, recheck for breath and circulation. If none, continue CPR until professional help arrives.

* AEDs (automated external defibrillators) are being widely deployed in public places. These smart defibrillators can be effectively used by nonmedical personnel after a few hours of training. The fast response that they allow can improve survival rates of victims.

Case History

Allen (1984) describes an instance where an unconscious electric shock victim was revived because

CPR was administered. Bill Longan, an electrician for a Texas Rural Electric Cooperative, went to hook up an irrigation pump near Munday, Texas, in July 1984. His wife, Karen, and two children went along. Ray Herring, the farm manager, met them at the pump.

While Bill was working on an overhead line from a plastic bucket, he was blinded by the sun and became entangled in a live 240-volt line. He collapsed in the bucket, unconscious. When Ray and Karen got the bucket down, Bill's lips were turning blue. They pulled him from the bucket and began CPR. Both Ray and Karen had taken CPR training in their communities. They continued CPR while trying to reach help (which never arrived) on the truck's CB radio.

After a while Bill began breathing on his own and eventually recovered consciousness. He was then driven to a hospital in Seymour, Texas, 25 miles away, and has completely recovered. Bill is alive because persons at the scene administered CPR.

Revival from fibrillation while CPR is being administered is unusual. As noted earlier, CPR may be required until medical personnel apply defibrillation procedures.

Worker Electrocution Statistics

Essentially every American worker is exposed to electricity every day. The National Institute for Occupational Safety and Health data show (NIOSH 1998):

- About 400 American workers are electrocuted each year and the average rate is about 0.4 electrocutions per 100,000 workers per year.
- Electrocution is the fifth leading cause of death and accounts for 7% of deaths.
- Electrocutions have decreased steadily over the years.
- About 44 electrocutions per year occur in agriculture/fisheries/forestry.
- The rate in these industries is 1.3 electrocutions per 100,000 workers per year. Only construction (2.4) and mining (2.2) have higher rates.
- Almost all deaths are from AC power, and two-thirds occur at high voltage (>600V).

Earlier analysis of electrical injuries on farms showed (Fletcher 1974):

- About 80% of injuries were fatal.
- Over 60% involved a high-voltage power line.
- Over 60% involved a grain elevator, grain auger, or irrigation equipment.
- Nearly half of the injuries involved use of improper or defective equipment.

Prevention of Shocks

Kline and Friauf (1954) list these basic causes of electrical shocks:

1. Equipment failure.
2. Human failure.
3. A combination of equipment failure and human failure.
4. A fortuitous combination of events so unlikely that even the most prudent of persons could hardly be expected to anticipate and guard against it.

Except in case 4, action can be taken to prevent the shocks from occurring.

Overhead Lines

Because many electrical injuries on farms involve high-voltage lines, proper caution in regard to contact of these lines is a very important prevention measure. Figure 3.3 is an example warning sign designed to be posted in locations where contact with an overhead line is possible.

Figure 3.4 shows common situations which can result in injuries with overhead lines.

Experience in Nebraska indicates that frequent reminders of the dangers of overhead lines are effective in preventing injuries. Mallie (1979) reports that in 1963, 7 deaths were reported in Nebraska due to contact with overhead lines. A safety program was

Fig. 3.3. Overhead lines warning sign.

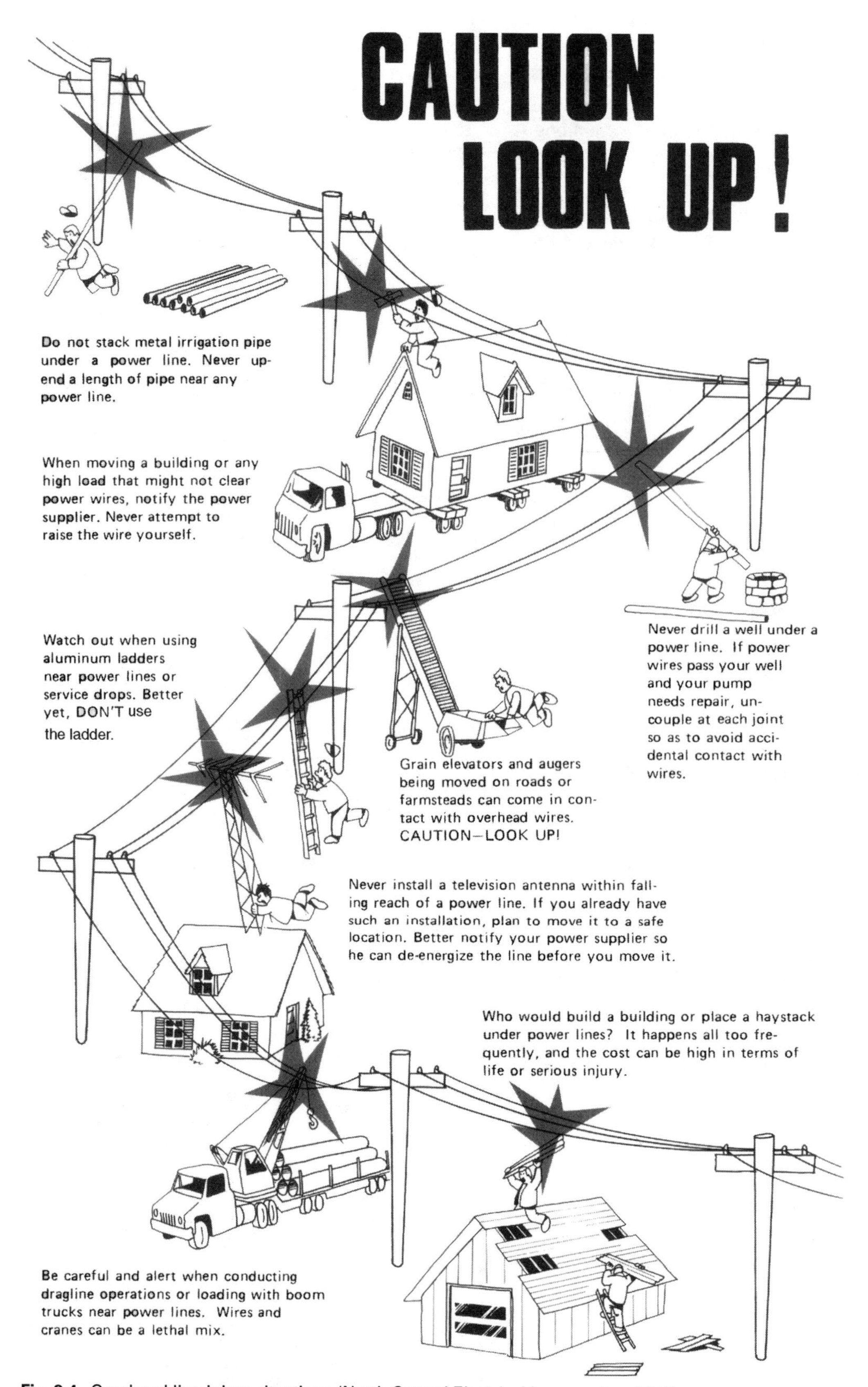

Fig. 3.4. Overhead line injury situations (North Central Electrical League circa 1976).

started which involved placement of decals as in Figure 3.2 on irrigation pumps, and radio and TV announcements. In 1964, one death was reported. The program continued and no deaths were reported in the following three years. Because of this success, the safety program was stopped. In 1968, two deaths were reported. Apparently, frequent reminders are needed to prevent the act without forethought which results in contact with an overhead line.

Working Around Electrical Equipment

Following these guidelines can reduce the likelihood of shock from electrical equipment:

1. Keep alert! Think about what you are doing and the possible danger involved.
2. Move slowly. Avoid sudden movements which could cause contact with an energized part.
3. Make sure a circuit is de-energized before working on the equipment.
4. Use equipment properly. Study the instructions.
5. Do not use equipment which may be dangerous due to defective grounding, wet conditions, poor insulation, or other factors.

Animal Shocks

The principles of shocks discussed for humans also apply to animals, and in fact some of the human electrocution parameters were established through tests on animals. Animals can feel shock effects at lower voltages because contact points are often through the mouth at a waterer and through bare feet to damp concrete or earth. In each case, skin resistance is very low. Low voltages called "stray voltages" can exist in livestock facilities and can cause severe problems for producers. Stray voltage is the topic of chapter 14.

References

Allen, Danita. 1984. He was turning blue. Successful Farming 82(12):34-AF.

Bete, Channing L. 1971. What everyone should know about electrical safety. Channing L. Bete Co., Inc. Greenfield, MA.

Cadick, John. 1994. Electrical safety handbook. McGraw-Hill Inc., New York, NY.

Dalziel, C. F. 1972. Electric shock hazard. IEEE Spectrum 9:41–50.

Dalziel, C. F., and W. R. Lee. 1968. Reevaluation of lethal electric currents. IEEE Transactions on Industry and General Applications 5: 467–476.

Fletcher, W. J. 1974. Where are the electrical accidents happening on the farm? ASAE paper 74-3501. American Society of Agricultural Engineers, St. Joseph, MI.

Kline, R. L., and J. B. Friauf. 1954. Electric shock, its causes and its prevention. Navships 250-660-42, Bureau of Ships, Navy Department, Washington, DC.

Kouwenhoven, W. B. 1949. Effects of electricity on the human body. Electrical Engineering 68: 199–203.

Kouwenhoven, W. B. 1969. Human safety and electric shock. In: Electrical safety practices, Monograph 112, pp. 91–97. Instrument Society of America, Pittsburgh, PA.

Mallie, D. J. 1979. The hazard of overhead power lines and irrigation pipe. Student project report. Agricultural and Biosystems Engineering Department, Iowa State University.

NIOSH. 1998. Worker deaths by electrocution. NIOSH publication 98-131. NIOSH, Cincinnati, OH.

North Central Electrical League. Circa 1976. Caution look up. NCEL, Minneapolis, MN.

OSHA. 1997. Controlling electrical hazards. OSHA 3075. U.S. Department of Labor, Washington, DC.

Thygerson, Oltan. 2001. National Safety Council First Aid and CPR, Fourth Ed. Jones and Bartlett Publishers, Sudbury, MA.

Problems

3.1. Explain why fuses and circuit breakers on branch circuits in farm buildings afford no personal protection against electrical shock.

3.2. An electric fencer is to be built which will give a 60-Hz shock of 0.05 s duration.

(a) How high can the shock current be and still not cause fibrillation in people?

(b) How high a voltage can be used without danger of exceeding this current?

3.3. What are the chances that an American working in agriculture/fishery/forestry will be electrocuted by high voltage during a 40-year working career? Express as 1 in (some number).

Chapter 4

THREE-PHASE CIRCUITS

Introduction

A 3-phase circuit uses three simultaneously energized conductors to carry power to a load in a very effective and economical system. This chapter will cover the basics of 3-phase circuits, 3-phase power measurement, and finally applications of 3-phase power. Unless otherwise noted, discussion assumes balanced 3-phase systems—that is, equal currents through and voltages between all three energized conductors.

Advantages of 3-Phase Power

Three-phase power offers these advantages over use of single-phase power:

- Power flow to a load is constant with time, whereas power flow in a single-phase circuit is pulsating.
- Three-phase motors are simpler and cheaper than comparable single-phase machines.
- Circuit conductors for 3-phase circuits can be smaller. A 3-phase system can carry a given quantity of power using 25% less conductor material than a single-phase system.

For these reasons, most electric power is generated as 3-phase, and most AC machines over 10 hp are 3-phase.

Types of 3-Phase Service

Generation of 3-Phase Electromotive Force

A single-phase AC generator can be constructed by placing a conductor loop on a shaft revolving in a magnetic field. As the loop turns, the conductor cuts lines of the magnetic field and an alternating voltage is produced between ends of the loop (see Figure 9.3). A 3-phase generator can be constructed by placing three separate loops on a single shaft, and mounting each loop so that it is at an angle of 120° to the other two.

Figure 4.1a shows simple generator loops, with their ends (output terminals) labeled aa′, bb′, and cc′. The voltage between any pair of output terminals is an AC waveform. However, since the loops are displaced 120° on the shaft, the waveforms of the voltages produced are displaced 120° in time from each other (Figure 4.1b). In other words, $e_{bb'}$ leads $e_{cc'}$ by 120°, but lags $e_{aa'}$ = by 120°. The waveforms are labeled with lowercase e to indicate instantaneous voltages. We can call each of these loops a "phase." Note at this point that the three phases are not interconnected in any way and no voltage exists between them.

Wye Connection

Now suppose that points a′, b′, and c′ are connected, and leads a, b, and c are brought out of the machine (Figure 4.2a). As illustrated in Figure 9.3, the conductors being brought out would each go through a slip ring and a brush to form a

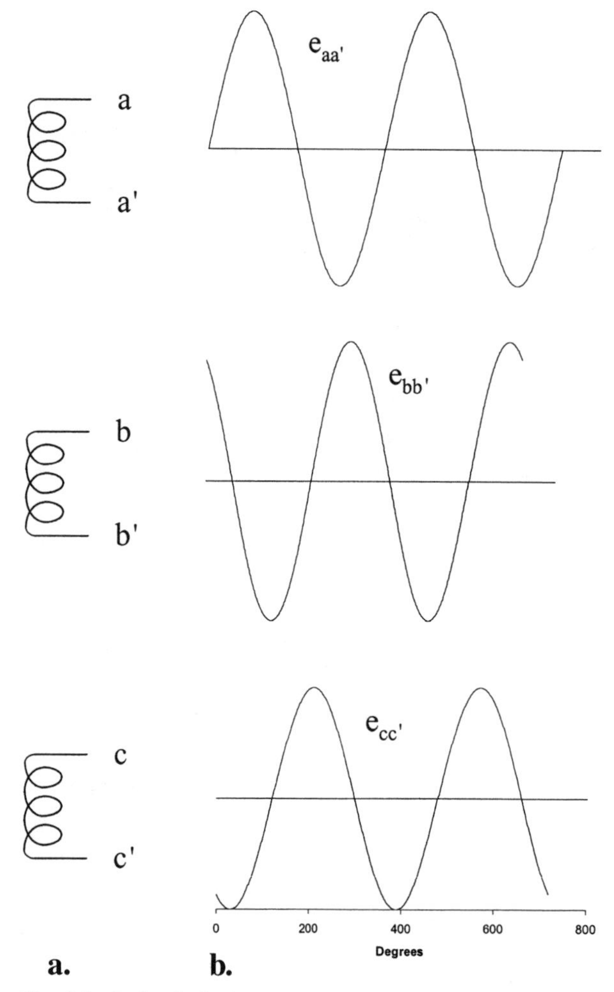

Fig. 4.1. A simple 3-phase generator.

conducting path to the rotating loop. If we call the common point n, the phasors are then arranged as shown in Figure 4.2b. The voltage phasors are seen to be mutually 120° out of phase. Assuming the standard counterclockwise rotation of phasors, and referencing on E_{na}, we see that E_{na} leads E_{nb} by 120°, and leads E_{nc} by 240°.

The circuit shown in Figure 4.2a can also be drawn in the configuration of a letter Y (Figure 4.3).

Voltages E_{na}, E_{nb}, and E_{nc} are called *phase voltages* because they are what would be measured across one phase winding of the generator. Assuming a balanced system where:

$E_{na} = E_{nb} = E_{nc} = E_p$ (the phase voltage), then, in polar notation,

$$\vec{E}_{na} = E_p \ \angle 0°$$
$$\vec{E}_{nb} = E_p \ \angle 240° \quad (\text{or } E_p \ \angle -120°)$$
$$\vec{E}_{nc} = E_p \ \angle 120° \quad (\text{or } E_p \ \angle -240°)$$

Also of interest are the voltages between leads a, b, and c. These are called *line voltages* since they are what would be measured between any two circuit conductors, or lines. It is necessary now to determine the relation between the phase and line voltages on a wye system. The line voltage, E_{ab}, between points a and b can be computed as follows:

$$\vec{E}_{ab} = \vec{E}_{an} + \vec{E}_{nb}$$

The subscript order of $\vec{E}_{an}$ assumes that the instantaneous polarity of the voltage has been reversed. That is,

$$\vec{E}_{an} = -\vec{E}_{na}$$

Figure 4.4 shows graphically how E_{ab} is computed. Figure 4.5 shows all phase and line voltages on a wye system.

Now we can numerically determine the relation between phase voltage (E_p) and line voltage (E_L) on a wye system. Assume:

$$E_p = E_{na} = E_{nb} = E_{nc} = 120 \text{ V}$$

Referring to Figure 4.4, let's compute the magnitude of $\vec{E}_{ab}$:

$$\vec{E}_{ab} = \vec{E}_{an} + \vec{E}_{nb}$$

Counting angles positive in the counterclockwise direction from reference;

$$\begin{aligned}
\vec{E}_{an} &= \qquad\qquad\qquad -120 + j0 \\
\vec{E}_{nb} &= 120 \cos 240 + j(120 \sin 240) \\
&= -60 - j103.92 \\
\overline{\quad} \\
E_{ab} &= -180 - j103.92 \\
&= 208\angle 210
\end{aligned}$$

therefore, $E_L = E_{ab} = 208$ V. E_L is the line voltage of the system.

The ratio of line to phase voltage is:

$$\frac{E_L}{E_P} = \frac{208}{120} = \sqrt{3}$$

thus, in a wye system,

$$E_L = E_p(\sqrt{3}) \qquad \textbf{(4.1)}$$

In Figure 4.5,

$$\vec{E}_{ab} = E_L \ \angle 210°$$
$$\vec{E}_{ca} = E_L \ \angle 330°$$
$$\vec{E}_{bc} = E_L \ \angle 90°$$

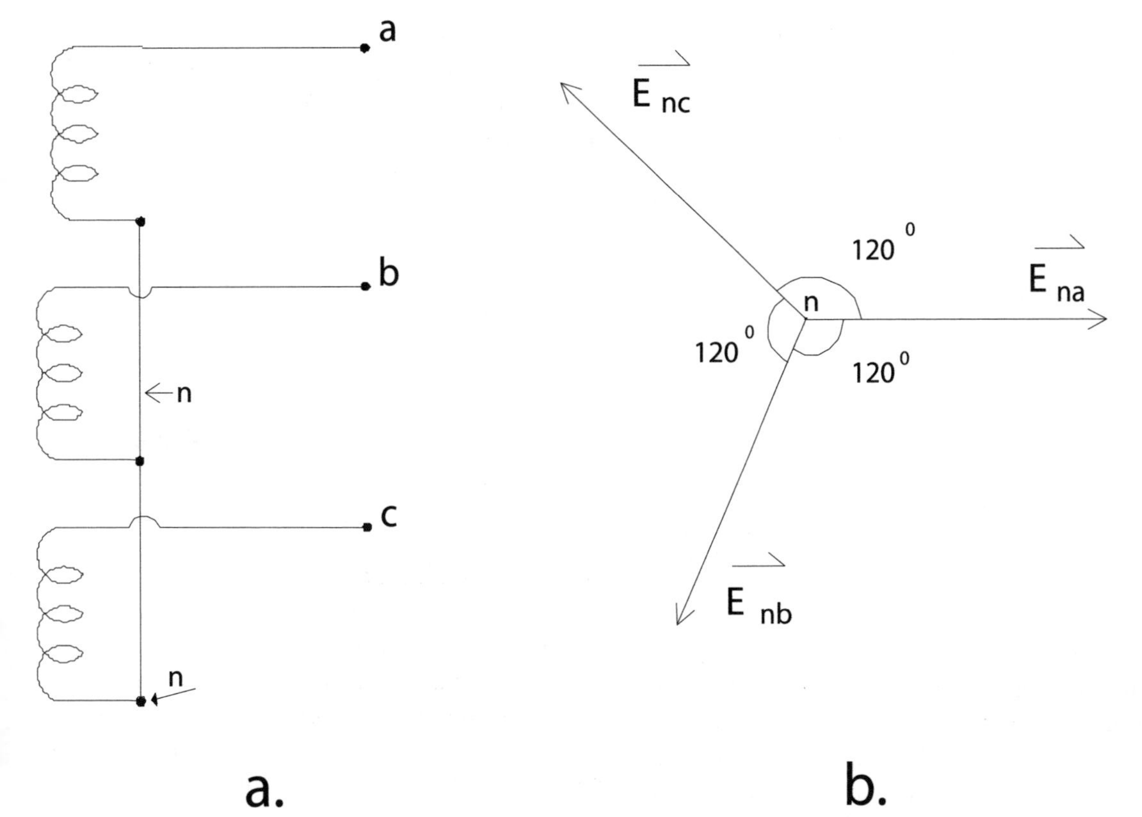

Fig. 4.2. Wye connection of windings.

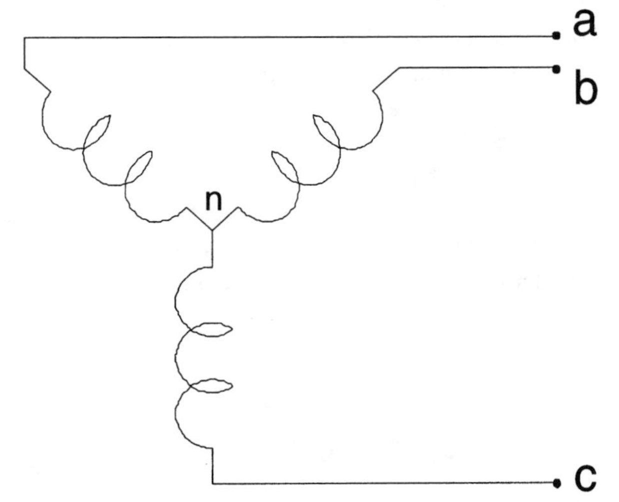

Fig. 4.3. Wye connection drawn to resemble the letter "Y."

If $\vec{E}_{na} = 120\angle 0°$, then $\vec{E}_{ab} = 208\angle 210°$, $\vec{E}_{ca} = 208\angle 330°$, and $\vec{E}_{bc} = 208\angle 90°$.

Common Wye Service

The type of service derived from a 3-phase distribution line is determined by the type of transformer used to step down distribution line voltage for a service connection. A common type of 3-phase service is the 120/208-V 3-phase wye service (Figure 4.6). The center of the wye is grounded, and a neutral conductor (n) brought along with the three energized conductors.

Loads can be connected to this service in several ways. A 3-phase wye load can be connected as illustrated to a line of the service and the neutral connected

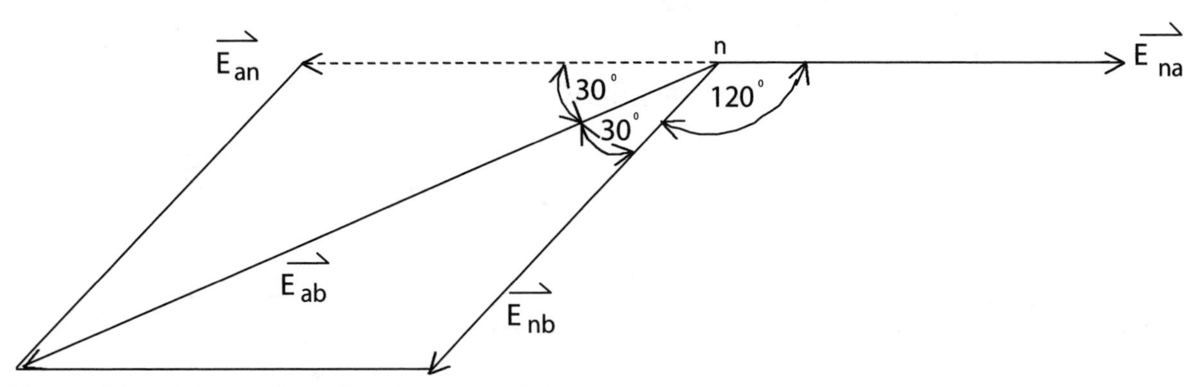

Fig. 4.4. Computation of a line voltage for a wye system.

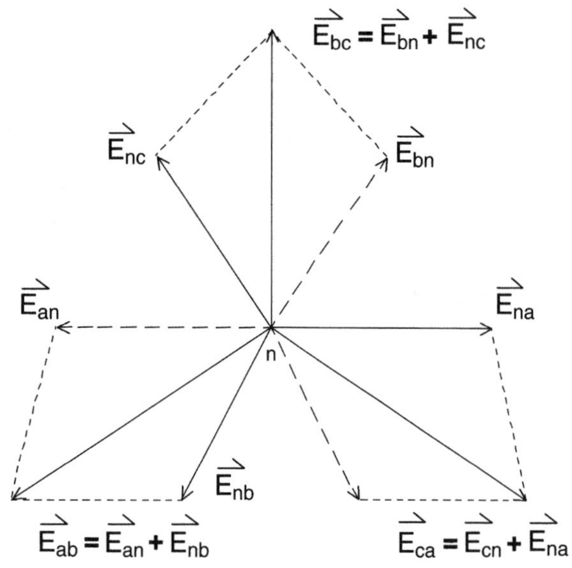

Fig. 4.5. Computation of wye line voltages.

to the center of the wye load. As will be seen later in this chapter, the neutral connection is not needed with a balanced wye load. Other types of loads that can be connected to the 120/208-V 3-phase wye service are illustrated in Figure 4.7. Note a neutral connection is not used for a 3-phase delta load or a 208-V single-phase load.

The neutral carries current only with the single-phase 120-V loads. A grounding conductor (not the neutral), originating at the wye center neutral of the service, may be extended to the frame of any of the load for equipment grounding.

Delta Connection

Suppose now the generator loops of Figure 4.1 are connected in series, as shown in Figure 4.8a. This circuit can be drawn to resemble a Greek letter delta (Δ) as shown in Figure 4.8b. For this reason it is called a delta connection. We now can drop the a′, b′, and c′ notations since these ends are connected to ends b, c, and a, respectively (Figure 4.8b).

By inspection of Figure 4.8, it can be seen that the voltage developed across a generator winding (a phase) also appears between two lines brought out from the delta. Thus, for a delta system,

$$E_L = E_p \quad \textbf{(4.2)}$$

The phase voltage is numerically equal to the line voltage. As in the case of the wye connection, the three voltages are mutually 120° out of phase.

Common Delta Service

A common type of delta service is 120/240-V 3-phase delta (Figure 4.9). As in the case of wye service, the configuration of voltages is determined by the way the secondary of the service transformer is connected. Available service voltages are shown in Figure 4.9. Figure 4.10 shows the types of loads which can be connected to this service, besides a delta 3-phase load.

The neutral is the center of one of the delta transformer windings. As with wye service, the neutral carries current only with the single-phase 120-V loads. A grounding conductor originating at the neutral may be extended to the frame of any of the loads.

Notice that no load is connected between point c and neutral. Phase c is, in the language of the electrician, the "wild leg." The voltage between c and n is 208 V, if the line voltage is 240 V. Voltages to neutral greater than 120 V cannot be used, so no load can be

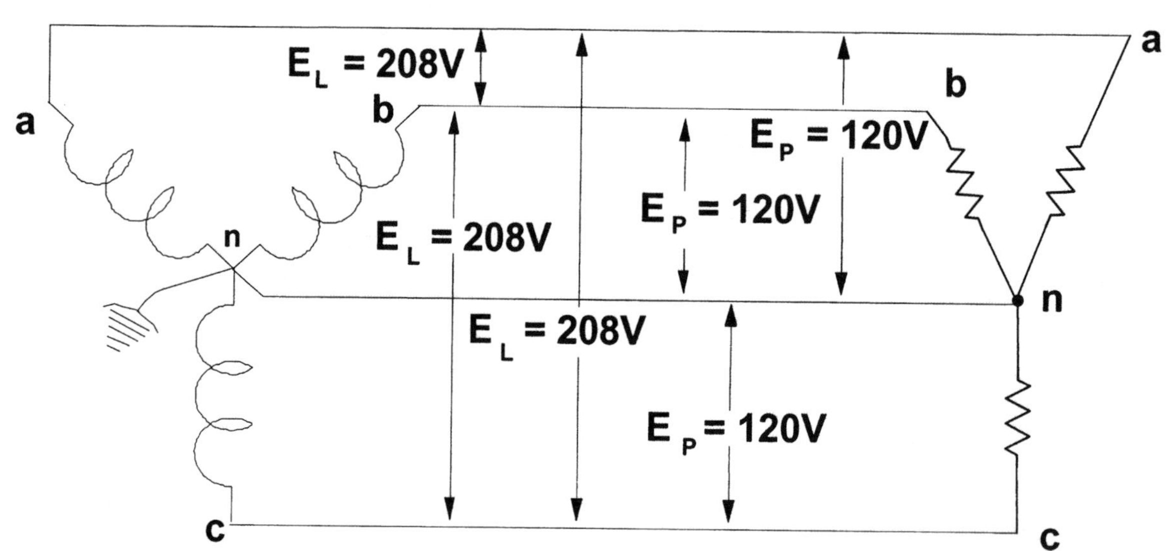

Fig. 4.6. 120/208-V 3-phase wye service, with 3-phase wye load connected.

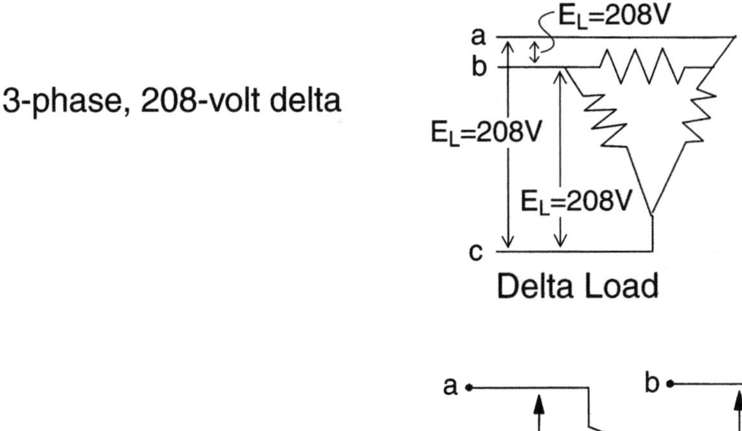

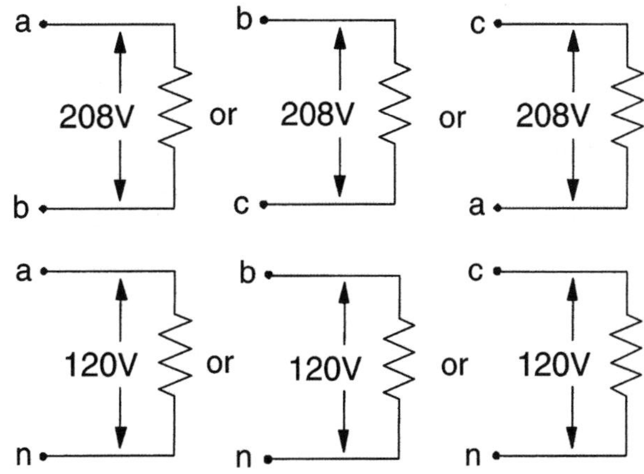

Fig. 4.7. Types of loads served from 120/208-V wye service, in addition to 3-phase wye load.

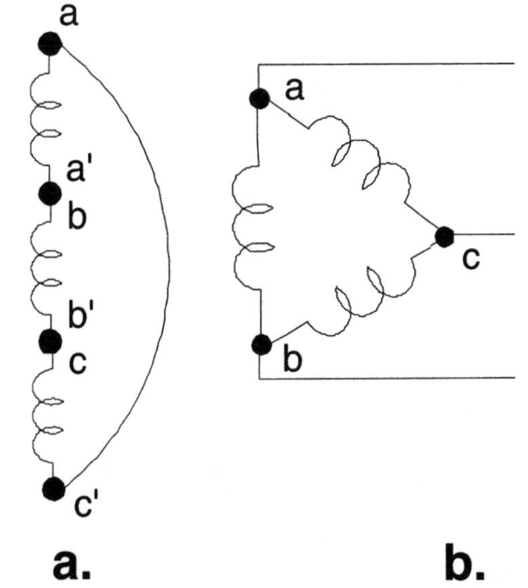

Fig. 4.8. Delta connection of generator loops.

connected between c and n. If this is done accidentally, and a 120-V load is energized to 208 V, the load will, most likely, burn out.

Three-Phase Loads

Three-phase load components can be connected in either a wye or a delta configuration. The type of service (wye or delta) does not dictate the load configuration. For example, a wye-connected load can be served from either a delta or a wye service. Transformer design determines a wye or delta service and line voltage of the service.

Balanced Wye Load

To illustrate power computations, a balanced, wye-connected 3-phase resistive load will be considered.

Example 4.1

Compute phase current, line current, individual load power, and total power for the balanced 3-phase load of Figure 4.11.

The current in any conductor between the transformer and load is a line current, designated as I_L. The current flowing through each branch of the wye

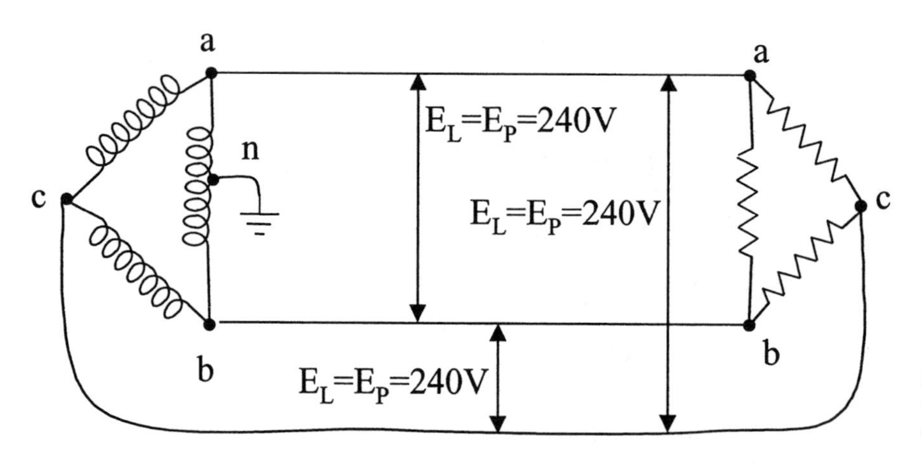

Fig. 4.9. 120/240-V 3-phase delta service, with data load connected.

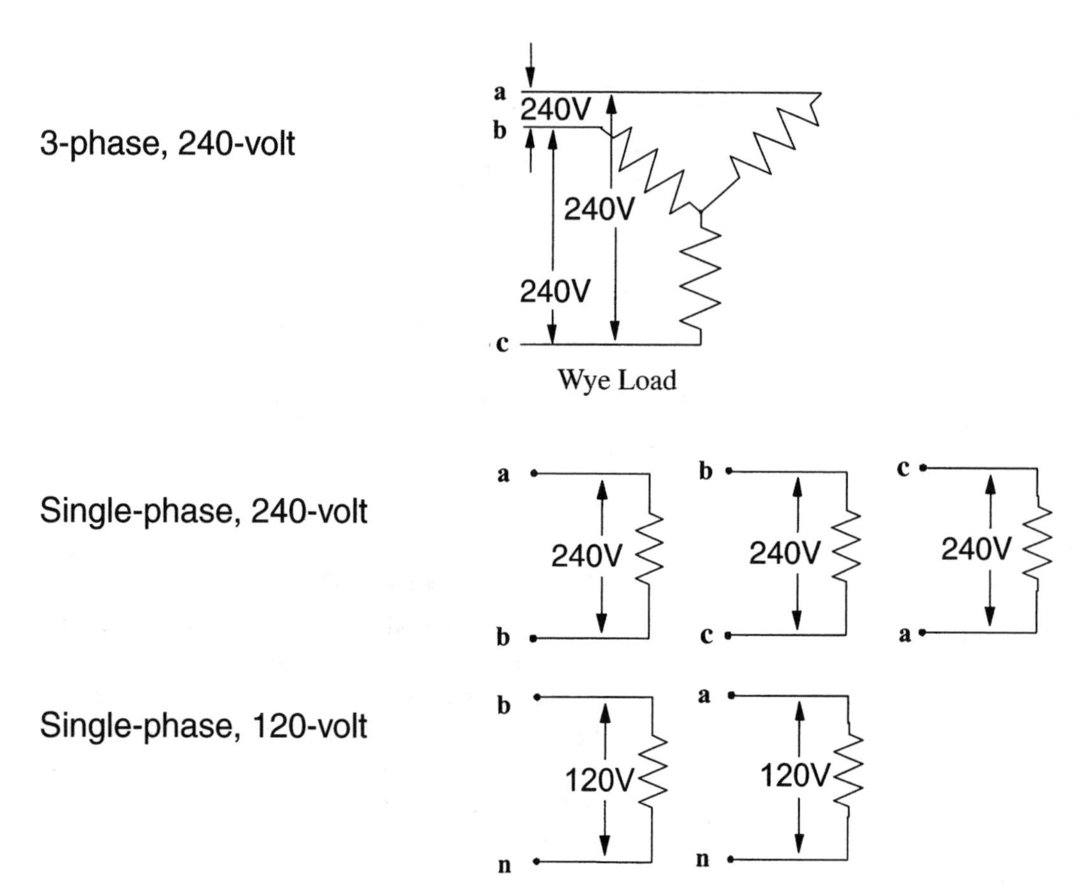

Fig. 4.10. Types of loads served from 120/240-V delta service.

load is a phase current, designated as I_p. Since there is no other connection to the conductor between the point where current is I_L and where it is I_p, the two must be equal. Thus, for a wye load,

$$I_p = I_L \tag{4.3}$$

With the wye load, it is convenient to look at one resistor at time (Figure 4.12). We can do this since there is no other connection to the conductor on the line side of the wye center.

The load of Figure 4.12 can be analyzed as a single-phase load:

$$I_{na} = \frac{120}{15} = 8 \text{ A} \quad \text{then,}$$

$$P_A = IE \cos \Theta = (8)(120)(1) = 960 \text{ W}$$

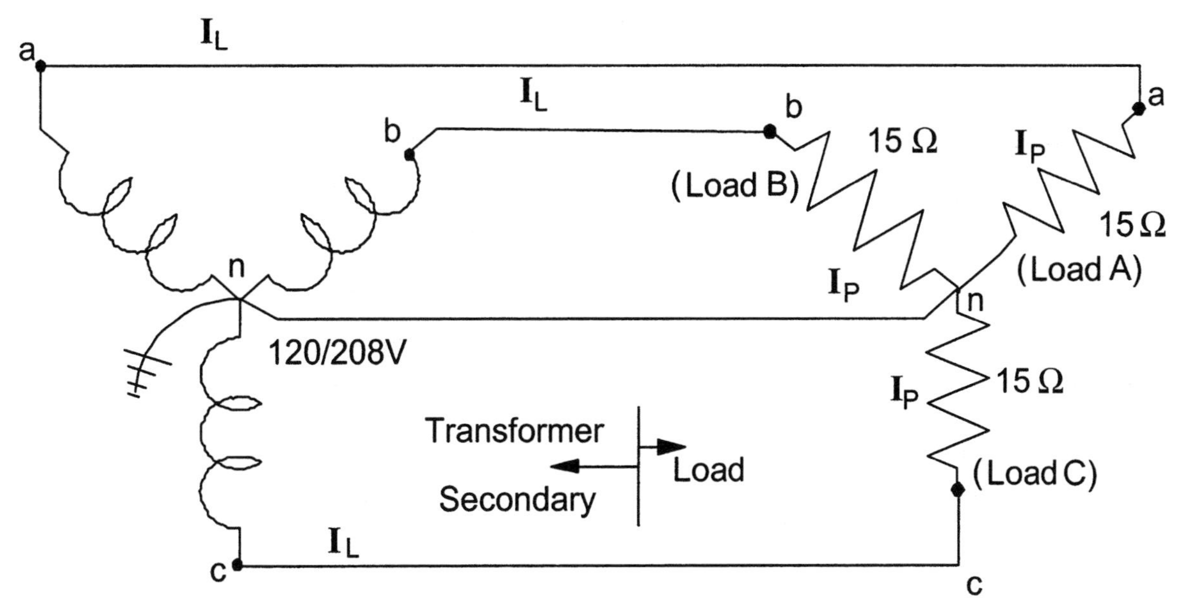

Fig. 4.11. Wye configured balanced resistive load.

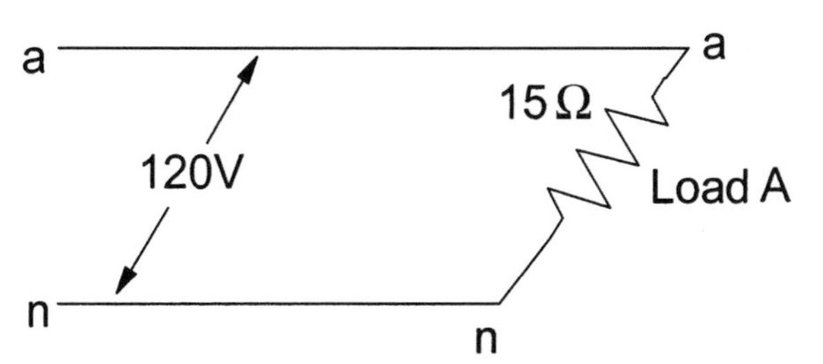

Fig. 4.12. Load A of the wye load of Figure 4.11.

Because the load of Figure 4.11 is composed of three identical resistors, each subjected to the same voltage, the total power is 3 times the power in load A:

$$3P_A = 3(960\ \text{W}) = 2{,}880\ \text{W}$$

Also, from the basic power equation:

$$P_{3\phi} = 3(I_P\, E_p \cos \Theta) = 3(8)(120)(1) = 2{,}880\ \text{W}$$

Generalized 3-phase equations are usually written in terms of line quantities, thus, we substitute for E_P and I_P using Equations 4.1 and 4.3:

$$P_{3\phi} = 3I_L \frac{E_L}{\sqrt{3}} \cos \Theta$$

or

$$P_{3\phi} = \sqrt{3}\,\cancel{\sqrt{3}}\ I_L\ \frac{E_L}{\cancel{\sqrt{3}}}\ \cos \Theta$$

or

$$P_{3\phi} = \sqrt{3}\ I_L\ E_L\ \cos \Theta \qquad \textbf{(4.4)}$$

The same power quantity results when this generalized power equation is used:

$$P_{3\phi} = \sqrt{3}\ (8)(208)(1) = 2{,}880\ \text{W}$$

Neutral Current in Balanced Wye Load

If the voltages of Figure 4.11 are designated in polar form, the current through the neutral can be computed:

$$\vec{E}_{na} = 120\angle 0°, \vec{E}_{nb} = 120\angle 240°, \vec{E}_{nc} = 120\angle 120°$$
$$Z_A = Z_B = Z_C = 15\angle 0°$$

The neutral current is the phasor sum of the current through the loads: $\vec{I}_n = \vec{I}_{na} + \vec{I}_{nb} + \vec{I}_{nc}$

$$\begin{aligned}
\vec{I}_{na} &= 8\angle 0° && = 8 + j0 \\
\vec{I}_{nb} &= 8\angle 240° = 8 \cos 240 + j8 \sin 240 \\
&\qquad = -4 && - j6.93 \\
\vec{I}_{nc} &= 8\angle 120° = 8 \cos 120 + j8 \sin 120 \\
&\qquad = -4 && + j6.93 \\
\overline{\vec{I}_n} &= && \overline{0 + j0}
\end{aligned}$$

The current to the center of a balanced wye load (in this case also the neutral) is seen to be zero. This conductor can, thus, be eliminated without affecting load operation.

Balanced Delta Load

Example 4.2

Find the phase currents, line currents, and total power of a delta load consisting of 15-Ω resistors forming a delta load and connected to a 120/208-V 3-phase service (Figure 4.13). Examining the circuit, these points are noted:

- The neutral is not extended to any point on the delta load. It may be connected to the frame of the load for equipment grounding.
- I_L may not be equal to I_p within the load since there are connection points at the corners of the delta.
- Each of the three resistors is subjected to E_L, which is 208 V in this case.

Looking at one resistor (Figure 4.14), we can compute I_p:

$$I_p = \frac{208}{15} = 13.867 \text{ A}$$

This is the current through the resistor. Note that we cannot say it is I_L since there is another connected conductor at each end of the load. For this load:

$$P = I_P E_L \cos\theta = (13.867)(208)(1) = 2{,}884.26 \text{ W}$$

Since there are 3 identical resistors, each with the same voltage across,

$$P_{3\phi} = 3\,(2884.26) = 8{,}652.78 \text{ W}$$

If the currents at any corner of the delta load (Figure 4.13) are added as phasors, it can be shown that:

$$I_L = I_P \sqrt{3} \quad \textbf{(4.5)}$$

Three-Phase Balanced-Load Equations

We now have the complete set of important 3-phase, balanced-load equations:

For wye loads: $I_L = I_p$ (4.3)
$E_L = E_p\sqrt{3}$ (4.1)

For delta loads: $I_L = I_p\sqrt{3}$ (4.5)
$E_L = E_p$ (4.2)

For wye or delta-balanced loads:
$P = \sqrt{3}\, I_L E_L \cos\Theta$ (4.4)

An example will illustrate use of these equations.

Example 4.3

A balanced resistive wye load is connected to a 3-phase source:

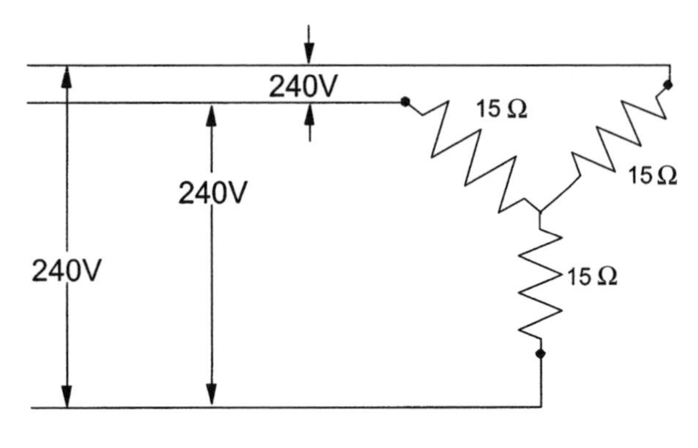

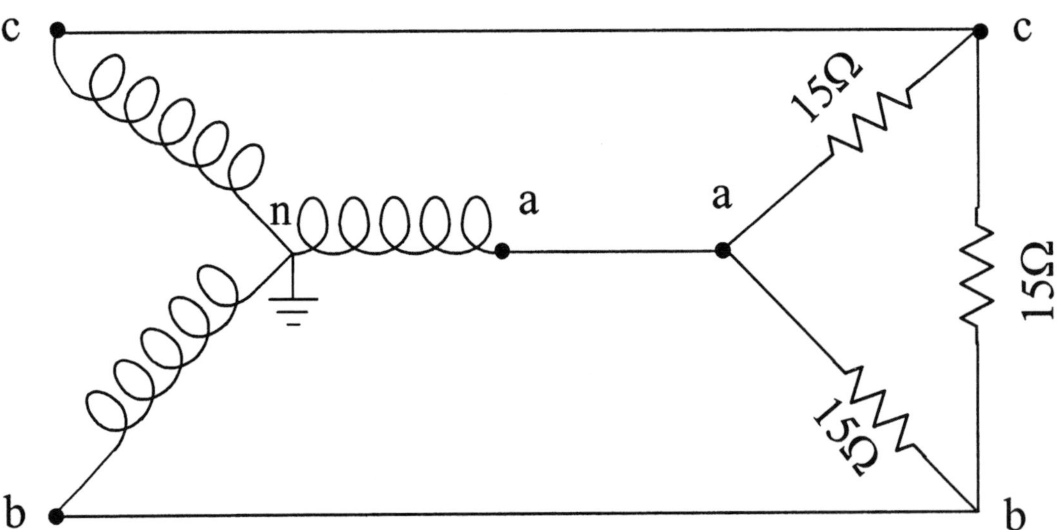

Fig. 4.13. Delta-configured balanced resistive load.

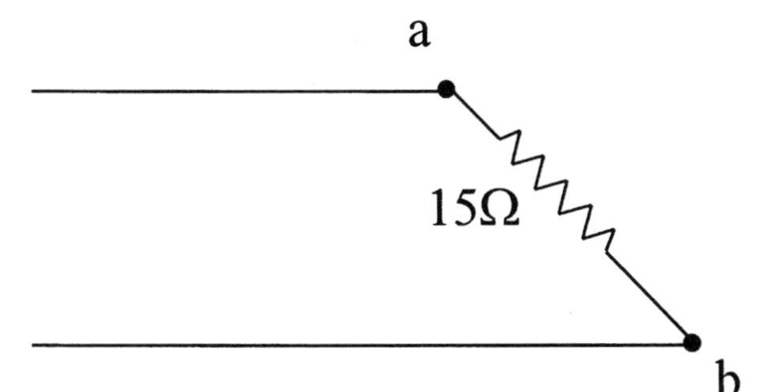

Fig. 4.14. Load A of the delta load of Figure 4.13.

a. What is I_L?

Each resistor sees $E_p = \frac{E_L}{\sqrt{3}} = 138.56$ V

$$E_p = \frac{240}{\sqrt{3}} = 138.56 \text{ V}$$

$$I_L = I_p = \frac{138.56}{15} = 9.238 \text{ A}$$

b. What power is used by each resistor?

$$P = I_p E_p \cos \Theta = (9.238)(138.56)(1) = 1{,}280 \text{ W}$$

c. What power is used by the whole wye load?

$$P_T = 3(1280) = 3{,}840 \text{ W}$$

or

$$P_T = \sqrt{3}\, I_L E_L \cos \Theta = \sqrt{3}\,(9.238)(240)(1) = 3{,}840 \text{ W}$$

d. What power would these 3 resistors draw if connected to the same source in a delta configuration?

$$I_p = \frac{240}{15} = 16 \text{ A}$$

$$I_L = \sqrt{3}\, I_p = 27.71 \text{ A}$$

$$P_T = \sqrt{3}\, I_L E_L \cos \Theta = \sqrt{3}\,(27.71)(240)(1) = 11{,}519 \text{ W}$$

e. What size resistors would be required to draw 3,840 W when delta-connected?

$$P = \frac{3840}{3} = 1{,}280 \text{ W}$$

$$I_p = \frac{1280}{240} = 5.333 \text{ A}$$

$$1{,}280 = I_p^2(R) = (5.333)^2 R$$

$$R = 45\ \Omega$$

f. What is the ratio of wye load resistance to delta load resistance for equivalent power?

$$\frac{R_D}{R_Y} = \frac{45}{15} = 3$$

Balanced Wye Load with Power Factor < 1

An example will illustrate computation on a system where the power factor is less than unity.

Example 4.4

A 3-phase wye-wound motor is connected to a supply with $E_L = 230$ V. The motor is modeled as an RL 3-phase circuit:

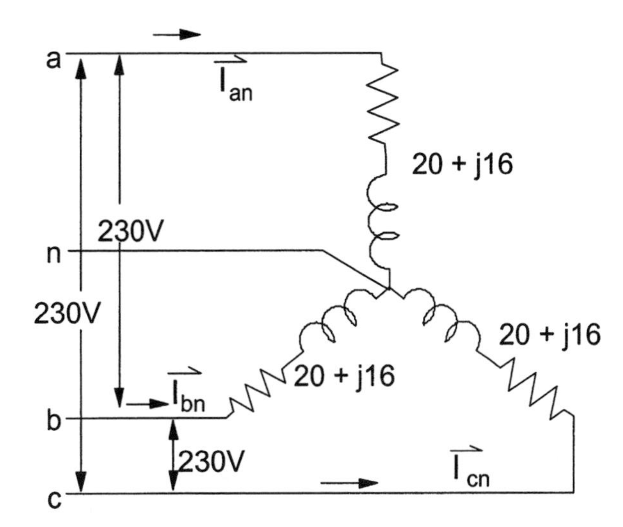

a. Compute the current in each conductor and draw a phasor diagram.

$$I_L = I_p = \frac{E_p}{Z} \qquad Z_a = Z_b = Z_c = Z$$

$$= 20 + j16 = 25.61\angle 38.66°$$

$$E_p = \frac{230}{\sqrt{3}} = 132.79 \text{ V}$$

Using the phase sequence of Figure 4.2,

$$\vec{I}_{an} = \frac{E_{na}}{Z} = \frac{132.79\angle 0°}{25.61\angle 38.66°} = 5.185\angle -38.66°$$

Likewise, $\vec{I}_{bn} = \dfrac{E_{nb}}{Z} = \dfrac{132.79\angle 240^\circ}{25.61\angle 38.66^\circ}$
$= 5.185\angle 201.34^\circ$

$\vec{I}_{cn} = \dfrac{E_{nc}}{Z} = \dfrac{132.79\angle 120^\circ}{25.61\angle 38.66^\circ}$
$= 51.85\angle 81.34^\circ$

Each phase current lags the corresponding phase voltage by 38.66°:

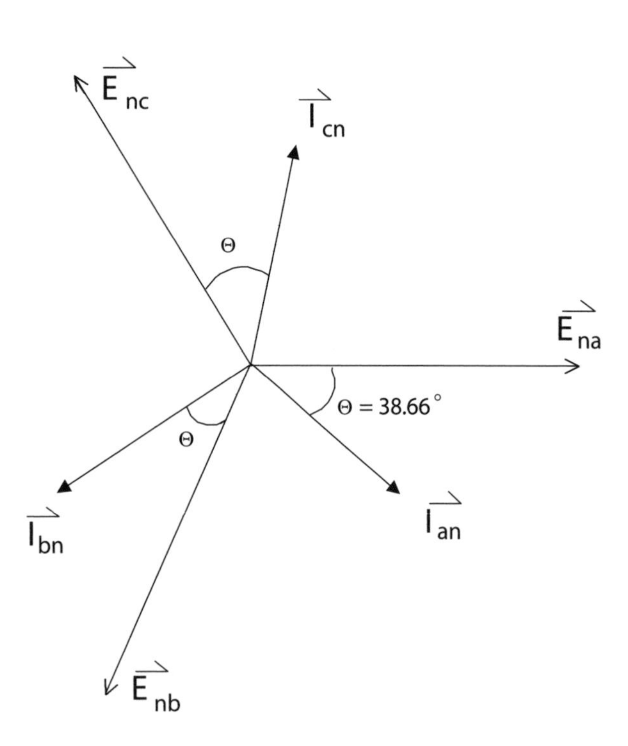

b. Compute the power used by the motor. Since the load is balanced, the generalized 3-phase power equation may be used:

$$P_T = \sqrt{3}\, I_L E_L \cos\Theta = \sqrt{3}\,(5.185)(230)(\cos 38.66)$$
$$= 1{,}612.9 \text{ W}$$

Note that the power factor angle used is the angle between the voltage across the RL load and the current through this load. It is not, in this case, the angle between I_L and E_L. It is left to the student to show that the sum of the currents to the load is zero and that the neutral carries no current.

Unbalanced Wye Load with Power Factor < 1

When the 3-phase wye load is unbalanced, computations are similar, except that the line currents are not equal and the neutral current is not zero. Again, an example will illustrate the procedure.

Example 4.5

A 4-wire 3-phase 120/208-V source is connected to the load shown.

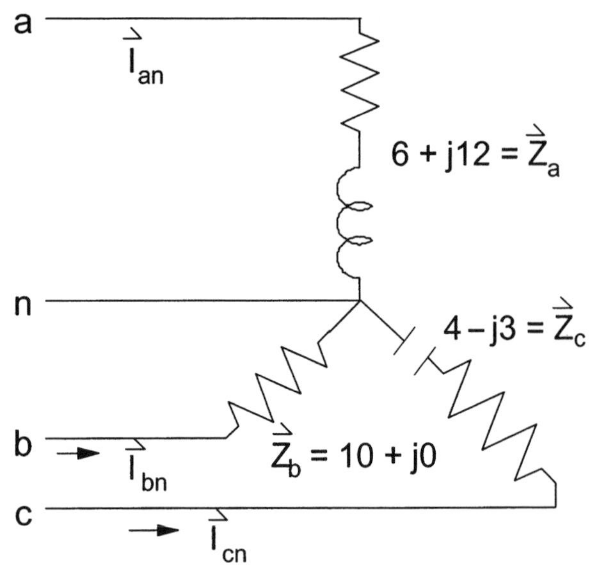

a. Compute line currents and neutral current.

$E_p = 120$ V

$\vec{E}_{na} = 120\angle 0^\circ, \vec{E}_{nb} = 120\angle 240^\circ,\ \vec{E}_{nc} = 120\angle 120^\circ$

$\vec{Z}_a = 13.42\angle 63.4^\circ, \vec{Z}_b = 10\angle 0^\circ,\ \vec{Z}_c = 5\angle -36.9^\circ$

$$\vec{I}_{an} = \frac{120\angle 0^\circ}{13.42\angle 63.4^\circ} = 8.94\angle -63.4^\circ = 4 - j7.99$$

$$\vec{I}_{bn} = \frac{120\angle 240^\circ}{10\angle 0^\circ} = 12\angle 240^\circ = -6 - j10.39$$

$$\vec{I}_{cn} = \frac{120\angle 120^\circ}{5\angle -36.9^\circ} = 24\angle 156.9^\circ = 22.0 + j9.42$$

$$\vec{I} = 20 = j8.96 = 21.9\angle -24.1^\circ$$

Shifting the phasor 180° gives I_n:

$$\vec{I}_n = 21.9\angle 165.9^\circ$$

b. Show currents and phase voltages on a phasor diagram.

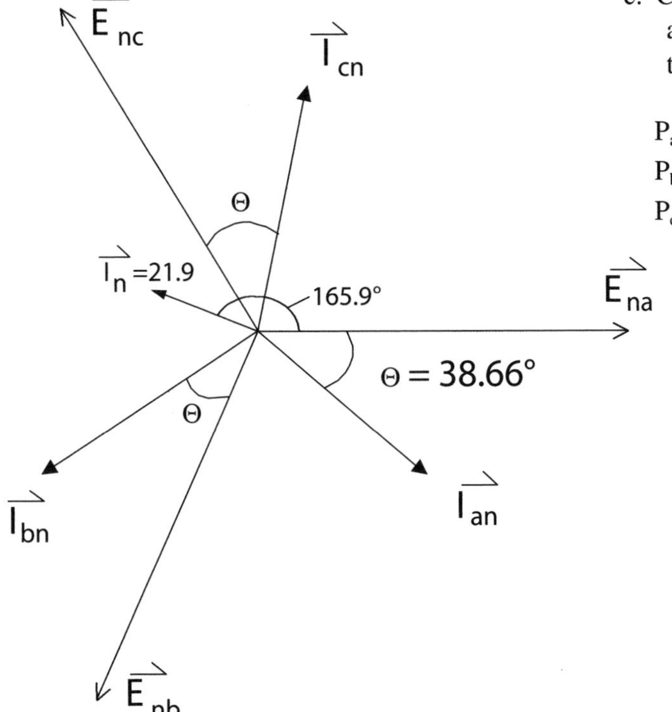

c. Compute power use. Since the load is unbalanced, we must sum the power to each of the three legs.

$$P_a = (120)(8.94) \cos 63.4° = 480 \text{ W}$$
$$P_b = (120)(12) \cos 0° = 1{,}440 \text{ W}$$
$$P_c = (120)(24) \cos 36.9° = 2{,}303 \text{ W}$$
$$P_T = 4{,}223 \text{ W}$$

Problems

4.1. Compute $\vec{E}_{na} + \vec{E}_{nb} + \vec{E}_{nc}$ for the wye voltages of windings shown in Figure 4.2b. Assume all three have an equal magnitude of 120 V.

4.2. A delta-connected balanced resistive load draws 40 kW when connected to a 240-V (line-to-line) 3-phase service.

(a) Compute I_L.

(b) Compute I_p.

(c) Compute resistor resistance.

4.3.

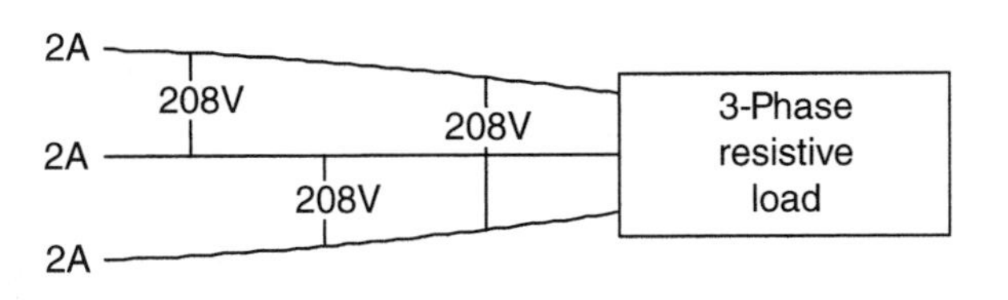

(a) What is the load power?

(b) What is the resistance of each resistor if the load is wye-connected?

(c) What is the resistance of each resistor if the load is delta-connected?

4.4. One of the hot wires to a 3-phase, 3-wire resistive heater is disconnected. What fraction of the original power will continue to be used if:

(a) The heater is wye-connected?

(b) The heater is delta-connected?

4.5. A 3-phase delta-connected load is composed of three 15-A, 3.6-kW resistive heaters, each having a resistance of R Ω . Compute I_P, I_L, E_P, E_L, and R.

4.6. A 3-phase wye-connected load is composed of three 30-A, 3.6-kW resistive heaters, each having a resistance of R Ω. Compute I_p, I_L, E_P, E_L, and R.

4.7.

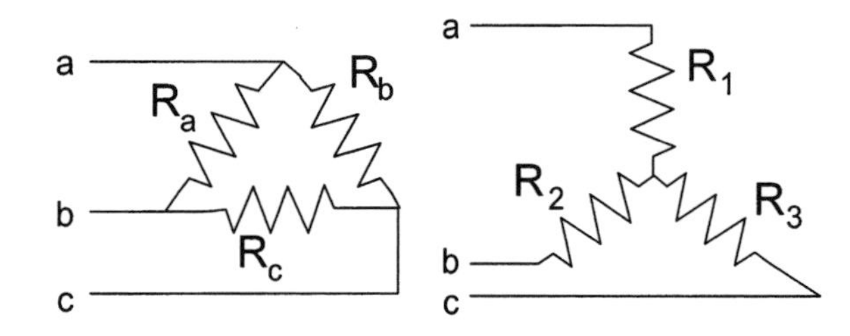

Resistors R_A, R_B, and R_C (not necessarily equal) connected in a delta are to be replaced by resistors R_1, R_2, and R_3 connected in a wye. The two circuits are to be equivalent when connected to a 3-phase source. Solve for R_1, R_2, and R_3, in terms of R_A, R_B, and R_C. *Hint:* Look at two terminals at a time, with the third unhooked.

4.8. A 3-phase, wye-wound motor draws 7 kW and 25.91 A when connected to a 120/208-V, 3-phase service. Assume the load is balanced. Model this motor as a series RL circuit and compute values for R and X_L.

4.9. Power is delivered to a load by a 3-phase 4-wire system with $E_L = 230$ V. The load is wye connected and consists of

$$Z_a = 30 + j0, Z_b = 20 + j20 \text{ and } Z_c = 20 - j20.$$

Compute I_a, I_b, I_c, and I_n and express in rectangular notation. Compute power in watts. Assume the phase sequence of Figure 4.2.

4.10. A farm is connected to a 4-wire 120/208-V 3-phase service. The loads (all wye-connected) are:

Phase A:	10 kW, power factor = 0.9 (lagging)
Phase B:	12 kW, power factor = 0.6 (lagging)
Phase C:	4 kW, power factor = 0.75 (lagging)

Compute each of the line currents, and the neutral current. Assume the phase sequence of Figure 4.2. Sketch a phasor diagram showing the phase voltages and all of the phase current.

4.11. Compute all ammeter readings, and power use for entire load.

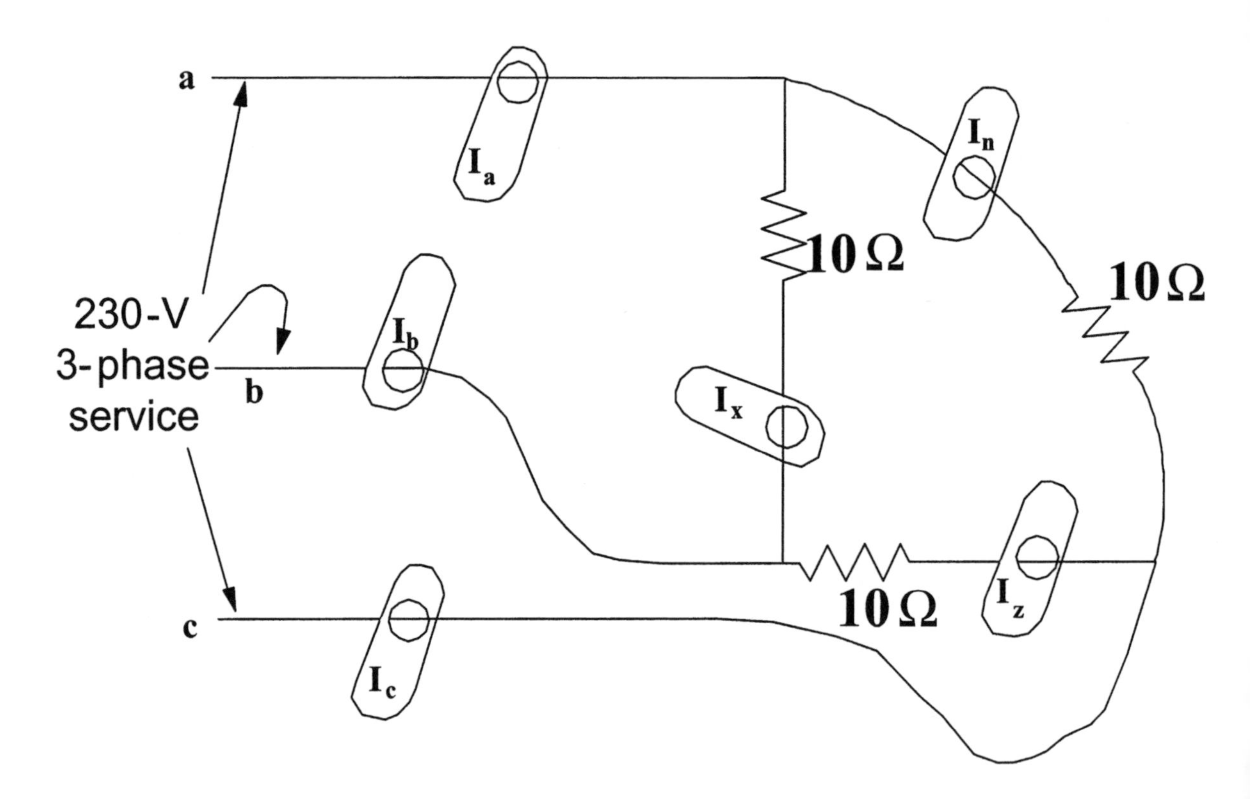

4.12. Assume a single-phase (120/208-V) subpanel is installed in a barn from a 120/208-V 3-phase wye service available on the farm. In chapter 1, it was said that 120-V loads should be balanced on a 3-wire service, resulting in zero current in the neutral wire. For the following circuit, calculate the current in the neutral wire. *Hint:* It is *not* zero.

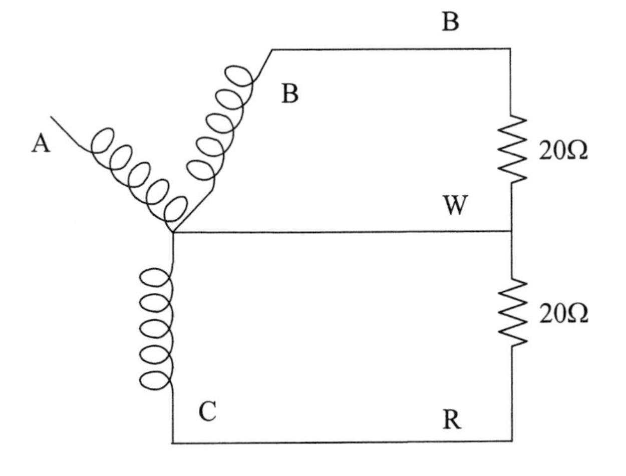

4.13.

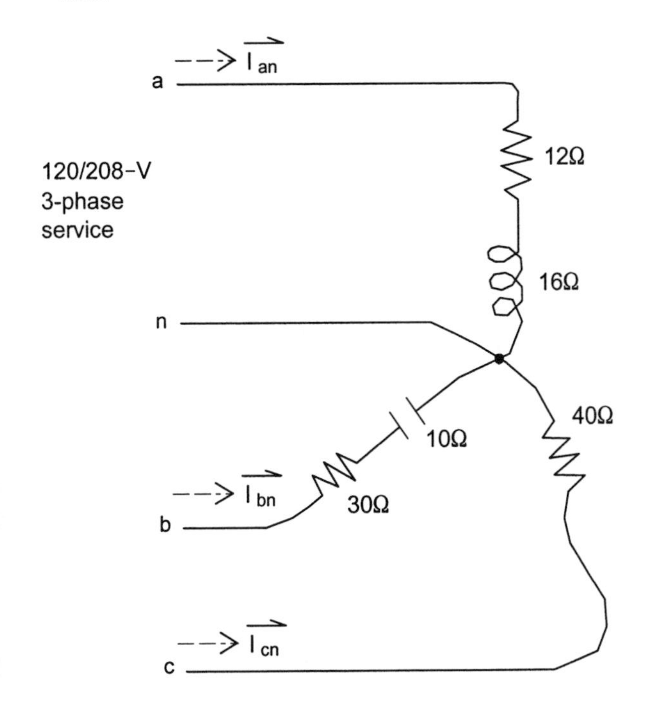

Assume the phase sequence of Figure 4.2. Compute neutral current and express in polar notation. Sketch a phasor diagram showing phase voltages and all currents.

4.14.

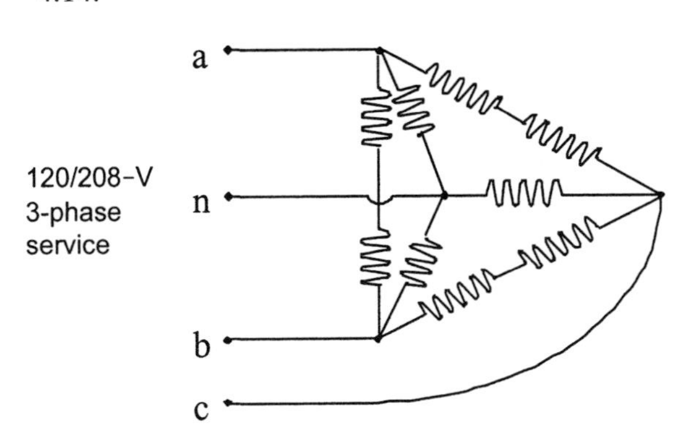

All resistors are 10 Ω.

Compute: Line current.
Total power use.

Chapter 5

CONDUCTOR DESIGN PRINCIPLES

Introduction

In chapter 1 we saw that electrons will flow through a piece of copper if electrons are scarce at one end and abundant at the other. Circuit analysis described in chapter 1 assumed that conductors carrying these electrons between the power source and load had no resistance. This assumption was appropriate for circuit analysis, but conductor resistance and other factors must be considered for circuit design. This chapter describes how to design real-world power circuits.

Conducting Materials

Conductors made of materials having atoms with loosely held outer-orbit electrons exhibit the least resistance to electron flow and are called *good conductors.* Most metals are in this category, being characterized by only one or two electrons in their outermost orbits.

The resistance of a length of conductor depends on the physical properties of the material, the length, and the cross-sectional area of the conductor:

$$R = \frac{\rho L}{A} \qquad \textbf{(5.1)}$$

where R = resistance of the conductor, ohms
ρ = resistivity of material, ohm-circular mil/ft
L = length of the conductor, ft
A = cross-section area of the conductor, circular mils or cmils

The circular mil area unit is defined as follows:

$$A = \text{cross-sectional area in cmils} = (\text{diameter in mils})^2$$
$$(1 \text{ mil} = 0.001 \text{ in.}) \qquad \textbf{(5.2)}$$

Notice that the higher the resistivity, the higher will be the resistance of a conductor having a given diameter and length. Resistivities of some common materials are listed in Table 5.1. When material cost, resistivity, and other properties are considered, copper and aluminum turn out to be the best metals for most conductor applications. Under special circumstances, all of these materials have been used as conductors.

TABLE 5.1 Resistivities of common conducting materials at 20°C

Material	Resistivity, ρ $\frac{\text{ohm cmil}}{\text{ft}}$
Silver	9.55
Copper	10.37
Gold	14.7
Aluminum	17.01
Tungsten	33.10
Platinum	66.9
Steel	100
Lead	129
Cast iron	360
Mercury	577

Example 5.1

Calculate the resistance of an aluminum rod 50 ft long and 0.25 in. in diameter.

$$(0.25\ \text{in.})\frac{1\ \text{mil}}{.001\ \text{in.}} = 250\ \text{mil}$$

$$A = (250\ \text{mil})^2 = 62{,}500\ \text{cmil}$$
$$= 62.5\ \text{kcmil, or thousands of circular mils}$$

The cross-sectional area of the rod is 62,500 cmil. Substituting into Equation 5.1, and using the resistivity of aluminum from Table 5.1, we get:

$$R = \frac{\rho L}{A} = \frac{17.01\ (\Omega\text{-cmil})}{\text{ft}} \frac{50\ \text{ft}}{62{,}500\ \text{cmil}}$$
$$= 0.0136\ \Omega$$

In the future, conductor designs may take advantage of superconductivity (decreased metal resistivity at low temperatures). The resistivity of some materials nearly reaches zero as their temperatures decrease to near absolute zero (−460°F). In the United States, Europe, and Japan, companies are racing to commercialize superconducting materials. One product will be a conductor capable of carrying over 100 times more current than conventional copper and aluminum conductors (American Superconductor 2001).

Insulating Materials

In addition to providing a low-resistance path for electron flow, a good conductor design must also employ a high-resistance covering or barrier to prevent electric current from leaving its intended path. The high resistance of this insulating material also comes about because of its atomic structure. In this case, the outer orbits are nearly full, and electrons may be drawn out of orbit only with difficulty.

The resistivity of insulating materials is very much greater than that of conductors. Common values are on the order of 10^7 Ω cmil/ft. The quality of an insulating material, however, is usually not measured by its resistivity since this property varies greatly from sample to sample, and also varies according to what voltage is impressed across the material.

Insulating materials can be compared by their dielectric strength (Table 5.2). This is the voltage per one-thousandth inch of thickness at which their resistance suddenly reaches a breakdown point and they become conductors.

Example 5.2 illustrates the meaning of dielectric strength.

TABLE 5.2 Dielectric strengths of some common insulating materials

Material	Dielectric strength, volts/mil
Air	80
Bakelite	500
Glass	200
Mica	2,000
Paraffined paper	1,200
Porcelain	750
Poly-vinyl chloride (PVC)	375

Example 5.2

A piece of bakelite 0.012-in. thick separates two conductors in a switch. Voltage between the two conductors is 240 V. Will the bakelite conduct current? From Table 5.2, the dielectric strength of bakelite is 500 V/mil.

$$\frac{500\ \text{V}}{\text{mil}}\ \frac{\text{mil}}{0.001\ \text{in.}}\ 0.012\ \text{in.} = 6{,}000\ \text{V}$$

If the voltage difference between the two conductors reaches about 6,000 volts, the bakelite separator will break down and become a conductor. Since 6,000 is 25 times the 240-V working voltage, the bakelite should effectively insulate the conductors from each other.

Conductor Size Designation

Conductor sizes in the United States are specified by an American Wire Gage (AWG) or kcmil size designation (column 1, Table 5.3). AWG gage num-

TABLE 5.3 Properties of conductors

Size	Area, cmil	Usual Number of Strands	Diameter each Strand, in.	Copper		Aluminum	
				Weight lb/1000 ft	Resistance* ohms/1000 ft	Weight lb/1000 ft	Resistance* ohms/1000 ft
AWG**							
14	4110	1	0.064	12.5	3.07	3.78	5.06
12	6530	1	0.081	19.8	1.93	6.01	3.18
10	10380	1	0.102	31.43	1.21	9.556	2.00
8	16510	1	0.128	49.98	0.764	15.20	1.26
6	26240	7	0.061	79.44	0.491	24.15	0.808
4	41740	7	0.077	126.3	0.308	38.41	0.508
3	52620	7	0.087	159.3	0.245	48.43	0.403
2	66360	7	0.097	205	0.194	62.3	0.319
1	83690	19	0.066	259	0.154	78.6	0.253
0	105600	19	0.074	326	0.122	99.1	0.201
00	133100	19	0.084	411	0.0967	125	0.159
000	167800	19	0.094	518	0.0766	157	0.126
0000	211600	19	0.106	653	0.0608	199	0.100
kcmil***							
250	250000	37	0.082	772	0.0515	235	0.0847
300	300000	37	0.090	925	0.0429	282	0.0707
350	350000	37	0.097	1080	0.0367	328	0.0605
400	400000	37	0.104	1236	0.0321	375	0.0529
500	500000	37	0.116	1542	0.0258	469	0.0424
600	600000	61	0.099	1850	0.0214	563	0.0353
700	700000	61	0.107	2160	0.0184	657	0.0303
750	750000	61	0.111	2316	0.0171	704	0.0282
800	800000	61	0.114	2469	0.0161	751	0.0265
900	900000	61	0.122	2780	0.0143	845	0.0235
1000	1000000	61	0.128	3086	0.0129	938	0.0212

* DC resistance at 75°C.
** American Wire Gage numerical designation.
*** kcmil = thousands of circular mils.

bers run from 40 to 0000. These gage numbers are like names and have no relation to the wire size. The AWG gages commonly used in electrical wiring are listed (14 through 0000). The corresponding diameters are listed in column 4. The AWG system applies only to non-ferrous metals.

The cross-sectional area in circular mils is listed for each wire size in column 2. Notice that for sizes larger than 0000, the size listed is the cross-sectional area in kcmil (thousands of circular mils). Conductors larger than AWG-8 are usually made up of several strands, rather than one solid cylinder, so that the conductor can be easily and repeatedly bent without breaking. The number of wires (strands) is listed in column 3 of Table 5.3. When the conductor is stranded, the area listed is the sum of the areas of the strands. Conductors smaller than AWG-6 may also be available in stranded form.

Example 5.3

What is the area of a conductor made up of 37 strands of 0.0900-in.-diameter wire?

$$0.0900\ \cancel{\text{in.}}\ \frac{\text{mil}}{0.001\ \cancel{\text{in.}}} = 90\ \text{mil}$$

$$A = (90\ \text{mil})^2 = 8100\ \text{cmil}$$

This is the area of each strand.

$$(8100)\frac{\text{cmil}}{\cancel{\text{strand}}}(37\ \cancel{\text{strand}}) = 299{,}700\ \text{cmil}$$

$$\cong 300{,}000\ \text{cmil} = 300\ \text{kcmil}$$

Circuit Conductor Design

In choosing conductors for specific applications, it is necessary to consider three factors:

- Environment in which the conductor will be placed.
- Amperage capacity (ampacity) of the conductor.
- Voltage drop.

Environment determines the type of insulation needed. Conductor size might be determined by any

TABLE 5.4 Insulated conductor and cable types commonly used in agriculture

Insulation Designation	Trade Name	Temperature, Rating	Description and Common Uses
THHN	Heat-resistant thermoplastic	90°C	Flame-retardant, heat-resistant thermoplastic insulated individual conductors. For use in conduit, dry, and damp locations. Available in several colors.
THWN	Moisture and heat-resistant thermoplastic	75°C	Flame-retardant, moisture- and heat-resistant, thermoplastic insulated individual conductors. For use in conduit, dry, and wet locations. Available in several colors.
NM*	Non-metallic sheathed cable	60°C	Two or three conductors (plus bare grounding conductor) in a moisture-resistant, flame-retardant non-metallic sheath. For use in normally dry locations. Cannot be imbedded in poured concrete or used as service entrance cable. Use in family dwellings not exceeding 3 floors above grade and other structures. Can be used exposed or concealed.
SE	Service entrance cable	75°C	Commonly 3 conductors in a flame-retardant, moisture-resistant covering. The neutral is braided around the two energized conductors. Type SE is used primarily between an above-ground point of attachment and the service entrance panel.
USE	Underground service entrance cable	75°C	Single conductors cabled into an uncovered assembly for direct burial as a feeder or branch circuit or service lateral. Covering is moisture resistant but not necessarily flame retardant, or protective against mechanical abuse.
UF*	Underground feeder cable	60°C	Two or three conductors (plus bare grounding conductor) with a flame-retardant moisture-, fungus-, corrosion-resistant covering for direct burial as a feeder or branch circuit. Also used for interior wiring in wet, dry, or corrosive locations. Use in livestock buildings. Cannot be exposed to direct sunlight unless label specifies "Sunlight Resistant." Cannot be used as service entrance cable. Cannot be embedded in poured concrete.
Multiplex (triplex, quadruplex)	Overhead feeder	90°C	Two or three insulated aluminum conductors wound around a bare stranded messenger which serves as a neutral, and supports the assembly. The messenger contains one steel strand for strength. For use as overhead feeders. Conductors are usually XHHW (Surbrook and Mullin 1985).

*NM and UF may be marked as NM-B and UF-B. This marking means the conductors within the cable are rated at 90°C. For the purpose of ampacity, the temperature rating of the cable remains 60°C.

one of the three factors. Each will be discussed, then a conductor design example will be presented.

Environment

The environment determines what kind of insulation the conductor needs. This is specified by the conductor or cable type designation. In most cases, this is printed on the insulation. Common conductor types, along with their uses and temperature ratings, are listed in Table 5.4. These are the types most commonly used in agricultural applications. Many more are available for particular applications.

It is possible that the environment will dictate conductor size for overhead spans. The National Electrical Code (NFPA 2002) rules are simple: Conductors for overhead spans shall not be smaller than AWG-10 copper or AWG-8 aluminum for spans up to 50 feet, and not smaller than AWG-8 copper or AWG-6 aluminum for longer spans unless supported by a messenger wire (NEC 225-6). Tensile strength limitation is the reason for this rule.

Selection of conductor type and insulation is very important since the life of the conductor or cable almost always corresponds to the life of the insulation. Conductors never wear out or run out of electrons as long as they are protected by insulation. Insulation, however, deteriorates over time. Insulation reacts with oxygen and compounds in its surroundings like ammonia and other manure gases, oil, gasoline, salts, and water. Some of these materials can also react with and corrode conductors if the insulation is broken. For example, buried aluminum cable rapidly corrodes and fails if its insulation is punctured. The rate of reaction increases with temperature and approximately doubles when temperature rises 10°C. Direct solar radiation also can speed up deterioration. In choosing insulation, the least expensive type which meets environmental requirements is the appropriate choice.

Ampacity

The ampacity of a conductor is its current-carrying capacity in amps. Ampacity depends on conductor resistance, the allowable operating temperature of the insulation, and the heat dissipation capability of the conductor. All conductors carrying current produce heat which must be transferred to the environment. Conductors which carry currents at or below their ampacities will operate cool enough to ensure a full useful life. Ampacities of several conductor/cable types are listed in Tables 5.5 and 5.7 (copper) and in Tables 5.6 and 5.8 (aluminum).

Several things can be noted from these tables. Refer to the ampacity section of Table 5.5:

- Ampacity increases with conductor size (AWG-14 UF can carry 15 amps, while AWG-4/0 UF can carry 175 amps).
- Ampacity for copper is higher than ampacity for aluminum in the same wire size (AWG-14 copper, UF [Table 5.5] will carry 15 amps but AWG-12 UF [Table 5.8] is required for 15 amps if the conductor is aluminum).
- Ampacity is higher for conductors which have higher temperature ratings. To carry 90 A, UF cable (60°) must be AWG-2, but USE cable (75°C) can be AWG-3 and THHN (90°C) can be AWG-4.

You might wonder, at this point, what the result would be if the ampacity of a conductor is exceeded. Consider an AWG-14 copper conductor, type THHN, in a conduit. From Table 5.7, its ampacity is 15 amps (read down column 4 to highest current which can be carried by AWG-14). If this conductor carried, say, 18 A for a long period of time, probably the only effect would be a significant decrease in the insulation life due to an increased operating temperature. If its current was increased to, say, 25 A, the insulation would probably become warm to the touch. At a current of 50 or 60 A, the insulation would begin to melt and smoke within a few seconds. The time factor is important. The ampacity can be exceeded for a very short time without any adverse effects. If, for example, the conductor supplies power for a motor, the ampacity can be exceeded while the motor starts. The specifics of motor conductor selection will be taken up in the section on motors circuits (chapter 10).

Derating Conductor Ampacity

Three situations may require table ampacity values to be reduced:

- Currents of loads operated continuously must be counted at 125% for ampacity purposes (NEC 210-19(a)). A continuous load is a load where the maximum current is expected to continue for 3 h or more (NEC 100).
- If a cable or conduit contains more than three current-carrying conductors, tabled ampacity values must be reduced. NEC article 310-15(b)(2)(a) lists adjustment factors.
- Most ampacity tables assume conductors are used in an ambient temperature of 30°C. At ambient temperatures above or below 30°C, tabled ampacity values are reduced or increased, respectively. NEC tables 310-16 and 310-17 contain ambient temperature correction factors.

Voltage Drop

Voltage drop occurs in the line as a result of resistance and current in the conductor. An example will help to explain the concept.

Example 5.4

In Figure 5.1, a line is strung from a pole, where we assume the voltage between the conductors to be constant at 120 V. The load draws a constant 50 A.

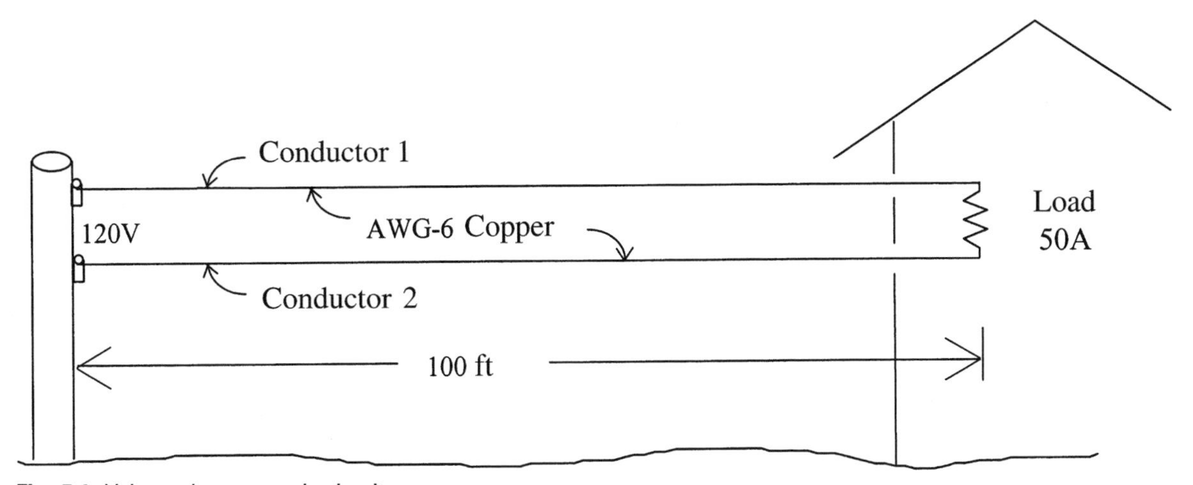

Fig. 5.1. Voltage drop example circuit.

TABLE 5.5 Minimum copper conductor sizes (AWG or mcmil) for 115–120 V branch circuits

	Ampacity Criterion							2% Voltage Drop Criterion																					
	In Air, Cable or Conduit							Length of Run, in Feet																					
				Direct Burial		Overhead in Air*		Compare size shown below with size shown to left of single vertical line. Use the larger size.																					
Load, in Amps	NM** UF**	RH,RHW TWH,TWHN THHW,SE	THHN	UF**	USE	Single	Triplex or Quad	20	40	60	80	100	125	150	175	200	225	250	275	300	350	400	450	500	550	600	650	700	750
5	14	14	14	14	14	10	8	14	14	14	12	12	12	10	10	8	8	8	8	8	6	6	6	6	4	4	4	4	4
7	14	14	14	14	14	10	8	14	14	12	12	10	10	8	8	8	8	6	6	6	6	4	4	4	4	3	3	3	2
10	14	14	14	14	14	10	8	14	12	12	10	8	8	8	6	6	6	6	4	4	4	4	3	3	2	2	1	1	1
15	14	14	14	14	14	10	8	14	12	10	8	8	6	6	4	4	4	4	3	3	2	2	1	1	1/0	1/0	1/0	2/0	2/0
20	12	12	12	12	12	10	8	12	10	8	6	6	6	4	4	4	3	3	2	2	1	1	1/0	1/0	2/0	2/0	3/0	3/0	3/0
25	10	10	10	10	10	10	8	12	8	8	6	6	4	4	3	3	2	2	1	1	1/0	1/0	2/0	2/0	3/0	3/0	4/0	4/0	4/0
30	10	10	10	10	10	10	8	12	8	6	6	4	4	3	2	2	1	1	1/0	1/0	2/0	2/0	3/0	3/0	4/0	4/0	4/0	250	250
35	8	8	8	8	8	8	8	10	8	6	4	4	3	2	2	1	1	1/0	1/0	2/0	2/0	3/0	3/0	4/0	4/0	250	250	300	300
40	8	8	8	8	8	8	8	10	6	6	4	4	3	2	1	1	1/0	1/0	2/0	2/0	3/0	3/0	4/0	4/0	250	300	300	300	350
45	6	8	8	6	8	8	8	10	6	4	4	3	2	1	1	1/0	1/0	2/0	2/0	3/0	3/0	4/0	4/0	250	300	300	350	350	400
50	6	8	8	6	8	8	8	8	6	4	4	3	2	1	1/0	1/0	2/0	2/0	3/0	3/0	4/0	4/0	250	300	300	350	350	400	400
60	4	6	6	4	6	8	6	8	6	4	3	2	1	1/0	2/0	2/0	3/0	3/0	4/0	4/0	250	300	300	350	350	400	500	500	500
70	4	4	6	4	4	8	6	8	4	3	2	1	1/0	2/0	2/0	3/0	3/0	4/0	4/0	250	300	300	350	400	500	500	500	600	600
80	3	4	4	3	4	6	4	6	4	3	1	1	1/0	2/0	3/0	3/0	4/0	4/0	250	300	300	350	400	500	500	600	600	600	700
90	2	3	4	2	3	6	4	6	4	2	1	1/0	2/0	3/0	3/0	4/0	4/0	250	300	300	350	400	500	500	600	600	700	700	750
100	1	3	3	1	3	4	4	6	4	2	1	1/0	2/0	3/0	4/0	4/0	250	300	300	350	400	500	500	600	600	700	700	750	800
115	1/0	2	2	1/0	2	4	3	6	3	1	1/0	2/0	3/0	4/0	4/0	250	300	300	350	400	500	500	600	600	700	750	800	900	900
130	2/0	1	2	2/0	1	3	2	4	2	1	2/0	3/0	4/0	4/0	250	300	350	350	400	500	500	600	700	700	750	900	900	IM	
150	3/0	1/0	1	3/0	1/0	1	1	4	2	1/0	2/0	3/0	4/0	250	300	350	400	400	500	500	600	700	750	800	900	IM			
175	4/0	2/0	2/0	4/0	2/0	2	1/0	4	1	2/0	3/0	4/0	250	300	350	400	500	500	600	600	700	750	900	IM					
200		3/0	3/0		3/0	1/0	2/0	4	1	2/0	3/0	4/0	300	350	400	500	500	600	600	700	750	900	IM						

Source: Hiatt (2000); reprinted with permission.

*Single conductors in overhead branch circuits must be at least AWG-12 copper or AWG-8 aluminum for spans up to 50 ft. For service conductors and for branch circuit spans greater than 50 ft, conductors must be at least AWG-8 copper or AWG 6 aluminum (NEC articles 225-6 (a) and 230-23).

**UF not permitted in sizes larger than AWG-4/0.

NM not permitted in sizes larger than AWG-2.

NM cannot be used in agricultural environments defined by NEC article 547.

In most applications, equipment will be rated at 60°C. Therefore, conductor ampacity must be that listed in the NM UF column (equivalent to 60°C) after all deratings are calculated.

TABLE 5.6 Minimum aluminum conductor sizes (AWG or mcmil) for 115–120 V branch circuits

Load, in Amps	Ampacity Criterion							2% Voltage Drop Criterion																					
	In Air, Cable or Conduit			Direct Burial		Overhead in Air*		Length of Run, in Feet. Compare size shown below with size shown to left of single vertical line. Use the larger size.																					
	NM** UF**	RH,RHW TWH,TWHN THHW,SE	THHN	UF**	USE	Single	Triplex or Quad	20	40	60	80	100	125	150	175	200	225	250	275	300	350	400	450	500	550	600	650	700	750
5	12	12	12	12	12	8	8	12	12	12	10	10	8	8	8	6	6	6	6	6	4	4	4	4	3	3	2	2	2
7	12	12	12	12	12	8	8	12	12	10	10	8	8	6	6	6	4	4	4	4	4	3	3	2	2	1	1	1	1/0
10	12	12	12	12	12	8	8	12	10	10	8	6	6	6	4	4	4	4	3	3	2	1	1	1	1/0	1/0	2/0	2/0	2/0
15	12	12	12	12	12	8	8	12	10	8	6	6	4	4	3	3	2	2	1	1	1/0	1/0	2/0	2/0	3/0	3/0	3/0	4/0	4/0
20	10	10	10	10	10	8	8	10	8	6	4	4	4	3	2	1	1	1	1/0	1/0	2/0	3/0	3/0	3/0	4/0	4/0	250	250	300
25	10	10	10	10	10	8	8	10	6	6	4	4	3	2	1	1	1/0	1/0	2/0	2/0	3/0	3/0	4/0	4/0	250	300	300	300	350
30	8	8	8	8	8	8	8	10	6	4	4	3	2	1	1/0	1/0	2/0	2/0	3/0	3/0	4/0	4/0	250	300	300	350	350	400	400
35	6	8	8	6	8	8	8	8	6	4	3	2	1	1/0	1/0	2/0	2/0	3/0	3/0	4/0	4/0	250	300	300	350	400	400	500	500
40	6	8	8	6	8	8	8	8	4	4	2	1	1	1/0	2/0	3/0	3/0	3/0	4/0	4/0	250	300	350	350	400	500	500	500	600
45	4	6	8	4	6	8	6	8	4	3	2	1	1/0	2/0	2/0	3/0	4/0	4/0	4/0	250	300	350	350	400	500	500	500	600	600
50	4	6	6	4	6	8	6	6	4	3	1	1	1/0	2/0	3/0	3/0	4/0	4/0	250	300	300	350	400	500	500	600	600	600	700
60	3	4	6	3	4	6	4	6	4	2	1	1/0	2/0	3/0	4/0	4/0	250	300	300	350	400	500	500	600	600	700	700	750	800
70	2	3	4	2	3	6	4	6	3	1	1/0	2/0	3/0	4/0	4/0	250	300	300	350	400	500	500	600	600	700	750	800	900	900
80	1	2	3	1	2	4	3	4	2	1	2/0	3/0	3/0	4/0	250	300	350	350	400	500	500	600	700	700	750	900	900	IM	
90	1/0	2	2	1/0	2	6	3	4	2	1/0	2/0	3/0	4/0	250	300	350	350	400	500	500	600	700	700	800	900	IM	IM		
100	1/0	1	2	1/0	1	4	2	4	1	1/0	3/0	3/0	4/0	300	300	350	400	500	500	600	600	700	800	900	IM				
115	2/0	1/0	1	2/0	1/0	3	1	4	1	2/0	3/0	4/0	250	300	350	400	500	600	600	600	700	800	900	IM					
130	3/0	2/0	1/0	3/0	2/0	2	1/0	3	1/0	2/0	4/0	250	300	350	400	500	500	600	600	700	800	900	IM						
150	4/0	3/0	2/0	4/0	3/0	1	2/0	3	1/0	3/0	4/0	300	350	400	500	600	600	700	700	800	900								
175		4/0	3/0		4/0	1/0	3/0	2	2/0	4/0	250	300	400	500	600	600	700	900	900	900									
200		5/0	4/0		5/0	2/0	4/0	1	3/0	4/0	300	350	500	600	600	700	800	IM	IM										

Source: Hiatt (2000); reprinted with permission.

*Single conductors in overhead branch circuits must be at least AWG-12 copper or AWG-8 aluminum for spans up to 50 ft. For service conductors and for branch circuit spans greater than 50 ft, conductors must be at least AWG-8 copper or AWG 6 aluminum (NEC articles 225-6 (a) and 230-23).

**UF not permitted in sizes larger than AWG-4/0.

NM not permitted in sizes larger than AWG-2.

NM cannot be used in agricultural environments defined by NEC article 547.

In most applications, equipment will be rated at 60°C. Therefore, conductor ampacity must be that listed in the NM UF column (equivalent to 60°C) after all deratings are calculated.

TABLE 5.7 Minimum copper conductor sizes (AWG or mcmil) for 230–240 V single-phase branch circuits

Load, in Amps	Ampacity Criterion: In Air, Cable or Conduit: NM** UF**	RH,RHW TWH,TWHN THHW,SE	THHN	Direct Burial: UF**	Direct Burial: USE	Overhead in Air*: Single	Overhead in Air*: Triplex or Quad	2% Voltage Drop Criterion — Length of Run, in Feet (Compare size shown below with size shown to left of single vertical line. Use the larger size.): 25	50	75	100	125	150	175	200	225	250	275	300	350	400	450	500	550	600	650	700	750	800
5	14	14	14	14	14	10	8	14	14	14	14	14	14	12	12	12	12	10	10	10	8	8	8	8	8	6	6	6	6
7	14	14	14	14	14	10	8	14	14	14	14	12	12	12	10	10	10	10	8	8	8	8	6	6	6	6	6	4	4
10	14	14	14	14	14	10	8	14	14	14	12	12	10	10	8	8	8	8	8	6	6	6	6	4	4	4	4	4	4
15	14	14	14	14	14	10	8	14	14	12	10	10	8	8	8	6	6	6	6	4	4	4	4	3	3	3	2	2	2
20	12	12	12	12	12	10	8	14	12	10	8	8	8	6	6	6	6	4	4	4	4	3	3	2	2	1	1	1	1
25	10	10	10	10	10	10	8	14	12	10	8	8	6	6	6	4	4	4	4	3	3	2	2	1	1	1/0	1/0	1/0	1/0
30	10	10	10	10	10	10	8	14	10	8	8	6	6	4	4	4	4	3	3	2	2	1	1	1/0	1/0	1/0	2/0	2/0	2/0
35	8	8	8	8	8	8	8	12	10	8	6	6	4	4	4	4	3	3	2	2	1	1	1/0	1/0	2/0	2/0	2/0	3/0	3/0
40	8	8	8	8	8	8	8	12	8	8	6	6	4	4	4	3	3	2	2	1	1	1/0	1/0	2/0	2/0	3/0	3/0	3/0	3/0
45	6	8	8	6	8	8	8	12	8	6	6	4	4	4	3	2	2	2	1	1	1/0	1/0	2/0	2/0	3/0	3/0	3/0	4/0	4/0
50	6	8	8	6	8	8	8	12	8	6	6	4	4	3	3	2	2	1	1	1/0	1/0	2/0	2/0	3/0	3/0	4/0	4/0	4/0	4/0
60	4	6	6	4	6	8	6	10	8	6	4	4	3	2	2	1	1	1/0	1/0	2/0	2/0	3/0	3/0	4/0	4/0	4/0	250	250	300
70	4	4	6	4	4	8	6	10	6	4	4	3	2	2	1	1	1/0	1/0	2/0	2/0	3/0	3/0	4/0	4/0	250	250	300	300	300
80	3	4	4	3	4	6	4	8	6	4	3	3	2	1	1	1/0	1/0	2/0	2/0	3/0	3/0	4/0	4/0	250	300	300	300	350	350
90	2	3	4	2	3	6	4	8	6	4	3	2	1	1	1/0	1/0	2/0	2/0	3/0	3/0	4/0	4/0	250	300	300	350	350	400	400
100	1	3	3	1	3	4	4	8	6	4	3	2	1	1/0	1/0	2/0	2/0	3/0	3/0	4/0	4/0	250	300	300	350	350	400	40	500
115	1/0	2	2	1/0	2	4	3	8	4	3	2	1	1/0	1/0	2/0	3/0	3/0	3/0	4/0	4/0	250	300	300	350	400	400	500	600	500
130	2/0	1	2	2/0	1	3	2	6	4	3	1	1/0	1/0	2/0	3/0	3/0	4/0	4/0	4/0	250	300	350	350	400	50	500	500	600	600
150	3/0	1/0	1	3/0	1/0	1	1	6	4	2	1	1/0	2/0	3/0	3/0	4/0	4/0	250	250	300	350	400	40	500	500	600	600	600	700
175	4/0	2/0	2/0	4/0	2/0	2	1/0	6	3	1	1/0	2/0	3/0	3/0	4/0	4/0	250	300	300	350	400	500	500	600	600	600	700	700	750
200		3/0	3/0		3/0	1/0	2/0	6	3	1	1/0	2/0	3/0	4/0	4/0	250	300	300	350	400	500	500	600	600	700	700	750	800	900
225		4/0	3/0	4/0	4/0	1/0	3/0	4	2	1/0	2/0	3/0	4/0	4/0	250	300	300	350	400	500	500	600	600	700	750	800	900	900	IM
250		250	4/0	250	250	2/0	4/0	4	2	1/0	2/0	3/0	4/0	250	300	300	350	400	400	500	600	600	700	750	800	900	IM	IM	
275		300	250	300	300	3/0	4/0	4	1	2/0	3/0	4/0	250	300	300	350	400	400	500	600	600	700	750	800	900	IM			
300		350	300	350	350	3/0	250	4	1	2/0	3/0	4/0	250	300	350	400	400	500	500	600	700	750	800	900	IM				
325		400	350	400	400	4/0	300	3	1/0	2/0	4/0	4/0	300	300	350	400	500	500	600	600	700	800	900	IM					
350		500	350	500	500	4/0	300	3	1/0	3/0	4/0	250	300	350	400	500	500	600	600	700	750	900	IM						
375		500	400	500	500	250	350	3	1/0	3/0	4/0	250	300	350	400	500	500	600	600	700	800	900	IM						
400		600	500	600	600	250	400	3	1/0	3/0	4/0	300	350	400	500	500	600	600	700	750	900	IM							

Source: Hiatt (2000); reprinted with permission.

*Single conductors in overhead branch circuits must be at least AWG-12 copper or AWG-8 aluminum for spans up to 50 ft. For service conductors and for branch circuit spans greater than 50 ft, conductors must be at least AWG-8 copper or AWG 6 aluminum (NEC articles 225-6 (a) and 230-23).

**UF not permitted in sizes larger than AWG-4/0.

NM not permitted in sizes larger than AWG-2.

NM cannot be used in agricultural environments defined by NEC article 547.

In most applications, equipment will be rated at 60°C. Therefore, conductor ampacity must be that listed in the NM UF column (equivalent to 60°C) after all deratings are calculated.

TABLE 5.8 Minimum aluminum conductor sizes for (AWG or mcmil) 230–240V single-phase branch circuits

	Ampacity Criterion							2% Voltage Drop Criterion																				
	In Air, Cable or Conduit							Length of Run, in Feet																				
Load, in	NM**	RH,RHW TWH,TWHN		Direct Burial		Overhead in Air*		Compare size shown below with size shown to left of single vertical line. Use the larger size.																				
Amps	UF**	THHW,SE	THHN	UF**	USE	Single	Triplex or Quad	25	50	75	100	125	150	175	200	225	250	275	300	350	400	450	500	550	600	650	700	750	800
5	12	12	12	12	12	8	8	12	12	12	12	12	12	10	10	10	8	8	8	8	6	6	6	6	4	4	4	4	
7	12	12	12	12	12	8	8	12	12	12	12	10	10	10	8	8	8	8	6	6	6	4	4	4	4	4	4	3	
10	12	12	12	12	12	8	8	12	12	12	10	8	8	8	6	6	6	6	6	4	4	4	4	3	2	2	2	2	
15	12	12	12	12	12	8	8	12	12	10	8	8	6	6	6	4	4	4	4	3	3	2	2	1	1	1	1/0	1/0	1/0
20	10	10	10	10	10	8	8	12	10	8	8	6	6	4	4	4	4	3	3	2	1	1	1	1/0	2/0	2/0	2/0	2/0	3/0
25	10	10	10	10	10	8	8	12	8	8	6	6	4	4	4	3	3	2	2	1	1	1/0	1/0	2/0	3/0	3/0	3/0	3/0	3/0
30	8	8	8	8	8	8	8	12	8	6	6	4	4	3	3	2	2	1	1	1/0	1/0	2/0	2/0	3/0	3/0	3/0	4/0	4/0	4/0
35	6	8	8	6	8	8	8	10	8	6	4	4	3	3	2	2	1	1	1/0	1/0	2/0	2/0	3/0	4/0	4/0	4/0	4/0	250	250
40	6	8	8	6	8	8	8	10	6	6	4	4	3	2	1	1	1	1/0	1/0	2/0	3/0	3/0	3/0	4/0	250	250	250	300	300
45	4	6	8	4	6	8	6	10	6	4	4	3	2	2	1	1/0	1/0	1/0	2/0	2/0	3/0	4/0	4/0	250	250	250	300	300	350
50	4	6	6	4	6	8	6	8	6	4	4	3	2	1	1	1/0	1/0	2/0	2/0	3/0	3/0	4/0	4/0	300	300	300	300	350	350
60	3	4	6	3	4	6	4	8	6	4	3	2	1	1/0	1/0	2/0	2/0	3/0	3/0	4/0	4/0	250	300	350	350	350	400	400	500
70	2	3	4	2	3	6	4	8	4	3	2	1	1/0	1/0	2/0	2/0	3/0	3/0	4/0	4/0	250	300	300	400	400	400	500	500	500
80	1	2	3	1	2	4	3	6	4	3	1	1	1/0	2/0	3/0	3/0	3/0	4/0	4/0	250	300	350	350	500	500	500	500	600	600
90	1/0	2	2	1/0	2	6	3	6	4	2	1	1/0	2/0	2/0	3/0	4/0	4/0	4/0	250	300	350	350	400	500	500	500	600	600	700
100	1/0	1	2	1/0	1	4	2	6	4	2	1	1/0	2/0	3/0	3/0	4/0	4/0	250	300	300	350	400	500	600	600	600	600	700	700
115	2/0	1/0	1	2/0	1/0	3	1	6	3	1	1/0	2/0	3/0	3/0	4/0	250	250	300	300	350	400	500	500	600	600	700	700	750	800
130	3/0	2/0	1/0	3/0	2/0	2	1/0	4	2	1	2/0	3/0	3/0	4/0	250	250	300	300	350	400	500	500	600	700	700	750	800	900	900
150	4/0	3/0	2/0	4/0	3/0	1	2/0	4	2	1/0	2/0	3/0	4/0	250	300	300	350	350	400	500	600	600	700	800	800	900	900	IM	
175		4/0	3/0		4/0	1/0	3/0	4	1	2/0	3/0	4/0	250	300	300	350	400	500	500	600	600	700	750	900	900	IM			
200		5/0	4/0		5/0	2/0	4/0	4	1	2/0	3/0	4/0	300	300	350	400	500	500	600	600	700	800	900	IM					
225		300	250		300	3/0	250	3	1/0	3/0	4/0	250	300	350	400	500	500	600	600	700	800	900	IM						
250		350	300		350	4/0	250	3	1/0	3/0	4/0	300	350	400	500	500	600	600	700	750	900	IM							
275		500	350		500	4/0	300	2	2/0	4/0	250	300	350	500	500	600	600	700	700	900	IM								
300		500	400		500	250	350	2	2/0	4/0	300	350	400	500	600	600	700	700	800	900									
325		600	500		600	300	400	1	3/0	4/0	300	350	500	500	600	700	700	750	900	IM									
350		700	500		700	300	500	1	3/0	250	300	400	500	600	600	700	750	900	900										
375		700	600		700	350	500	1	3/0	250	350	400	500	600	700	750	800	900	IM										
400		900	700		900	400	600	1	3/0	300	350	500	600	600	700	800	900	IM											

Source: Hiatt (2000); reprinted with permission.

*Single conductors in overhead branch circuits must be at least AWG-12 copper or AWG-8 aluminum for spans up to 50 ft. For service conductors and for branch circuit spans greater than 50 ft, conductors must be at least AWG-8 copper or AWG 6 aluminum (NEC articles 225-6 (a) and 230-23).

**UF not permitted in sizes larger than AWG-4/0.

NM not permitted in sizes larger than AWG-2.

NM cannot be used in agricultural environments defined by NEC article 547.

In most applications, equipment will be rated at 60°C. Therefore, conductor ampacity must be that listed in the NM UF column (equivalent to 60°C) after all deratings are calculated.

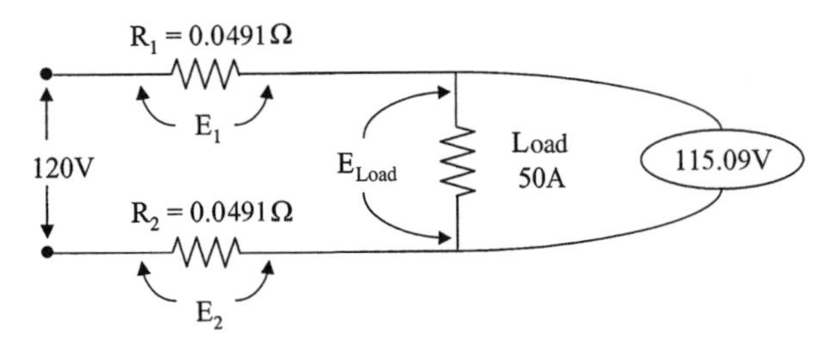

Fig. 5.2. Voltage drop example with conductor resistance.

We might first check the ampacity. In Table 5.3 we see that a simple AWG-6 copper conductor strung in air (60°C) can safely carry up to 80 A. This conductor will not overheat since it is operating well below its ampacity value.

Now we need to calculate the resistance of each of the conductors. From Table 5.3 we read a resistance of 0.491 Ω/1000 ft. Each of our conductors is 100 ft long, and each has a resistance of:

$$\frac{0.491\ \Omega\ (100\ \text{ft})}{1000\ \text{ft}} = 0.0491\ \Omega$$

We can represent Figure 5.1 by the series circuit in Figure 5.2.

Here we are using "perfect" conductors, and showing the resistance of each real conductor as a 0.0491-Ω resistor. Being a series circuit, the voltages add around the circuit. Thus:

$$120 = E_1 + E_{Load} + E_2$$

Since we know the current through R_1 and R_2 we can compute E_1 and E_2:

$$E_1 = I_1 R_1 = (50)(0.0491) = 2.455\ \text{V}$$
$$E_2 = I_2 R_2 = (50)(0.0491) = 2.455\ \text{V}$$

Now we can solve the previous equation for E_{Load}.

$$E_{Load} = 120 - 2.455 - 2.455 = 115.09\ \text{V}$$

We see that because of the resistance of the conductors, a voltage drop of 4.91 V has occurred between the pole and the load, and the load sees 115.09 V, not 120 V. A voltmeter in Figure 5.2 reads this voltage. This is often expressed as % voltage drop:

$$\%\ \text{voltage drop} = \frac{120 - 115.09}{120}(100)$$

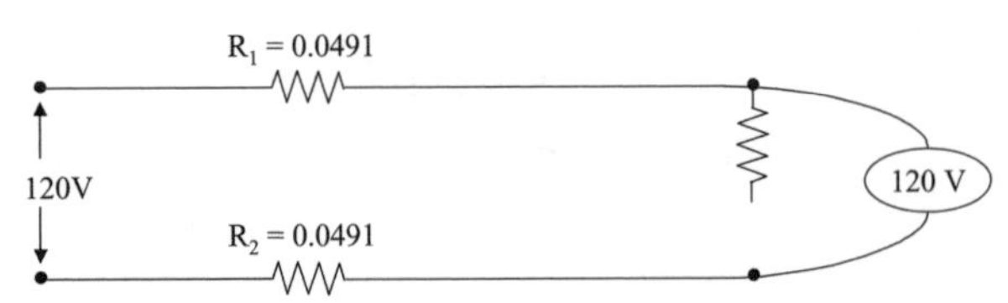

Fig. 5.3. Circuit of Example 5.2 with load turned off.

$$= \frac{4.91}{120}(100) = 4.09\%$$

As a result of this voltage drop, load performance will be degraded. If it is a resistive heater, heat output will decrease more than 8%, since power use is proportional to the square of voltage. If it is an incandescent lamp, usable light output will drop about 10%. If the load is a motor, its starting torque will be decreased, and it will need to draw a higher current to deliver its rated output. There is also a power loss in the lines:

$$P_{loss} = I_{line} E_{drop} = I_{line}(E_1 + E_2)$$
$$= 50(4.91) = 245.5\ \text{W}$$

An important principle is illustrated in Figure 5.3. If the load current is zero, the voltage drop is zero and a voltmeter at the end of the circuit reads 120 V.

We see that excessive voltage drop is undesirable. Looking at the factors influencing voltage drop, it is apparent that voltage drop can be lessened by:

1. Decreasing the load.
2. Using larger conductors.
3. Placing the load nearer to the pole.
4. Using a conductor material having a lower resistance.
5. Using a higher voltage.

The approach of using a higher voltage is especially effective. In Example 5.4, the power is delivered at 120 V: P = (50 A)(120 V) = 6000 W. Voltage drop is 4.91 V, and power loss is 245.5 W.

If voltage is increased to 240 V, the same power can be delivered with a current of 25 A: P = (25)(240)=6,000 W. Voltage drop, which is proportional to I, is cut in half: $E_{drop} = 2(25)(0.0491) = 2.455$ V. Line loss, which is proportional to I^2, is cut to one-fourth: $P_{loss} = (25A)^2\ (0.0491 + 0.0491) = 61.4$ W.

Availability of transformers to raise or lower AC voltages allows this approach to be used to raise system voltages for transmission over long distances, and subsequently lower them to safe values at the point of use (See Line Power, chapter 1).

Designing for Acceptable Voltage Drop

Because all conductors regardless of size do have resistance, we cannot eliminate voltage drop in a line carrying current. However, the circuit can be designed for a specified, acceptable voltage drop. The National Electrical Code states that these drop limits will provide "reasonable efficiency of operation" (NEC 210-19, FPN No. 4):

3% maximum voltage drop in branch circuit at farthest power outlet

5% maximum voltage drop in the feeder and branch circuit combined

Branch circuit designs for 2% voltage drop are common. It is necessary to meet ampacity requirements for safety. It is desirable to meet voltage drop guidelines for proper load performance. If total conductor cost is considered (ownership cost plus operating cost), the least-cost conductor size may be larger or smaller than the size which follows the code suggestion. This is discussed more in the section on conductor economics.

The smallest conductor size allowed for building wiring by the NEC is AWG-14 copper, and AWG-12 for aluminum or copper-clad aluminum (NEC 310-5). An equation for calculating the conductor size needed for a specified voltage drop can be derived by shifting some terms in Equation 5.1:

$$R = \frac{\rho L}{A}$$

Solving for A: $$A = \frac{\rho L}{R}$$

From Ohm's Law: $$R = \frac{E_{drop}}{I_{line}}$$

Substituting for R: $$A = \frac{\rho I_{line} L}{E_{drop}}$$

L is the total length of conductor. It is more convenient to use a one-way length in the equation. Call it ℓ:

$$2\,\ell = L$$

From Table 5.1, the resistivity of copper is 10.37 Ω-cmil/ft.
Then

$$A = \frac{(10.37)\, I_{line}\, 2\ell}{E_{drop}} = \frac{(20.74)\, I_{line}\, \ell}{E_{drop}}$$

The equation as commonly used for copper is:

$$A = \frac{(22)\, I_{line}\, \ell}{E_{drop}} \qquad \text{(for copper)} \qquad \mathbf{(5.3)}$$

The value of the constant is raised to 22 to account for line inductance, which has a slight current limiting effect.

For aluminum, a factor of 1.6 is used to account for aluminum's higher resistivity:

$$A = \frac{(1.6)\,(22)\, I_{line}\, \ell}{E_{drop}} \qquad \text{(for aluminum)} \qquad \mathbf{(5.4)}$$

An example problem illustrating the use of these equations follows.

Example 5.5

Design a feeder line to a building with these specifications:

Load: 78 A @ 230 V
Aluminum conductors to be strung overhead
One-way length: 273 ft
Voltage drop: 2%

Environment—First, consider the environment. From Table 5.4, we see that multiplex (triplex) is the proper choice. Triplex is available only in aluminum and has a temperature rating of 75°C. Considering ampacity we refer to Table 5.8, column 8. AWG-3 conductor is the smallest size which will carry 78 A.

To find the size to satisfy the voltage drop specification, we use Equation 5.4. To provide for connections, 10% should be added to the length:

273 feet
+ 27 feet (10% of 273)
300 feet

$$E_{drop} = (.02)\,(230) = 4.6\text{ V}$$

$$A = \frac{(1.6)(22)(78)(300)}{4.6} = 179{,}061\text{ cmil}$$

Referring to Table 5.3, the smallest conductor which has at least 179,061 cmil size is AWG-0000.

Since ampacity requires AWG-3 and voltage drop requires AWG-0000, the AWG-0000 must be chosen. There will be 3 conductors strung to the building. The sizes calculated are for the two energized conductors. Because extreme unbalance in load is unlikely, the neutral can be assumed to carry a much lower current, and may be specified 2 sizes smaller. However, triplex is usually assembled with the bare neutral the same size as the energized conductors. For this job we need:

300-ft AWG-0000 triplex (aluminum)

Tables 5.5, 5.6, 5.7, and 5.8 can be used to size conductors for the voltage drop criterion. In this case, we use Table 5.8 and find once again AWG-0000 to satisfy voltage drop. (In some cases, due to rounding procedures, the wire size calculated may not exactly agree with that specified in the table).

For this example, let's check what voltage the load sees with the circuit as designed. This calculation can be done two ways:

Circular mil method—The AWG-0000 conductor has an area of 211,600 cmil. Solving Equation 5.3 for E_{drop}:

$$E_{drop} = \frac{1.6(22)(I_{line})(\ell)}{A} = \frac{(1.6)(22)(78)(300)}{211{,}600}$$

$$E_{drop} = 3.89 \text{ V}$$

With the load on, the load sees 230 V − 3.89 V = 226.11 V. Since 3.89 V < 4.6 V, the design voltage drop criterion has been met.

Resistance method—From Table 5.3, the resistance of 0000 aluminum conductor is 0.100 ohms/1000 ft. Conductor resistance is:

$$2(300 \text{ ft})\left(\frac{0.100\ \Omega}{1000 \text{ ft}}\right) = 0.060\ \Omega$$

We must use two times the 300-ft run length because current flows through both conductors. Using Ohm's Law:

$$E_{drop} = I_{line} R_{line} = (78\text{A})(0.060\ \Omega) = 4.68 \text{ V}$$

Either calculation method is suitable for use. The difference in calculated values is due to slight disagreements between tabled resistance values and Equation 5.3 constants.

Example 5.6

Example 5.5 used a one-way length of 300 ft, which is relatively long. This example uses the same set of conditions, except for a one-way length of 30 ft instead of 300 ft.

Design a feeder line to a building with these specifications:

Load: 78A @ 230 V
Aluminum conductors to be strung overhead
One-way length: 27 ft
Design voltage drop: 2%

Environment

Aluminum triplex (same as Example 5.5)

Ampacity

AWG-3 (same as Example 5.5)

Voltage drop

27 ft
+ 3 ft (10% of 27)
30 ft

$$E_{drop} = (0.2)(230) = 4.6 \text{ V}$$

$$A = \frac{(1.6)(22)(78)(30)}{4.6} = 17{,}906 \text{ cmil}$$

Referring to Table 5.3, the smallest conductor which has an area of at least 17,906 cmil is AWG-6 with an area of 26,240 cmil.

In this example, ampacity requires AWG-3, and voltage drop requires AWG-6. The AWG-3 size must be used. Although voltage drop would be less than 2% if AWG-6 conductors were used, they cannot safely carry 78 A. Voltage drop tends to dictate conductor size for longer circuits, while ampacity dictates for shorter circuits. What's long and what's short? In this example, distances between 70 and 88 ft require AWG-3 for voltage drop, the same as the minimum ampacity sizes. For distances over 88 ft, voltage drop determines size. For distances under 70 ft, ampacity dictates.

Example 5.7

Design a branch circuit conductor to be run inside a dry farm building to meet these specifications:

Load: 35 A @ 230 V
Copper conductors to be fastened to wall
One-way length: 18 ft
Design voltage drop: 2%
Dry location

Type NM will be the cheapest conductor type to use, considering the different types on Table 5.4. From Table 5.7, ampacity dictates AWG-8.
For voltage drop:

$$\begin{array}{r} 18 \text{ ft} \\ +1.8 \text{ ft} \\ \hline 19.8 \text{ ft} \end{array} \quad \text{(use 20 ft)}$$

$$E_{drop} = 4.6 \text{ V}$$

$$A = \frac{(22)(35)(20)}{4.6} = 3345 \text{ cmil}$$

Table 5.3 says size AWG-14 is needed. In this case, ampacity dictates, and AWG-8 must be used. Thus, we will need 20 ft of AWG-8/3 copper type NM with ground. If there are no 115-V loads, a neutral conductor is not needed and AWG-8/2 with ground can be used, with the white conductor marked at its ends with black tape or marking pen to indicate it is an energized conductor.

These two examples suggest that for short distances, ampacity will determine size, and for long distances, voltage drop will be the determining factor. Since "short" and "long" are indefinite, both ampacity and voltage drop should be checked.

Estimating Total Cost of Electrical Conductors

The larger the conductor installed in an electrical system, the greater the initial cost. This higher initial cost can fool individuals having electrical systems installed into thinking they can save money by installing the smallest allowable conductors. This section presents a procedure for finding the size which will minimize total conductor cost.

Conductor Costs

The total annual cost of a conductor can be calculated as the sum of the annual ownership cost and the annual line losses:

$$C_A = C_C\,(CRF) + C_L \tag{5.5}$$

where C_A = total annual conductor cost, \$/yr
C_C = total first cost of the installed conductor, \$
CRF = capital recovery factor, decimal
C_L = cost of line losses, \$/yr

C_C is the turnkey cost of the conductor and includes purchase price, installation, and any other changes. CRF (Equation 5.6) is the fraction of C_C which, if paid in equal increments over the life of the conductor, will have the original cost plus compound interest all paid by the end of its life.

$$CRF = \frac{i(1 + i)^n}{(1 + i)^n - 1} \tag{5.6}$$

where i = interest rate, decimal
n = expected life of conductor, years

The interest rate is the rate at which funding can be secured from any selected source.

Yearly Line Losses

The yearly line losses of any given conductor can be calculated using the following formula:

$$C_L = \frac{R_W \times I^2 \times H_A \times C_E}{1000}$$

where C_L = line losses, \$/yr
R_W = conductor resistance, ohms
I = average load current, amps
H_A = time of load use, hours/yr
C_E = electrical energy cost, \$/kWh

An example will illustrate use of the procedure.

Example 5.8

A 120-V branch circuit is 45.5 ft (one way) and carries 40 A for 500 h/yr. The circuit is in steel conduit; the load does not operate continuously.

The minimum copper THHN conductor size for ampacity is AWG-8.

TABLE 5.9 Conductor costs for Example 5.6

R_W	Conductor size, AWG	Conductor cost, $/ft	Installed circuit cost, $	Installed circuit (cost)(CRF), $/yr	C_L $/yr	C_A
.307	14	.0338	7.13	.67	19.65	20.32
.193	12	.045	9.50	.90	12.35	13.25
.121	10	.067	14.13	1.33	7.74	9.07
.0764	8	.119	23.61	2.23	4.89	7.12
.0491	6	.17	35.87	3.39	3.14	6.53
.0308	3	.314	66.25	6.25	1.97	8.22
.0245	2	.41	86.51	8.17	1.57	9.74

Assume:

- 100 ft of conductor is required.
- Electrical energy costs $0.08/kWh.
- Wire life is 20 yr.
- Cost of capital is 7%.
- Total installed circuit costs is 210% of conductor cost.

Table 5.9 shows calculated values for Example 5.6. The right column lists values of C_A, the annual conductor cost. Minimum annual conductor cost occurs for AWG-6 conductor. Smaller conductors have excessive line loss, while larger conductors have high ownership costs.

If load use is reduced to 50h/yr, the lowest cost is for AWG-10. If load use is increased to 5,000 h, the lowest cost is for AWG-2.

References

American Superconductor. 2001. Electric power applications of superconductivity. American Superconductor, Westborough, MA. Available at www.amsuper.com

Hiatt, R. S. (ed). 2000. Agricultural wiring handbook, 12th ed. National Food and Energy Council, Inc., Columbia, MO.

NFPA. 2002. National Electrical Code 2002. National Fire Protection Association, Inc., Quincy, MA.

Surbrook, T. C., and R. C. Mullin. 1985. Agricultural Electrification. South-Western Publishing Co., Cincinnati, OH.

Problems

5.1. (a) An AWG-10 conductor has a diameter of 0.102 in Table 5.3. What diameter steel wire would be required to have the same resistance as an AWG-10 copper conductor of the same length? (State diameter in inches.)

(b) Describe a specific application where steel is used as a circuit conductor.

5.2. (a) About what voltage can be put across a layer of plastic electrical tape (PVC .007-in. thick)?

(b) From the electrical insulation standpoint is it necessary to wrap more than one layer in farmstead wiring applications? Explain.

5.3. (a) What voltage is available at the load with the circuit as shown?

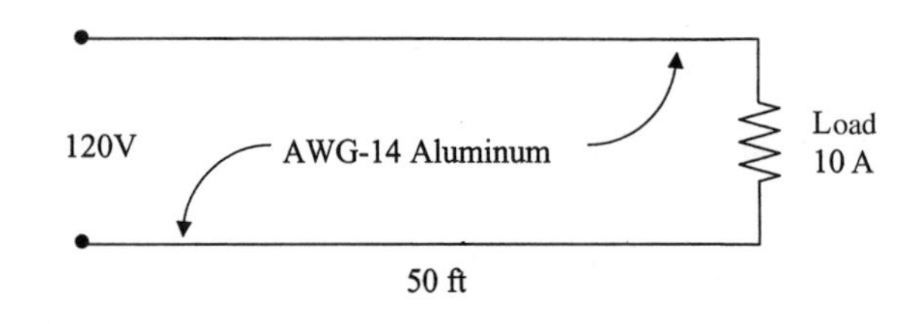

(b) What is the percent voltage drop?

(c) What voltage is available at the load end of the circuit with the load disconnected?

5.4. A branch circuit is to be installed underground to supply a 240-V 2.4-kW tank heater 220 ft away (measured distance one way). Design a conductor for this branch circuit. Specify conductor or cable type, size, material, design voltage drop, length needed. (Refer to page 1.31 in MWPS-28 for a discussion of underground wiring.)

5.5. A branch circuit is to be installed inside a dry farm building to supply a 120-V 3.6-kW heater. The one-way distance is measured as 60 ft. Select a cable for this branch circuit. Specify cable length, conductor type, material, design voltage drop, conductor size required for ampacity, conductor size required for voltage drop, and conductor size to be used.

5.6. Specify conductors for the feeder circuit described and calculate the voltage available at the load with the load on. Allow 10% of the length for connections. Choose a design voltage drop.

Load: 70 A, 230 V available at meter pole. The meter pole is 150 ft from the service entrance switch in the building. The feeder is to be run underground.

5.7. A branch circuit is to be run 95 ft (*including* connections) to a 4.6-kW, 240-V electric water heater inside a dry farm building. Wiring is to be done as cheaply as possible. A 2% design voltage drop and copper conductors are to be used. Assume the service entrance supplies a constant 240 V.

(a) Specify the conductor or cable type to be used.

(b) What color will the conductors be?

(c) What size of conductor does ampacity dictate for this circuit?

(d) What size of conductor does voltage drop dictate?

(e) What size should be used?

(f) With the heater turned off, what voltage is present at the load end of the circuit?

(g) With the heater turned on, exactly what voltage is present across the load if the conductor from (e) is used?

5.8. A single-phase, 5-hp motor is the only load operating on a 240-V branch circuit. The motor, being fully loaded, draws 28 A and operates at a power factor of 0.7. The conductor is AWG-3 copper, with a 600-ft run (including connections). The panel supplies a constant 240 V. Compute the phase angle at the panel, and the voltage across the motor. The load consists of the motor and the conductor resistance. *Hint:* Conductor voltage drop is in phase with the motor current.

5.9. A 230-V, AWG-8, copper, 3-wire feeder is exactly 100-ft long and serves a building containing these loads: 20 A at 230 V, 6 A at 115 V (black to white), 12 A at 115 V (red to white). All the loads are resistive. Compute the total power loss in the feeder, and the energy loss per month if the loads are operated 60 h per month.

Chapter 6

ELECTRICAL SAFETY

The topic of electrical safety involves safety for persons and animals, and also safety for electrical equipment and facilities. In some cases, principles are pertinent to both areas. This chapter includes information on several topics important to electrical safety.

Underwriters Laboratories, Inc.

Many purchased products (electrical, as well as non-electrical) have the familiar UL mark visible somewhere (Figure 6.1). Underwriters Laboratories, Inc. is a not-for-profit product safety testing and certification organization. When a product has the UL mark, it means the manufacturer has submitted samples of the product to UL for evaluation, and UL has found that the samples meet UL's safety requirements. Selecting electrical products with UL marks can help answer questions about the relative safety among competing product brands, especially when the brand names involved are not well known. Absence of a UL mark should raise concerns about an electrical product.

UL evaluates over 18,000 types of products made by over 61,000 manufacturers and operates in countries all over the world. Some products have marks from similar organizations in other countries. The CSA mark, for example, comes from Canadian Standards Association International in Canada.

Grounding

The term *ground* refers to a conductive path to the earth. By *grounding* a tool or electrical system, a low-resistance path to the earth is intentionally created. Grounding is a method of protecting persons from electrical shock. Grounding also enhances safety by providing a low-resistance path to earth, which limits voltage imposed by lightning, line surges, or unintentional contact with higher voltage lines. *System grounding* refers to the components which reliably connect the electrical system neutral to earth through a low-resistance path. *Equipment grounding* provides a reliable low-resistance path between conducting, non-current carrying components (such as equipment frames) and the system ground.

Equipment Grounding

The most common electric service in the United States is the 3-wire, 120/240-V single-phase service where the transformer secondary winding center point is grounded (see Figure 1.25). This grounded neutral is then extended on through the system for safety. The concept of equipment grounding is illustrated by the circuits of Figure 6.2.

The figure shows two 120-V branch circuits extending to two electrical machines with conducting frames. Switches and receptacles are omitted for simplicity. The machine frames are not in contact with the earth. Machine a is powered through a 2-wire circuit and the frame is not safety grounded. This lack of a ground connection

Fig. 6.1. UL Mark.

may violate NEC rules (NFPA 2002). Machine b is identical to machine a, except the branch circuit includes a green or bare conductor extending from the neutral buss in the panel to the frame of the machine. Assume that an accidental connection occurs between the hot circuit conductor and the machine frame, on each machine. Possible causes for such a connection include:

- Conductor insulation rubbed on the machine frame until the insulation wore through.
- A sharp object accidentally cut through the insulation.
- The installer stripped too much insulation off the conductor.
- A rodent chewed through the insulation.

Since the frame of machine a is not grounded, this accidental connection, by itself, causes no change in load operation and the machine will operate normally. However, the machine frame is now energized to 120 V, with respect to ground.

When a person standing on the ground contacts the machine frame, the person's body closes the circuit and the person is likely to be shocked. If the person's resistance is at the minimum of 500 Ω (see Table 3.1), Equation 3.2 predicts that shock current will reach 240 mA. This is in the fibrillation range (see Table 3.2) and death can result. This electrocution, described in a NIOSH study, apparently occurred in a similar scenario:

> Refrigeration technician performing maintenance tasks electrocuted when he contacted the improperly grounded refrigeration unit of a walk-in cooler at a restaurant (Case 91-32, NIOSH 1998).

Machine b is safety-grounded by a bare or green grounding conductor connected to the machine frame

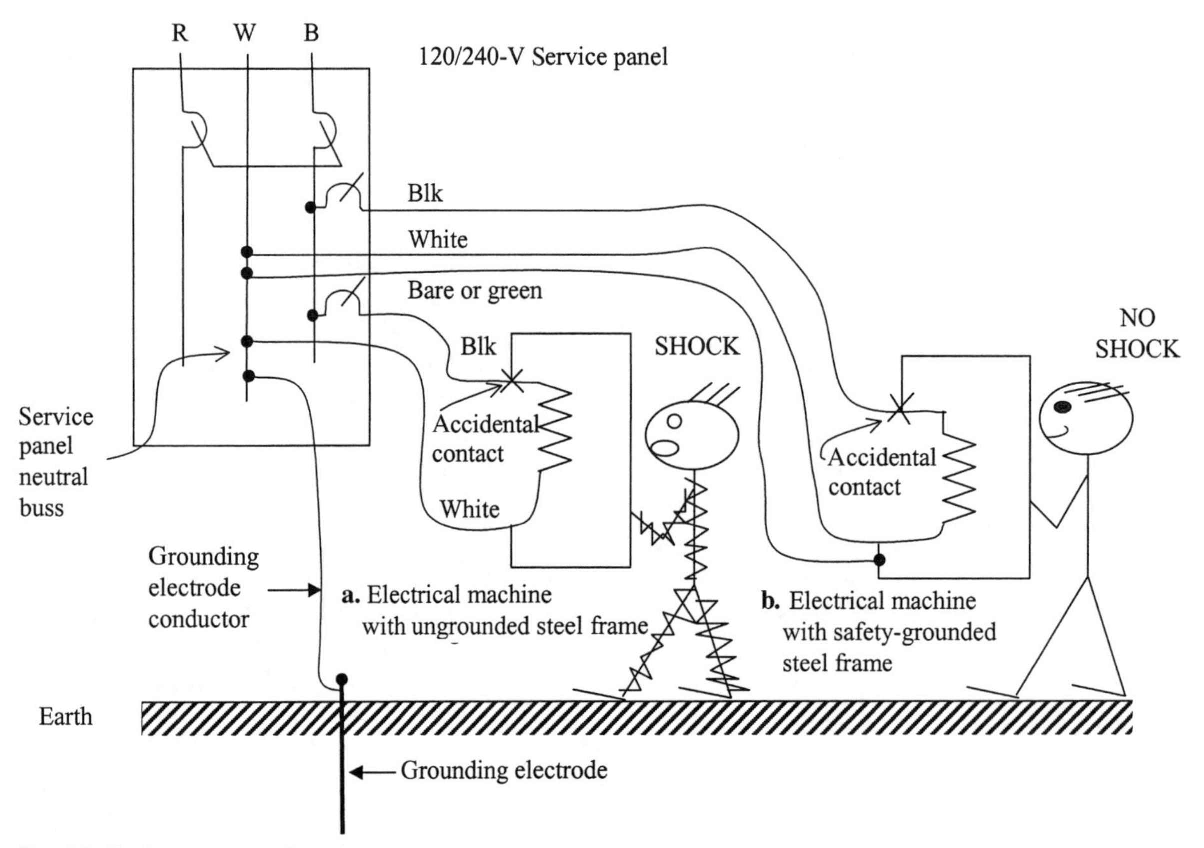

Fig. 6.2. Equipment grounding.

and extended to the neutral buss in the service panel. When the accidental contact occurs, a circuit is closed and one of two things will happen:

- If there is good contact (that is, the contact produces a low-resistance path from the machine frame to the black conductor), a short circuit occurs and the branch circuit overcurrent device opens.
- If there is only poor contact (that is, the path to the black conductor has high resistance), a leakage current too small to open the overcurrent device will flow through the grounding conductor. This leakage current may continue to flow indefinitely.

In either case, the machine frame remains at near ground potential and the person is not shocked. The voltage between the machine frame and ground is not high enough to produce a shock when a person completes the circuit.

Equipment Grounding Rules

In general, the NEC requires exposed non-current carrying metal parts of cord-and-plug-connected, and permanently wired electrical equipment, to be grounded. The many associated rules and exceptions are found in NEC Article 250 (NFPA 2002). The grounding conductor is usually the green or bare conductor carried along with the circuit conductors, as illustrated in Figure 6.2b. Plugs and receptacles are designed with terminals included for the grounding conductor. The most common is the NEMA (National Electrical Manufacturers Association) 5-15R and 5-15P 2-pole, 3-wire plug and receptacle (Figure 6.3). NEMA lists this set as 2-pole, 3-wire. The two poles are the current-carrying blades. The grounding electrode is counted as a wire but not as a pole. NEC requires use of grounding type receptacles virtually everywhere (NEC 210-7). The principles illustrated here are employed in many types of equipment grounding systems.

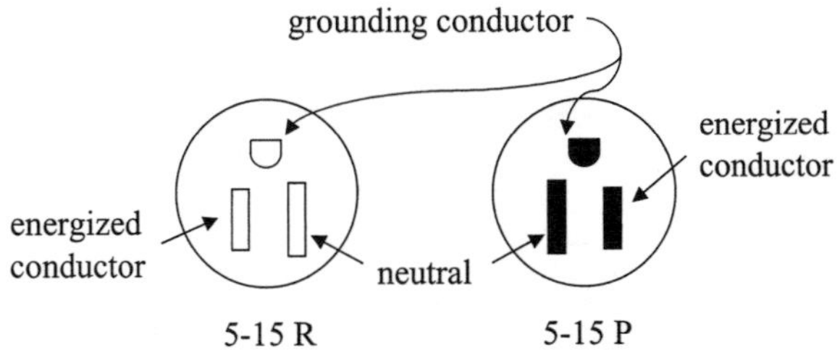

Fig. 6.3. NEMA 15-A, 125-V receptacle (left) and plug (right).

System Grounding

The basic approach to system grounding consists of connecting the main service neutral and the service panel neutral in each building to earth by means of a grounding electrode, which is typically an 8-ft-long copper or copper-clad steel rod driven into the earth. A grounding electrode conductor connects the grounding electrode and the neutral (Figure 6.2). There are many additional code rules related to system grounding. Refer to NEC Article 250, Hiatt 2000 Section 29, and MWPS 1992 for system grounding rules and installation procedures. Local codes may contain additional system grounding requirements.

Double Insulation

Double insulation is an alternative to equipment grounding that has been in use in the United States since the late 1960s. A double-insulated tool or appliance has only two conductors in its connection cord (Figure 6.4). There is no grounding conductor and there is no grounding electrode in the plug.

To be called *double insulated*, the tool or appliance must have a superior insulation system. It has normal electrical insulation required to make it function, and in addition, all electrical parts (motor, brushes, switches) are surrounded by additional insulation, or air space. Exposed parts are either non-conducting, or if conducting, are isolated from electrical parts by a

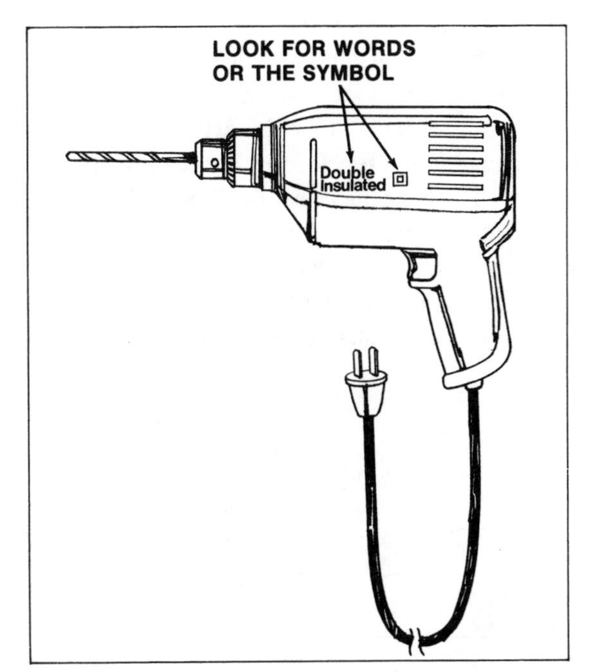

Fig. 6.4. Double-insulated 115-V, 3/8-in portable drill. (John Deere Publishing, Moline, IL; reprinted with permission.)

non-conducting link. For example, the drive trains of double-insulated drills contain a non-conducting link so that there is no conducting path from the chuck to the motor rotor. Double-insulated tools and appliances are usually exempt from the NEC safety grounding rule (NEC 250-114).

Double insulation eliminates some problems inherent with safety-grounded tools and appliances. For example, if you are holding the metal frame of a safety-grounded appliance, your body is well grounded, and simultaneous contact with some energized object can result in a shock which would not have occurred in the absence of safety grounding. Double-insulated shavers have, however, caused electrocutions which would have been unlikely with safety grounding (Dalziel 1972). Electrocutions have occurred when victims reached for energized cord-powered double-insulated shavers that had fallen into wash basins or toilet bowls.

Ground-Fault Circuit Interrupters

A ground-fault circuit interrupter (GFCI) is a device capable of sensing a ground fault on a branch circuit and then opening the circuit when the ground fault current exceeds a certain value for a certain period of time. A ground fault occurs when there is a closed circuit from an energized conductor to ground or earth. For example, if a person's body completes a circuit from an energized conductor to ground, current through the body constitutes a ground fault. If the circuit is equipped with a GFCI, it should be able to sense the ground fault and open the circuit fast enough to avoid a shock that can cause harm.

Operation of a GFCI can be explained by reference to Figure 6.5. Black and white circuit conductors extend through the center of a donut-shaped differential transformer. When any kind of a normal load is operating, the instantaneous directional sum of currents through the donut is always zero. For example, if at some instant 10 A is flowing toward the load through the black wire, there is simultaneously 10 A flowing away from the load in the white wire. The instantaneous sum is zero, and the differential transformer secondary current is zero. Whenever a ground fault occurs, ground fault current carried by the black conductor is flowing through the differential transformer, but then flowing through the ground fault to earth and not through the neutral circuit conductor. This creates a current imbalance at the differential transformer and results in a current in its secondary, which feeds into a solid-state circuit board. When the imbalance (which is the absolute value of the instantaneous current difference between I_W and I_B) exceeds the *trip value* (about 5 mA), the circuit board signals a circuit breaker to open the black conductor. This all happens in about .025 sec, which is fast enough so that if the ground fault current is flowing through a person's body, the person probably will not go into fibrillation.

Some other factors related to ground fault interrupters:

- The GFCI detects only ground faults. If a person is not in contact with the earth because he or she is wearing rubber-soled shoes or is standing on a dry wood floor and is being shocked by touching the black conductor and the white conductor, a GFCI will not open the circuit because there is no ground fault.
- When a GFCI trips, only the black (energized) conductor is opened.
- A GFCI obtains power needed to operate from the circuit it serves.
- Every GFCI has a test button. Pressing the test button causes a 5- to 7-mA current to flow through the differential transformer in the black conductor, but around the differential transformer on the return. The GFCI reacts as if there is a ground fault, and trips. Manufacturers recommend testing GFCI operation by pressing the test button monthly.

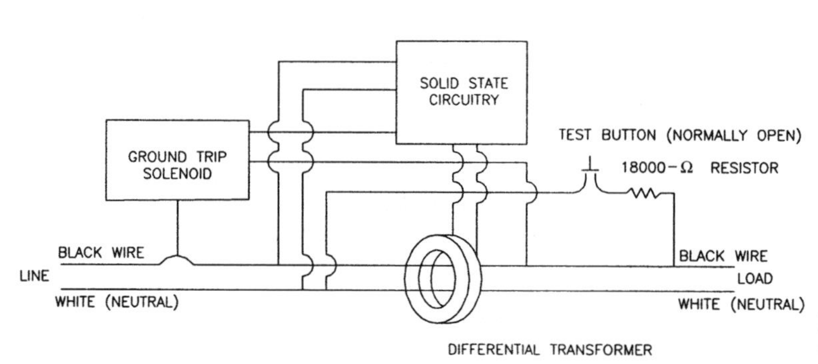

Fig. 6.5. Ground fault circuit interrupter (GFCI) operation principles.

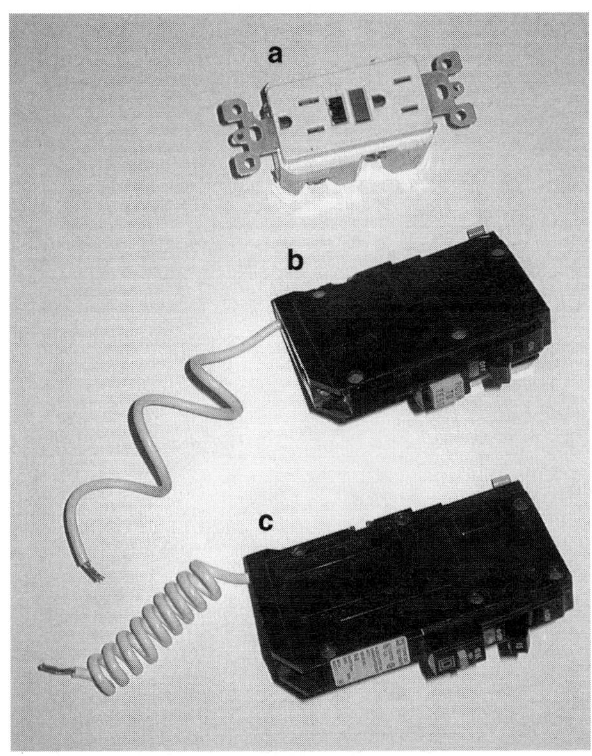

Fig. 6.6. Ground fault and arc fault circuit interrupters: a. Receptacle type GFCI; b. Circuit breaker type GFCI; c. Circuit breaker type AFCI.

- GFCIs are also used for equipment protection. Such GFCIs will have different time-trip current values than class A GFCIs for personal protection, which are discussed here.

GFCI Types

The two common types of GFCIs are the receptacle type (Figure 6.6a), which combines a duplex receptacle and a GFCI, and the circuit breaker type (Figure 6.6b), which combines circuit breaker and GFCI. Operating characteristics are similar for each of these types. The receptacle-type GFCI is designed to provide ground fault protection to all downstream receptacles on the same circuit. Circuit breaker-type GFCIs cost five to eight times as much as receptacle types and are less popular. GFCIs are sometimes included with other equipment such as drop cords and portable hair dryers.

GFCI Characteristics

UL requires that a Class A GFCI be capable of opening a circuit at least as fast as the time predicted by Equation 6.1 (Beausoliel and Meese 1976).

$$T = \left(\frac{20}{I}\right)^{1.43} \qquad \textbf{(6.1)}$$

The Dalziel fibrillation line (Equation 6.2) is also shown.

$$T = \frac{13{,}456}{I^2} \qquad \textbf{(6.2)}$$

where T = time, s
I = fault current, mA

Figure 6.7 includes the line defined by Equation 6.1. An example will illustrate use of the figure.

Example 6.1

An equipment failure has caused a fault current of 90 mA to flow in a branch circuit equipped with a GFCI. Refer to Figure 6.7.

a. What is the maximum GFCI trip time allowed by UL, at this fault current level? Reading down from the UL requirement line, at 90 mA, T ≅ 0.11 s. Solving Equation 6.1: T = 0.12 s.
b. What is the maximum time an adult can endure this shock current without danger of fibrillation? Reading down from the electrocution threshold line, T ≅ 1.7 s. Solving Equation 6.2: T = 1.7 s.
c. What is the typical trip time if this circuit has an additional 15 A load? Reading down from the 15-A load line T ≅ 0.03 s.
d. What is the typical trip time if this circuit has no additional load? Reading down from the 0-A load line, T ≅ 0.015 s.

GFCI Use

GFCIs are used for protection of personnel in areas where moisture or some other factor can increase the likelihood of a ground fault shock. The NEC requires that GFCI protection be installed on all receptacles on 120-V, 15- and 20-A circuits in these areas in dwelling units and other buildings (NEC 210-8):

- Bathrooms
- Accessible receptacles in garages and other non-habited buildings having a floor at or below grade and used for storage or work
- Accessible outdoor receptacles
- In crawlspaces at or below grade
- In unfinished basements
- On kitchen and wet bar countertops
- On rooftops
- In locations such as livestock buildings equipped with equipotential planes (NEC 547-9)

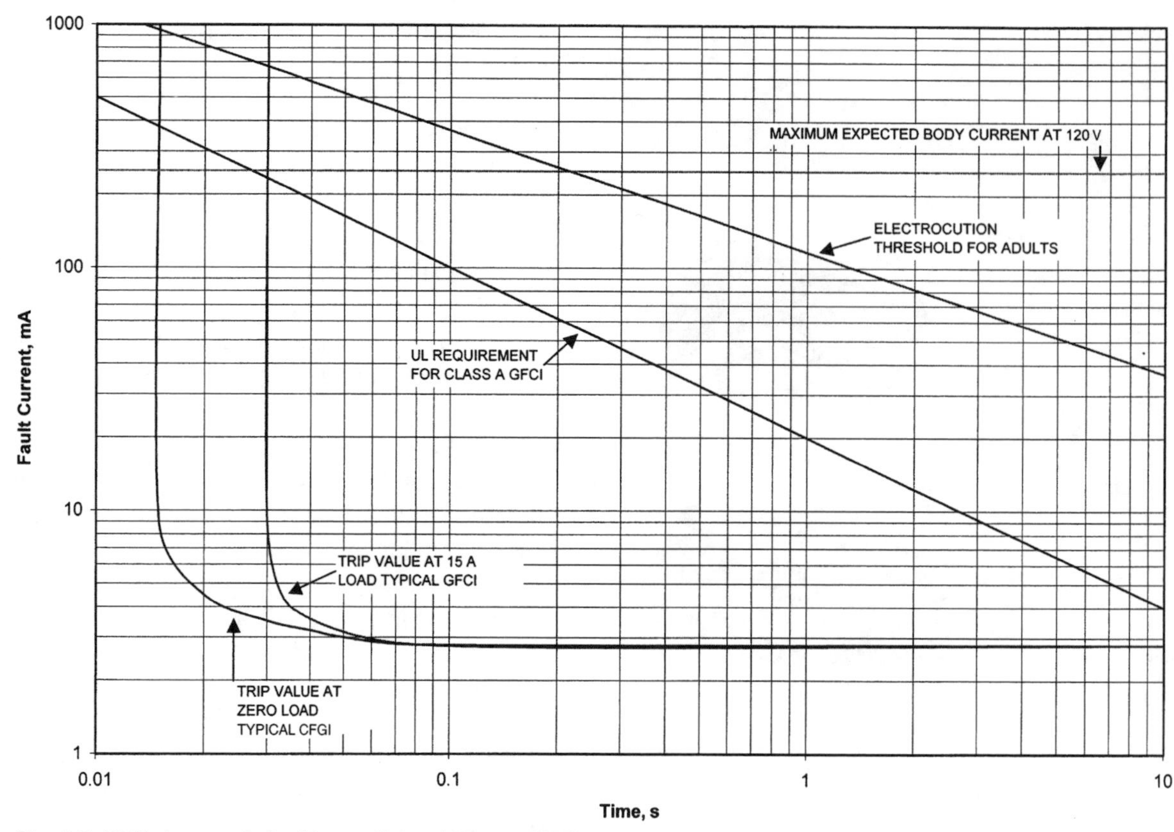

Fig. 6.7. GFCI characteristics (Beausoliel and Meese, 1976).

- For receptacles serving water pumps or other loads serving a pool [NEC 680-6 (a)(1)]
- For receptacles within 20 ft of the inside walls of a pool or fountain [NEC 680-6 (a)(3)]

Lighting outlets in close proximity to pools are required to be GFCI protected [NEC 680-6 (b)(1)].

Refer to the NEC for exceptions to, and further explanations of, these rules. The NEC does not prohibit use of GFCIs anywhere, so they can be applied in other areas where there is a perceived ground fault shock hazard.

Although GFCI receptacles, and other receptacles served by GFCIs, are equipped for 3-wire grounding plugs, they work equally well for 2-wire loads. GFCIs are also available for 240-V circuits.

Nuisance Tripping

Nuisance tripping of a GFCI occurs when current leakage, and not a ground fault, causes the GFCI to trip. The 5-mA trip current is so low that some loads may regularly have current leakage values which exceed this value. Examples of loads which may cause nuisance tripping:

- Certain brands of personal computers
- Circuits over 100 ft long
- Certain gaseous discharge lamps

Damp conditions increase likelihood of nuisance tripping. If an appliance trips the GFCI when plugged in (especially older, 3-wire power tools), this may indicate a faulty appliance which should not be used.

Arc-Fault Circuit Interrupters

Arc-fault interrupters (AFCIs) are specialized circuit breakers that can open branch circuits when a fault on the circuit is causing arcing. The arcing can occur when insulation breakdown causes a short circuit and allows contact and current flow between the hot and neutral conductors of the circuit.

Arcing caused by such faults may not result in currents high enough or long enough to cause conventional circuit breakers to open; however, they may cause fires. Residential bedrooms are likely locations for such fires (Engel 2002). Beginning in 2002, NEC 210 (B) requires AFCI protection on branch circuits supplying 15-A or 20-A outlets in dwelling unit bedrooms (NFPA 2002).

Circuit breaker-AFCI combination units are similar in appearance to circuit breaker GFCIs (Figure 6.6c). AFCI test buttons are blue, to distinguish them from yellow GFCI test buttons.

Overcurrent Protection

The purpose of overcurrent protection is to open a circuit when the current is high enough, for a sufficient length of time, to damage or shorten the life of the insulation on the circuit conductor. Circuit overcurrent device ratings should always be less than or equal to the conductor ampacities, except for rare instances, such as electric motor circuits, where another device protects the conductor. Since circuit ratings are typically 15 or 20 A or more, overcurrent devices seldom, if ever, provide personal protection.

Overcurrent Devices

The common overcurrent devices are the fuse and the circuit breaker. Symbols are shown in Figure 6.8.

Operation of Fuses

A fuse is a device which can be inserted in an electrical circuit for overcurrent protection. A fuse contains a fusible link through which the current flows. "Fusible" means capable of being liquefied by heat. The fusible link is commonly made of zinc or copper. A fuse is said to "blow" when it opens. Figure 6.9 shows the action of a simple fuse carrying an overload current.

During overload, the fuse is carrying a current which exceeds its rating, but no more than 10 times its rating. First, the center of the narrow section approaches the molten state (Figure 6.9a). Then, the center portion drops away and an arc is established across the gap, and is extinguished (Figure 6.9d). This opens the circuit. The time for all of this to occur may be a fraction of a second, or many minutes, depending on fuse rating, circuit current, and many other factors. Under short-circuit conditions (currents more than 10 times the device rating), the entire center section of the fuse is rapidly heated to a temperature high enough to volatilize it (Figure 6.10). Arc-extinguishing filler in the fuse cools and condenses the vapor. This stops the arc and opens the circuit.

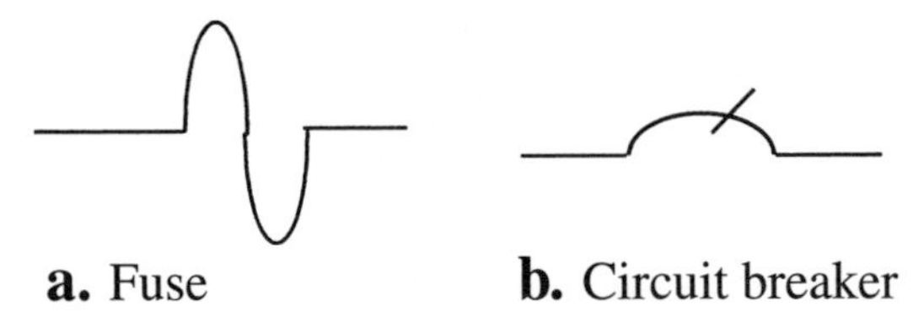

Fig. 6.8. Overcurrent device symbols.

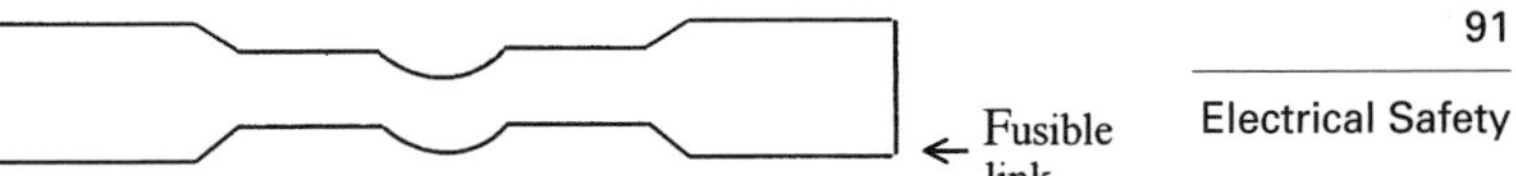

a. Center portion is overheated and sagging.

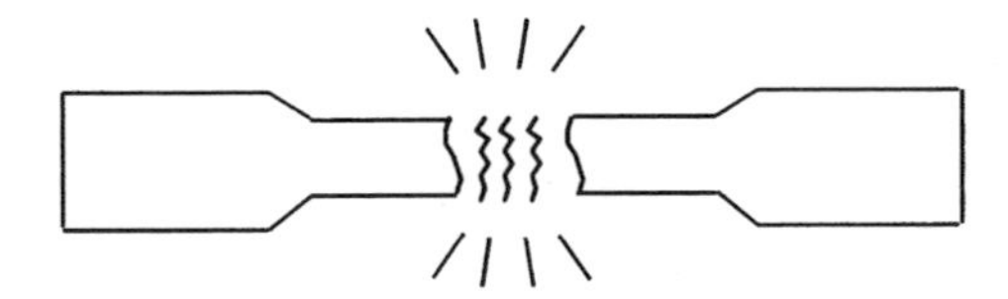

b. Center portion has melted and dropped away. An arc has been established across the gap.

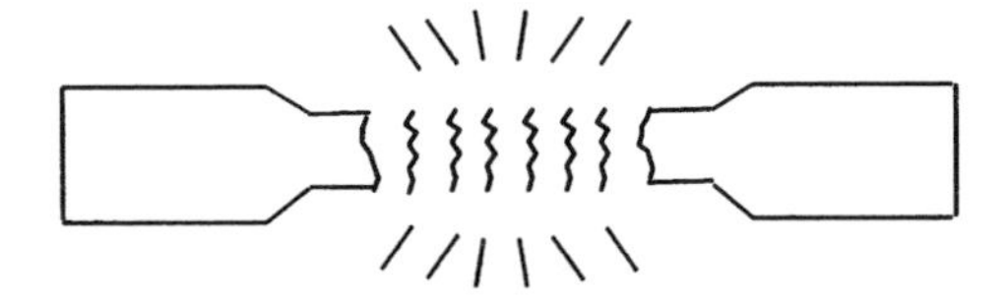

c. Heat of the arc burns back metal at each end.

d. Arc is extinguished because gap became too wide. The circuit is open.

Fig. 6.9. Operation of a simple fuse during an overload.

Dual Element Fuses

A dual element or "time-delay" fuse has two elements in series and either element can open the circuit. Figure 6.11a shows the elements, which consist of a fusible link in series with a spring-loaded soldered junction. Under short-circuit conditions, the fusible link vaporizes to open the circuit (Figure 6.9b). Under long-time overload conditions, solder at the junction heats up and melts, allowing the spring to separate the parts and open the circuit (Figure 6.11c). The time required to open the circuit is longer than for a

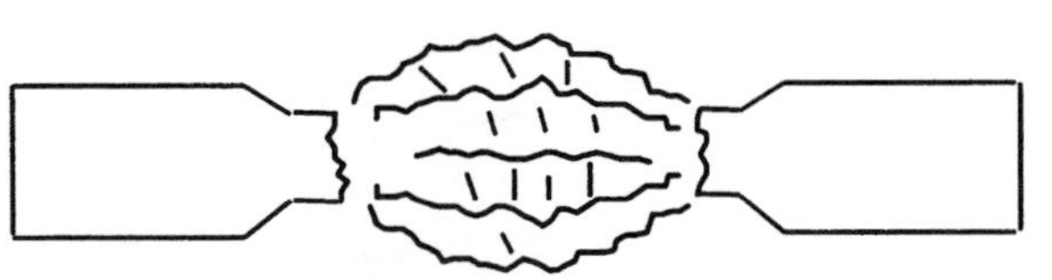

Fig. 6.10. Operation of a simple fuse during a short circuit.

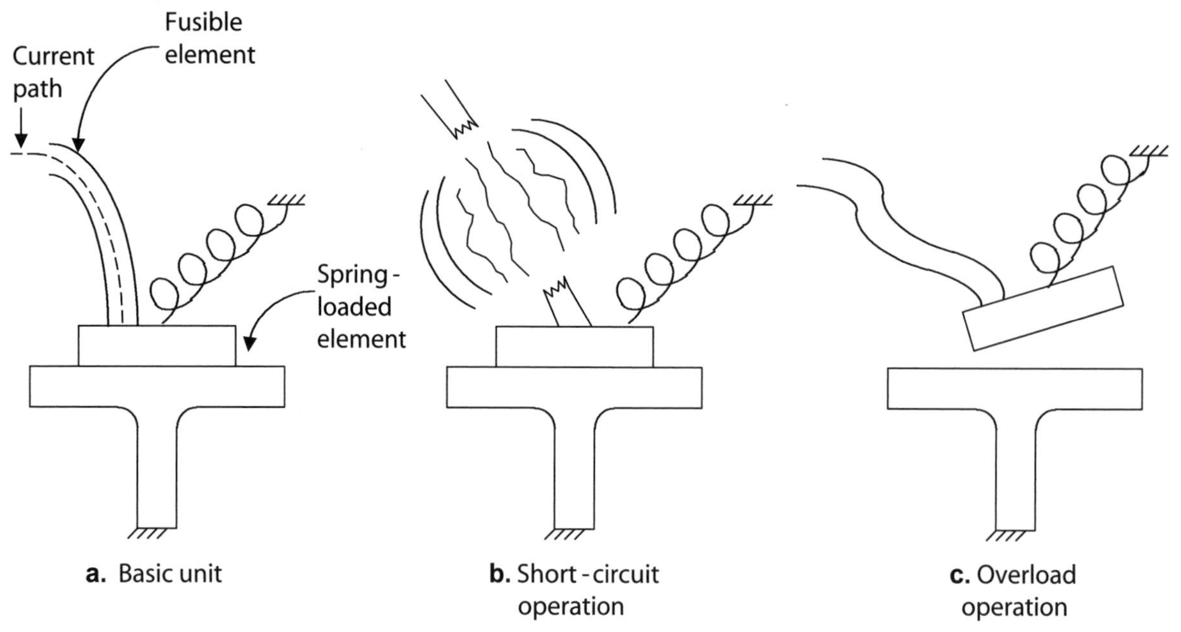

Fig. 6.11. Dual-element fuse operation.

conventional fuse. This added time can allow a motor to start without blowing the fuse needlessly.

Circuit Breakers

A circuit breaker is a mechanical overcurrent device. The active element inside which causes it to trip can be thermal, magnetic, or a combination of thermal and magnetic. Action of a thermal circuit breaker is illustrated in Figure 6.12. The thermal element is a bimetallic strip consisting of two metals having different coefficients of thermal expansion. As the element heats up due to circuit current, the strip bends toward the side with the lower coefficient of thermal expansion. This permits the spring to pull apart the contacts and open the circuit. Since the tripping action depends on a heat buildup over time, this type may be too slow to open under short-circuit conditions.

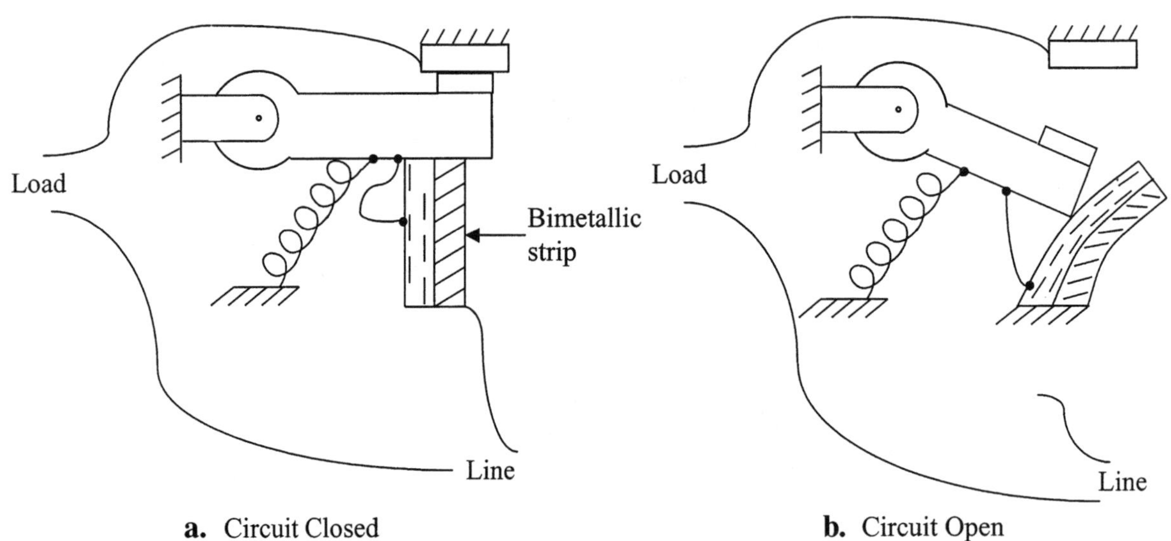

Fig. 6.12. Operation of a thermal circuit breaker.

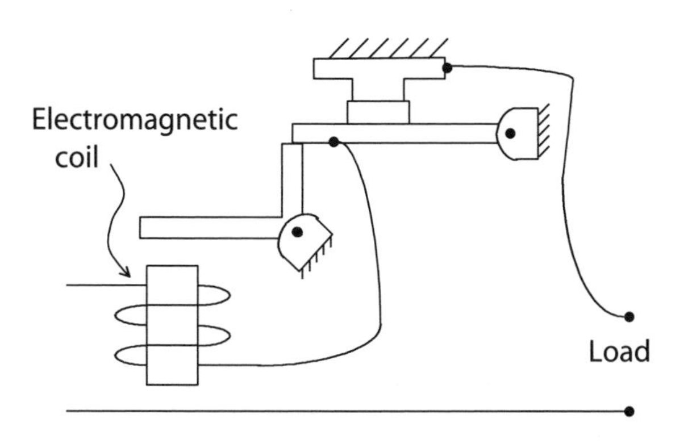

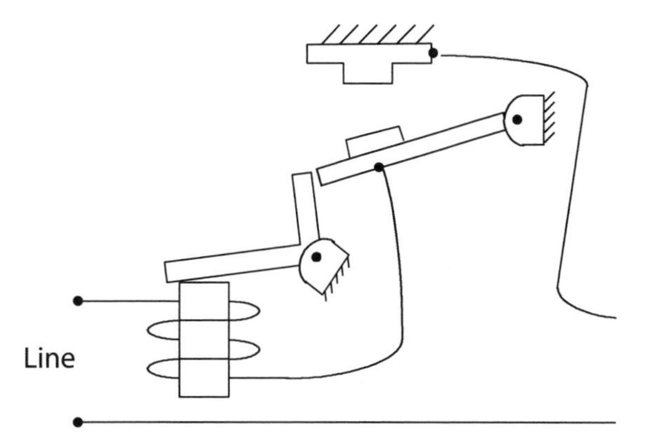

a. Circuit closed **b.** Circuit open

Fig. 6.13. Operation of a magnetic circuit breaker.

Figure 6.13 shows the operation of a magnetic circuit breaker. The active element is an electromagnetic coil. Force exerted by the coil increases as current increases. At a certain current value, coil force pulls down the trip lever allowing the spring to separate the contacts and open the circuit. The magnetic device may tend to trip too fast during overload condition, causing a nuisance trip of the breaker. The fast action of the magnetic breaker for short-circuit conditions and the slow response of the thermal-type breaker are combined in the thermal-magnetic breaker, which contains both types of elements. Thermal-magnetic mechanisms are used in standard circuit breakers.

Overcurrent Device Characteristics

Figure 6.14 is a graph of typical time-trip characteristics for a standard (inverse-time) circuit breaker. "Inverse time" means the trip time varies inversely, or decreases as current increases. An example illustrates use of the characteristics curve.

Example 6.2

A 10-A inverse-time breaker is carrying 60 A. How long will it take for the breaker to open the circuit?

$$\frac{60\ \not{A}}{10\ \not{A}} = 6 \text{ times its rating or } 600\% \text{ of rating}$$

The breaker is carrying 600% of its rating. Reading up to the line of Figure 6.14 at 600% of rated current, the trip time is about 1.8 s.

Comparing Fuses and Circuit Breakers

For many agricultural applications, either fuses or circuit breakers are available for overcurrent protection. Most new applications use circuit breakers. Table 6.1 lists some advantages of each device.

Circuit breakers in the service entrance panel are inherently safer than fuses in the service entrance panel. Properly installed circuit breakers have all potential contact points fully enclosed, whereas when a fuse is removed, there is a possible shock hazard. This shock hazard is similar to the hazard present when changing a light bulb, and requires care.

Since the circuit breakers are fully enclosed, they are less prone to tampering. When a circuit breaker trips, the circuit can be easily reset. However, when fuses blow, there is a possibility of replacing the fuse with a higher rated fuse. Replacing a 15-A fuse with a 20-A fuse may allow the load to operate, but the increased current can cause a fire hazard. Completely bypassing a fuse, such as placing a penny behind the fuse, can result in tragedy. Such tampering is unlikely but still possible with circuit breakers.

Circuit breakers offer more flexibility for controlling circuits. In addition to providing overcurrent protection, they can be used as a switch to control a load under certain circumstances. A common appli-

TABLE 6.1 Circuit breaker–fuse comparison

Circuit Breaker	Fuse
Can be used as a switch	Lower first cost
Less prone to tampering	Higher reliability
Easier to re-close circuit	Usually higher in current interrupting capacity
Dead-front box	

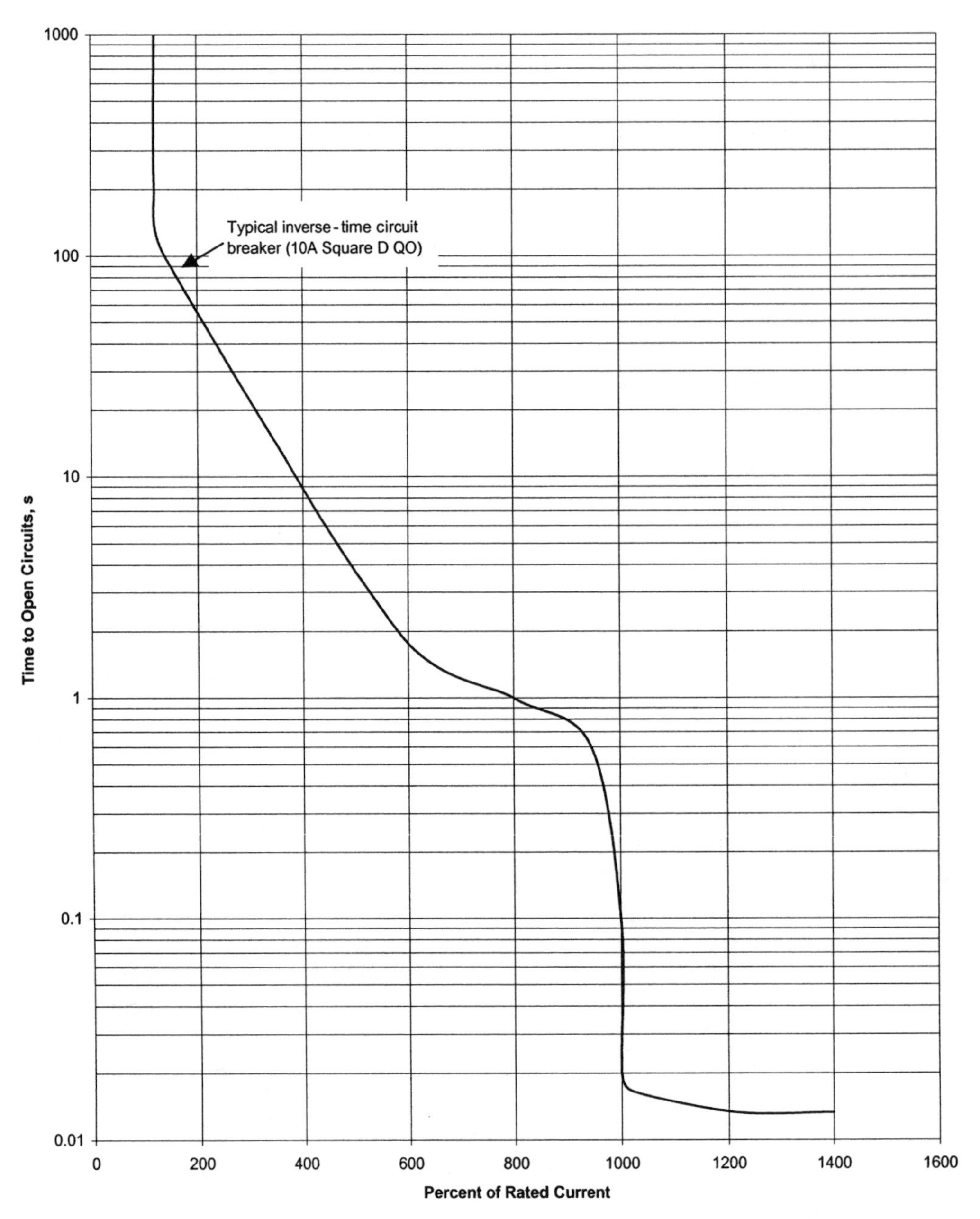

Fig. 6.14. Qo characteristic curve: typical overcurrent device characteristics. (Square D Inc., Palatine, IL; reprinted with permission.)

cation of using a circuit breaker to control a motor is a silo unloader in a cattle-feeding operation. If fuses are used for overcurrent protection, a device such as a knife switch is required to control the motor.

Circuit breakers can become inoperative if used in some corrosive or wet environments. This results in an unsafe situation. Exercising the mechanism periodically by turning it off and on is a good practice.

References

Beausoliel, Robert W., and William J. Meese. 1976. Survey of ground-fault circuit interrupter usage for protection against hazardous shock. National Bureau of Standards, U.S. Department of Commerce, Washington, DC.

Dalziel, C. F. 1972. Electric shock hazard. IEEE Spectrum 9: 41–50.

Engel, J. 2002. Arc-fault circuit interrupters: Bringing a new level of electrical protection into the home. NEC Digest (February-March): 44–49.

Hiatt, R. S. 2000. Agricultural wiring handbook, 12th ed. National Food and Energy Council, Inc. Columbia, MO.
MWPS. 1992. Farm buildings wiring handbook, 2nd ed. Midwest Plan Service, Iowa State University, Ames, IA.
NFPA. 2002. National electrical code 2002. National Fire Protection Association, Quincy, MA.
NIOSH. 1998. Worker deaths by electrocution. NIOSH publication 98-131. NIOSH, Cincinnati, OH.

Problems

6.1. A circuit breaker–GFCI combination was installed in a residence service panel. (This is an actual case that happened in Ames.) The GFCI was installed to protect an outdoor receptacle and a bathroom receptacle. When loads were plugged into the bath receptacle only, they operated normally. When any load was plugged into the outdoor receptacle, the GFCI tripped and power to bath and outdoor receptacles was shut off even though no ground fault existed. Draw the circuit as incorrectly wired.

6.2. As a person removes a 120-V plug from a wall receptacle, the person's right thumb and index finger simultaneously contact the neutral and energized terminals of the plug before their contact with the receptacle is broken.

(a) Scenario I: The receptacle is not GFCI-protected and the person is standing on a wet concrete floor. Explain what is likely to happen, and why.

(b) Scenario II: The receptacle is GFCI-protected and the person is standing on a wet concrete floor. Explain what is likely to happen, and why.

(c) Scenario III: The receptacle is not GFCI-protected and the person is standing on a dry rubber mat. Explain what is likely to happen, and why.

(d) Scenario IV: The receptacle is GFCI-protected and the person is standing on a dry rubber mat. Explain what is likely to happen, and why.

6.3. As a person removes a 120-V plug from a wall receptacle, the person's right index finger contacts the energized terminal of the plug before its contact with the receptacle is broken.

(a) Scenario I: The receptacle is not GFCI-protected and the person is standing on a wet concrete floor. Explain what is likely to happen, and why.

(b) Scenario II: The receptacle is GFCI-protected and the person is standing on a wet concrete floor. Explain what is likely to happen, and why.

(c) Scenario III: The receptacle is not GFCI-protected and the person is standing on a dry rubber mat. Explain what is likely to happen, and why.

(d) Scenario IV: The receptacle is GFCI-protected and the person is standing on a dry rubber mat. Explain what is likely to happen, and why.

6.4. An adult person is drawing a 200-mA hand-to-hand ground fault shock on a 15-A branch circuit having no other load.

(a) How long can the person endure this shock before fibrillation is likely?

(b) What is the maximum trip time allowed if the GFCI is to meet the UL standard for Class A GFCIs?

(c) What is the typical trip time for a GFCI under this condition?

6.5. The branch circuit overcurrent device on a motor circuit is a standard circuit breaker rated at 10 A.

(a) If the motor draws 80 A for 2 sec, as it starts, will the breaker open before the motor is started?

(b) How long can the motor take to start without opening the breaker if it draws 30 A during starting?

6.6. How long will it take for a 10-A standard breaker to open:

(a) If it is carrying 10 A?

(b) If a short circuit causes it to carry a current of 120 A?

Chapter 7

FARM ELECTRICAL SERVICE

A modern farmstead uses electrical power extensively in every building since productivity is so closely tied to use of electrical equipment. The system to supply this electric power needs to possess characteristics of safety, adequate capacity, expandability, and cost efficiency.

Safety is achieved by compliance with the current National Electrical Code (NFPA 2002), or applicable local code. Insurance companies may also have specific requirements related to safety, which they require as a condition of insurability.

Adequate capacity means that components are large enough to supply power needed at every location on the farmstead. A system with *expandability* is designed so that capacity can be increased in the future, with minimum cost and disruption.

A *cost-efficient* system considers ownership cost and operating cost of components and is designed for minimum total cost over the life of the system. The procedure to do this for a branch circuit was explained in chapter 5.

The steps in designing a farmstead electrical system are:

- Compute demand and specify service equipment for each farm building.
- Locate metering/distribution point on the farmstead.
- Compute capacity of farm service.
- Design service conductors for each building and for the farm.

Building Demand and Service Equipment

The procedure for computing building demand and selecting service equipment will be explained by means of examples.

Example 7.1

Compute building demand and select service equipment for a farm shop. Electrical equipment in the shop is specified, along with its current at 230 V:

	Amps (at 230 V)
Air compressor, 5 hp (28 A × 1.25)	35
Hydraulic pump, 5 hp	28
Drill press, 2 hp	12
Lathe, 2 hp	12
Vent fan, ½ hp	4.9
Grinder, ⅓ hp (7.2 @ 115 V)	3.6
LP furnace, ⅓ hp (7.2 @ 115 V)	3.6
Water heater 4,500 W	20
Welder	35
12 lighting outlets (1.5 A each @ 115 V)	9
20 duplex receptacles (1.5 A each @ 115 V)	15
Total connected load :	178.1

Notes:

- Full-load currents of motors are listed in Table 10.6.
- Following NEC 430.24, the full-load current of the largest motor, or one of the largest, is taken at 125% to allow extra capability for motor starting.
- For motors which will be operated at 115 V (grinder and LP furnace), list the current at 230 V.
- Lighting outlet demand is assumed to be 1.5 A per outlet at 115 V (NEC 220.3). This can be counted as 0.75 A at 230 V because when two 115-V outlets are balanced between red and black conductors, together they add 1.5 A of demand to the 230-V service. Three-wire circuits are discussed in chapter 1.
- Duplex receptacles are also assumed to contribute 0.75 A each to demand at 230 V (NEC 220.3).

The total connected load for the shop is 178 A at 230 V. Since it is very unlikely that all these loads will ever be on simultaneously, Table 7.1 provides a rational approach to computing the highest likely load to ever be needed.

First, we need to list shop loads that will operate "without diversity," that is, are likely to operate at one time. For the shop, we decide this will include the air compressor (35 A), water heater (20 A), and LP furnace (3.6 A), which start automatically, and lights (9 A) for a total of 68 A. This load is entered with a demand factor of 100%. The other loads will tend to be on selectively and individually. Taking the next 60 A at 50% adds 30 A. This leaves 178 − 68 − 60 = 50 A at 25%, which is 13 A. The sum, as shown on Table 7.1, is 111 A. We have estimated that the building demand for the shop is 111 A. The service panel and main circuit breaker need to be rated for at least 111 A. Common available sizes are 30, 60, 100, 150, 200, 225, 300, 400, 600, and 800 A. A 150-A panel and a 150-A main circuit breaker are chosen.

TABLE 7.1 Method for computing farm loads for other than dwelling unit (NEC Table 220.40)

Load at 230 V (A)	Demand factor (%)	Farm shop (A)
Loads expected to operate without diversity, but not less than 125% full load current of the largest motor and not less than the first 60 A of load	100	68
Next 60 A of all other loads	50	30
Remainder of other load	25	13
	Building demand:	111

Conclusion—The farm shop building demand is 111 A, and a 150-A panel and main circuit breaker are specified.

Location of Metering/Distribution Point

A central power distribution point is commonly used on farmsteads. This approach has some advantages, compared to distribution from the house or from one of the farm buildings. The meter, placed at the central distribution point, is easy to get to. In the case of a fire in one building, service to the other buildings is not disrupted. Services to new buildings can be added with minimum disruption. If the service distribution point is the load center (defined by use of amp × feet moments as shown in the following example) the investment in conductor is minimized. If overhead conductors are used, the presence of several conductors over the central part of the farmstead may be a disadvantage.

Example 7.2

Find the load center for the farmstead shown in Figure 7.1. This farmstead consists of the farm shop discussed in Example 7.1, plus three additional buildings and a well pump. For information on electrical needs of specific farm buildings and the farm residence, refer to the *Agricultural Wiring Handbook* (Hiatt 2000).

To find the load center, the calculation procedure is as follows:

Building	Demand, amps	X Distance, ft	X × Demand, amp ft	Y Distance, ft	Y × Demand, amp ft
Shop	111	100	11,100	210	23,310
Hog finishing	85	225	19,125	250	21,250
Residence	150	100	15,000	40	6,000
Well pump	10	175	1,750	10	100
Grains system	150	320	48,000	100	15,000
Sum of building demands	506		Total 94,975		Total 65,660

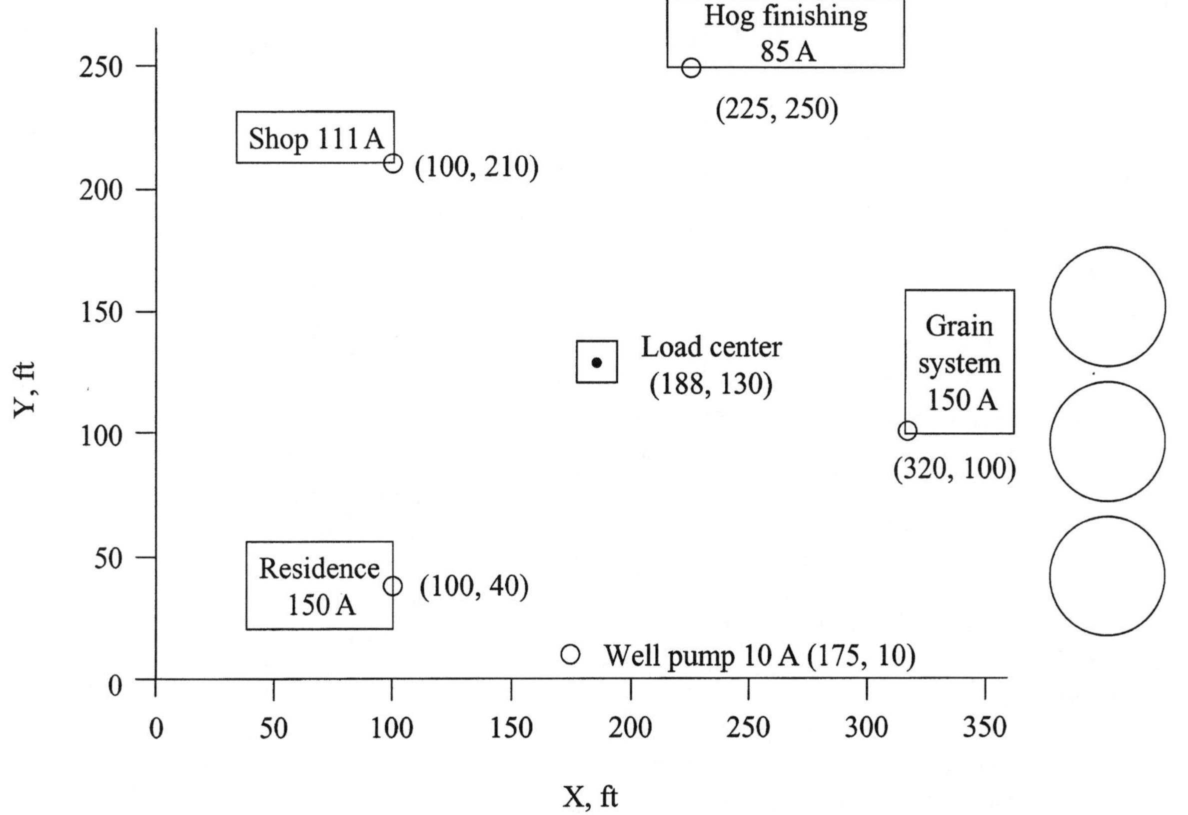

Fig. 7.1. Farmstead for Example 7.2.

$$\text{load center x distance} = \frac{94{,}975\ \text{A ft}}{506\ \text{A}} = 188\ \text{ft}$$

$$\text{load center y distance} = \frac{65{,}600\ \text{A ft}}{506\ \text{A}} = 130\ \text{ft}$$

Conclusion—The load center of the farm is at point x = 188 ft, y = 130 ft. Considerations such as landscaping, topography, or trees may dictate moving the distribution point slightly from the calculated load center. This is not a problem, since most of the advantages of using this procedure will still be true. Selection of the actual distribution point should be done in consultation with the electric power company.

Capacity of Farm Service

NEC article 220.41 defines a procedure for calculating a total farm load, taking into account diversity among the various loads. The procedure is explained in Example 7.3.

Example 7.3

Calculate the total farm load for the farm defined in Example 7.2. Table 7.2 shows the calculation procedure.

Conclusion—The total farm load is 443 A. This is a rational upper limit of the demand load likely for this farm. Note that it is substantially lower than the total of the building demands (506 A).

Designing Service Conductors

With distances and demand loads defined, it is now possible to design the service conductors extending from the load center of the farm to the individual buildings. These conductors are called "service laterals" if they are run underground and "service drops" if they are run overhead. The procedure is explained in Example 7.4.

Example 7.4

Design the service conductors for the farmstead system described in Example 7.3. Farmstead service conductors can be designed in several ways:

1. *Transformer on Meter Pole, all Overhead Conductors*—This is the least cost alternative, but there will be six sets of overhead conductors connected to the meter pole, and one set will be bare, high-voltage conductors. Traffic under these

TABLE 7.2 Total farm load calculation (NEC 220.41)

Load	Building Demand, amps	Demand Factor	Demand, amps
Residence Other loads	150	100%	150
Largest load			
grain center	150	100%	150
2nd largest load			
shop	111	75%	83
3rd largest load			
hog finishing	85	65%	55
Sum of remaining loads			
well pump	10	50%	5
Sum of building demands	506 amps	Total farm load	443 amps

conductors means there is a possibility of entanglement and injury. Some may think these conductors are unsightly.

2. *Transformer on a Distribution Line Post, all Overhead Conductors*—Cost is higher, but safety is improved because no high-voltage conductors pass over the farm yard, and since all conductors are triplex, all energized conductors are insulated.
3. *All Underground, with Transformer at Load Center*—With this alternative, high-voltage conductors are extended underground to the farm load center. There, the service transformer is placed on a concrete pad above ground, along with the watthour meter. Service laterals are run underground to each building. This alternative costs at least twice as much as the previous overhead designs, but the hazard of overhead entanglement is eliminated. Also, the possibility of poles and lines coming down in wind and ice storms is eliminated. Virtually all new residential suburban services are designed this way. Service laterals may be subject to rodent damage. Nicks in the insulation can let in moisture and lead to rapid conductor failure. Service lateral failures are difficult to locate and expensive to repair.

The underground system will be chosen for this example. The conductors extending from the power line to the transformer are high voltage, and will not be sized here.

On the secondary (low-voltage) side of the transformer, a system capable of supplying the total farm load of 443 A needs to be installed. A schematic is shown in Figure 7.2.

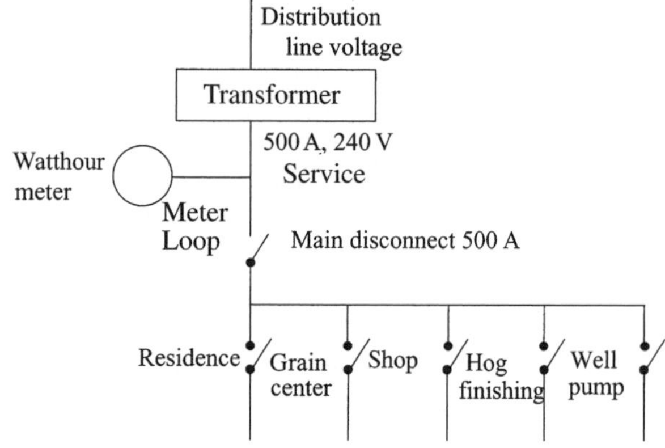

Fig. 7.2. Farm distribution system for Example 7.3.

We can now go through the conductor design procedure from chapter 5. Considering the environment (direct burial), UF and USE cable are suitable. However, only USE is approved for service conductors. We choose 3-conductor USE and assume it will be copper.

Conductor design numbers are summarized in Table 7.3. We'll start with the residence. We can calculate distance to the load center using coordinates on Figure 7.1:

$$D = \sqrt{(188-100)^2 + (130-40)^2} = 126 \text{ ft}$$

Unless better data are available, we increase this distance by 10% to allow for connections and other extra distances. For 150 A, copper USE, direct

TABLE 7.3 Farmstead service laterals for Example 7.4

Load	Building Demand, amps	Distance, ft	Ampacity Size, AWG	Voltage Drop Size, AWG	Size to Use, AWG
Residence	150	139	1/0	2/0	2/0
Shop	111	131	2	1/0	1/0
Hog finishing	85	138	3	1/0	1/0
Grain system	150	149	1/0	2/0	2/0
Well pump	10	133	14	10	10

burial, the minimum size for ampacity is AWG-1/0 (Table 5.7, left section). For 2% voltage drop, and 139 ft, enter the right section of Table 5.7. The size (for 150 ft) is AWG-2/0. Therefore, the 2/0 size is chosen. A similar procedure is used for the other loads, and the larger of the ampacity and voltage drop sizes is specified for use.

For wiring procedures and details not discussed here, refer to the *Agricultural Wiring Handbook* (Hiatt 2000) and to the *Farm Buildings Wiring Handbook* (MWPS 1992).

References

Hiatt, R. S. 2000. Agricultural wiring handbook, 12th ed. National Food and Energy Council, Inc., Columbia, MO.
MWPS. 1992. Farm buildings wiring handbook, 2nd ed. Midwest Plan Service, Ames, IA.
NFPA. 2002. National electrical code 2002. National Fire Protection Assn., Inc. Quincy, MA.

Problems

7.1. Calculate total connected load and building demand for a grain center having these loads: Eight receptacles, 12 lights, one 7.5-hp fan, four 5-hp fans, one 3-hp conveyor, three 1-hp conveyors, one 2.5-kW heater. All lights, two receptacles, the large fan, one 5-hp fan, and all conveyors are likely to operate without diversity.

7.2. A wiring system is being designed for the farmstead below. Because of staggered scheduling of loads, there is diversity among the three identical poultry houses.

(a) Locate the load center.

(b) Calculate the total farm demand.

(c) Specify sizes of all circuit breakers at the load center, assuming there is one main disconnect, plus a disconnect for each building.

(d) Specify conductor for each service lateral. Specification should include conductor insulation type, number of conductors, material, size, length.

(e) Specify panel/main breaker amperage rating for each building.

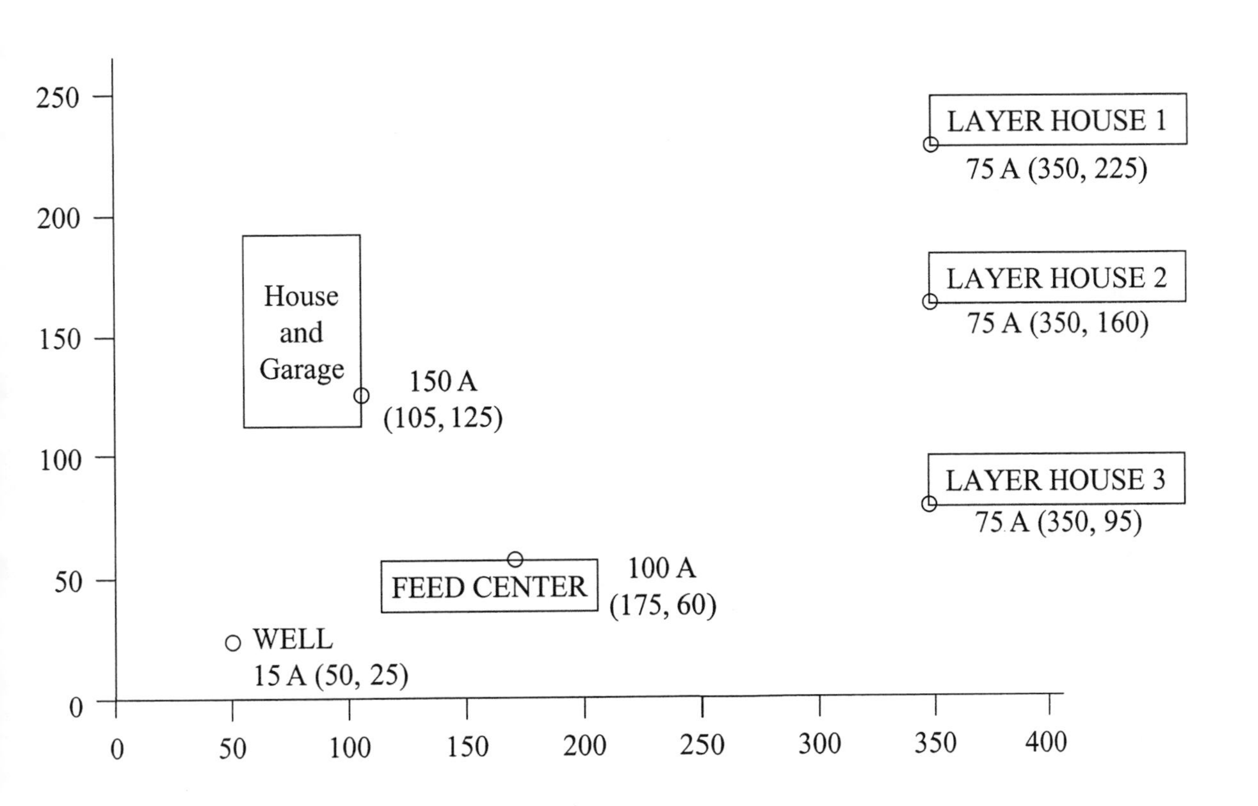

Chapter 8

ELECTRICAL CONTROLS

"Power is nothing without control." The author of this statement probably did not have electric power in mind when this quotation was written, but it still holds true. Electricity is mostly useless and dangerous without proper control. Starting with the basic switch, we'll discuss several control devices and circuits in common use with electric power.

Switches

The basic electrical switch (Figure 8.1) was discussed in chapter 1. It is convenient to refer to the moving part as the *pole* and the stationary part as the *throw.*

Open switch = Pole not contacting throw, and no complete circuit through the switch.

Closed switch = Pole contacting throw and a complete circuit present through the switch.

Note that this terminology is opposite to the terminology used with valves for fluids, where an open valve will allow liquid flow and a closed valve prevents flow.

Switches are rated by a maximum voltage they are insulated for, and for the maximum current they are capable of switching. For example, common room light switches usually have a 300-V, 15-A rating. Sometimes switches are rated in hp for use on electric motor circuits. This is discussed in more detail in chapter 10. Also, some switches have different current ratings depending on whether they are used with AC or DC.

Four-Letter Switch Notation

Switches are often designated as to function by four initials (Table 8.1). To derive the four-letter designation:

- Count the total number of poles (SP for single pole, DP for double pole, etc.).
- Count the number of throws *each pole* can contact (ST for single throw, DT for double throw, etc.).
- Put together the form letters (SPST, for example).

In the table, notice that the *multipole switches* have a vertical dotted line crossing both poles. This line means that all the poles are mechanically connected so that they all operate together. Notice that the table describes two types of DPDT switches. The single-phase transfer switch is the more conventional design. The 4-way switch is a special case, since the two poles share two throws.

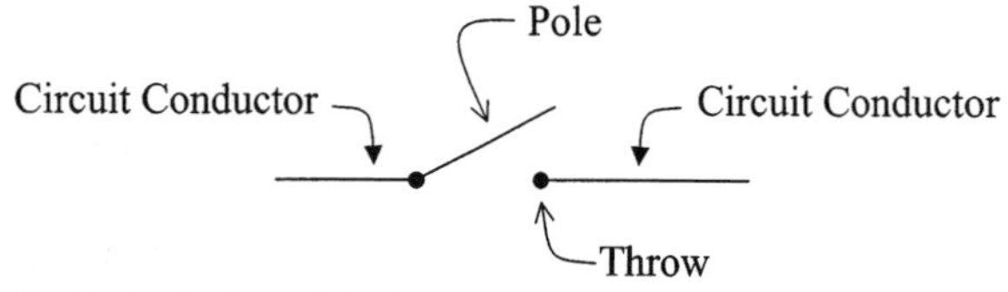

Fig. 8.1. Basic switch (single pole, single throw).

Lighting Circuits

Combinations of switches can be connected to control lighting circuits when control is needed at multiple locations. The switches are explained in Table 8.1, and Figure 8.2 shows some examples. In the figure:

- One lamp is shown, although multiple lamps in parallel can be switched.
- Only the rudimentary circuits are shown, with the lamp beyond the switches. Circuits for other lamp locations can be figured out with the shown circuits as the starting point.
- Wire colors are not shown. In actual use, a continuous white neutral conductor should extend to the

TABLE 8.1. Four-letter switch notation

Switch designation	Circuit	Common name	Common uses
SPST		Single-pole switch, S_1	Light circuit
SPDT		3-way switch, S_3	2-switch light circuit
DPST		Double-pole switch	230-V, single-phase switch box
DPDT	position 1 position 2	4-way switch, S_4	Light circuits with ≥ 3 switches
DPDT		Single-phase transfer switch	Standby generator circuit
TPST		3-phase load switch	3-phase switch box

S
Gold
Hot
N
a. Switched from one location
Silver

S_3
S_3
Hot
travelers
N
b. Switched from two locations

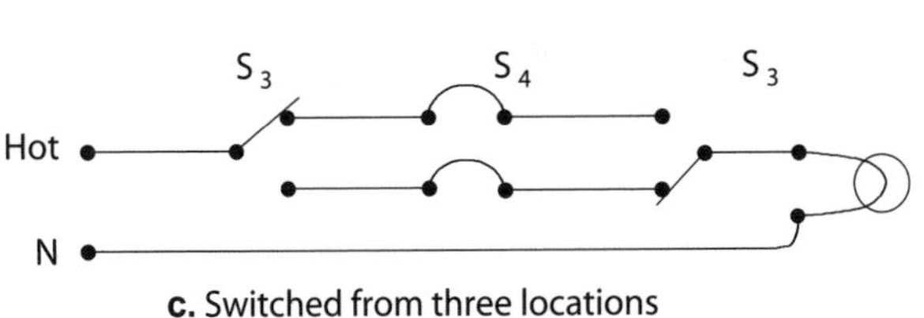

c. Switched from three locations

Fig. 8.2. Lamp-switching circuits.

silver screw of the lamp. Other non-neutral conductors should be colored, or coded for hot at the ends, if white. To code for hot, black tape is wrapped around both ends of the white conductor.

In circuit a, the lamp is switched from one location. The switch is placed in the hot conductor so that a continuous neutral is maintained. The figure shows that the hot conductor is connected to the gold screw on the lamp fixture, and the neutral (white) conductor is connected to the silver screw. When this convention is followed, lamp threads are connected to neutral and the lamp base is energized.

Circuit b allows control of the lamp from either switch, regardless of the position of the other switch. If a room or building has two doors, having a switch at each door adds convenience. This arrangement is also used at the ends of stairs and hallways. The energized conductors between the switches are called "travelers."

Circuit c shows that placing a 4-way switch between the two 3-way switches allows control from three locations. More than one 4-way switch can be inserted between the two 3-way switches to allow switching from any number of locations.

Electromechanical Relays

The electromechanical relay is a device which uses the force and motion of an electromagnet to change the position of an electrical switch. It is old technology but still used in thousands of applications to accomplish electrical control. To learn the concept, refer to Figure 8.3 where a 100-A load is switched by an SPST switch located some distance from the load and the power source. In Figure 8.3a, control is accomplished using only an SPST switch. This arrangement works adequately, but it may have some disadvantages:

- Conductors extending to the switch must carry the load current.
- The switch must be able to handle load current and load voltage.

Use of an electromechanical relay (Figure 8.3b) can result in a less-expensive and handier system. When the relay is used, the system has two circuits: a coil circuit and a load circuit. The coil circuit extends to the control switch. When the control circuit is closed, a low current (< 0.5 A) flows through the coil, producing a force which pulls the load switch closed against the force of a spring. The load switch remains closed as long as the control switch is closed. When the control switch is opened, the coil is de-energized and the spring pulls the load switch open, shutting off the load.

When the relay is used, the heavy load circuit extends only between the power source and the load. The load contacts of the relay must be capable of handling the load current and voltage. The coil circuit needs to be able to handle only the coil current. In the

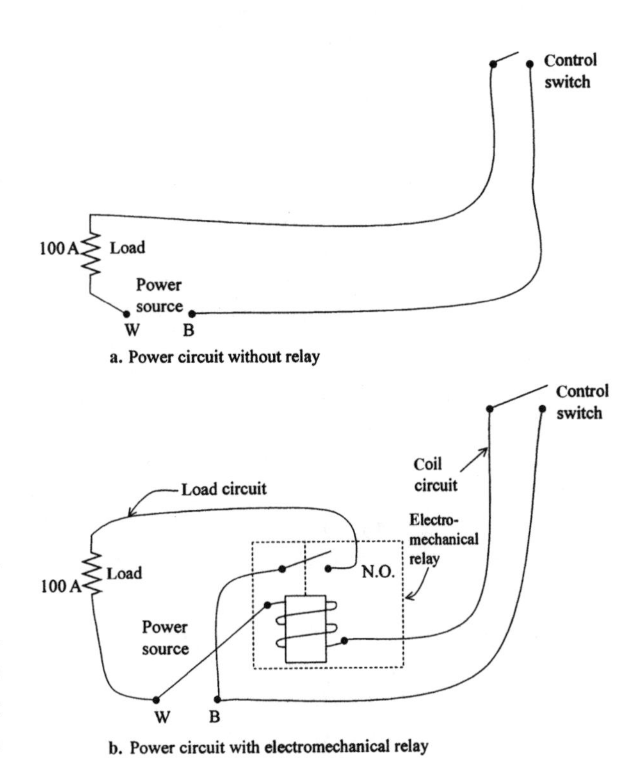

Fig. 8.3. Electromechanical relay.

circuit illustrated, coil voltage is the same as load voltage. In practice, the coil voltage can be different from the load voltage since it is a separate circuit. Safer, lower coil voltages of 12 or 24 volts are common. Relay circuits similar to that of Figure 8.3b are used in many applications.

Refrigerators

The control switch is a thermostat in the refrigerator. The heavy load is the compressor. The relay eliminates the need of running the compressor current up to the thermostat.

Automobile Lights

DC systems in automobiles use relay circuits for the heavy headlight load. The light switch on the dash then handles only the relay coil current.

Normally Open and Normally Closed

The relay load switch of Figure 8.3b is marked NO, which stands for normally open. Relays are pictured in normal position, that is, with no voltage applied to the coil. If the relay has NO contacts, the contacts will be *open* with *no voltage* to the coil, and *closed* when the coil is *energized.* If the relay contacts are NC (normally closed), the contacts are *closed* when the coil is *de-energized*, and *open* when the coil is *energized.* Using NC instead of NO contacts will reverse the logic of load operation. This is illustrated in Example 8.1.

Example 8.1

Sketch the circuits for a system in which a 115-V load is controlled by an electromechanical relay so that when an SPST control switch is open, the load is on, and when the control switch is closed, the load is off. Assume coil voltage is 115 V.

To obtain the operating logic specified, an NC relay with a 115-V coil is employed. In Figure 8.4, the load is on when the control switch is off and is off when the control switch is on.

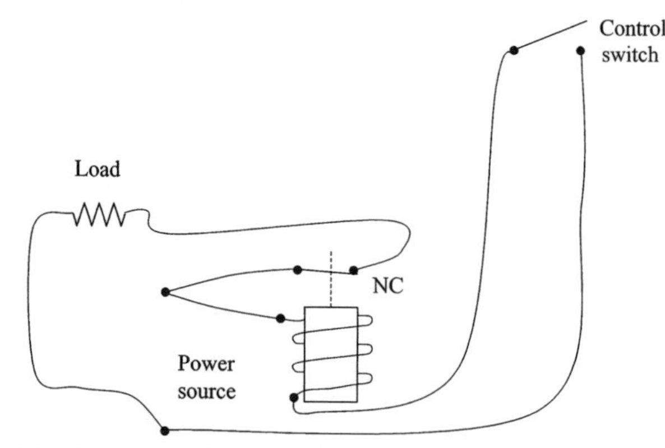

Fig. 8.4. Circuit for Example 8.1.

Numerous relay configurations are available. Some have multiple poles. Some have poles equipped with both NO and NC contacts. Many load current and coil voltage combinations can be obtained. Different enclosures are available, to meet the needs of the environment.

Solid-State Relays

Solid-state relays provide functions similar to those of electromechanical relays, but they have no moving parts. Coil and contact functions are performed by solid-state electronic devices such as transistors or silicone-controlled rectifiers. For some applications, they are superior to electromechanical relays. See chapter 13 for further discussion of solid-state relays.

Ladder Diagrams

A ladder diagram is a type of electrical circuit diagram in which conductors and circuit components are arranged to resemble the vertical side rails and horizontal rungs of a ladder. Specifying a circuit by means of a ladder diagram often makes the circuit much easier to comprehend and simpler to draw.

Ladder Diagrams of Electromechanical Relay Circuits

Figure 8.5 illustrates how to describe a relay circuit by means of a ladder diagram. The ladder diagram is constructed by first drawing the vertical ladder rails, representing the neutral and energized conductors. We'll use the convention of placing the neutral rail (W) on the right and the energized rail (B) on the left. Each complete circuit of the relay circuit system is a rung on the ladder. Table 8.2 shows common symbols used for ladder diagrams. It is important to note that location of the various components on a ladder diagram bears no relation to where they are on the actual circuit. For example, in Figure 8.5b, relay coil RC-1 (the circle) is not placed close to the NO contacts it controls, as it is in Figure 8.5a. Therefore, coils and the contacts which they control need to be labeled the same way so it is always clear which coil controls which contacts.

NO vs. NC Contacts

The on/off operating logic of a load can be reversed with a relay by changing from one type of

TABLE 8.2 Control circuit symbols

	Normally open (NO)	Normally closed (NC)
Push button		
Limit switch		
Temperature-actuated switch		
Flow switch		
Level or float switch		
Contacts controlled by relay coil		
Time delay relay Delay begins when coil energized		
Delay begins when coil de-energized		

Contacts are drawn showing their inactive normal state. When a switch is actuated or a relay energized, the contacts it controls change state. Normally open (NO) contacts close and normally closed (NC) contacts open.

Coil, —○— .

Motor, —(M)— .

contacts to the other type of contacts. This is illustrated in Example 8.2.

Example 8.2

The circuit of Figure 8.5 is changed by using a relay having NC contacts instead of NO contacts. Sketch the ladder diagram and describe the action of the circuit. Figure 8.6 shows the ladder diagram to be the same as Figure 8.5b, except the relay contact is NC. The operating logic of the circuit is reversed by this change. When the control switch is open, the coil is de-energized and the load is on. When the control switch is closed, the relay coil is energized and the load turns off.

Magnetic Motor Starter

A magnetic motor starter (MMS) is a widely used device for controlling an electric motor by means of an electromechanical relay (Figure 8.7). In addition to the switching function of the relay, an MMS also contains motor overload and undervoltage protection systems. The relay is actually a 4PST type. Three of the poles (x,y,z) are for load switching. The fourth (auxiliary contact) is needed for the coil circuit. This will be explained later. There are three load poles because three are used for 3-phase motors. Even though one pole or two poles are used for single-phase 120-V or 240-V motors respectively, there is such a predominance of 3-phase motors that MMS manufacturers may build only 3-pole types.

In series with each load pole is a heater (interlocking C symbol). The heater is selected according to motor full-load current, and each heater carries motor current. In close proximity to each heater is an open-on-rise thermostatic switch designed to open if its heater is too hot due to excessive motor current. These are the overloads.

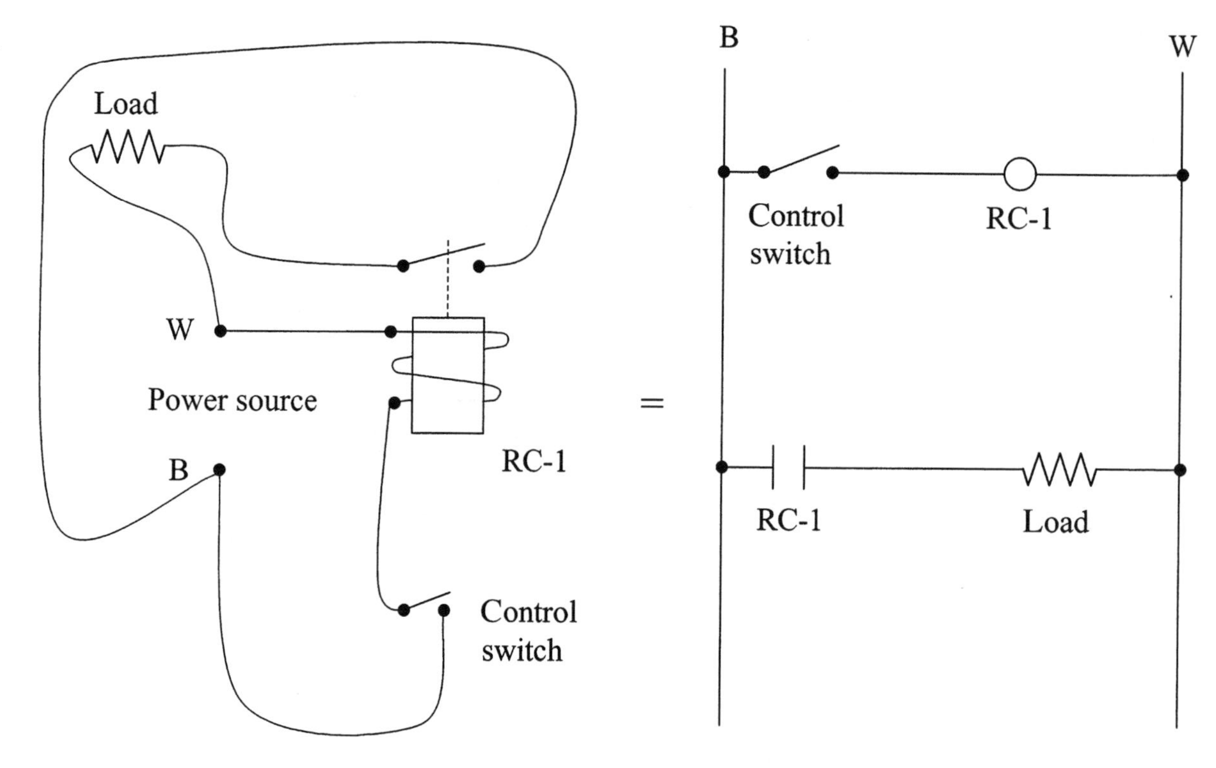

Fig. 8.5. Ladder diagram of an electromechanical relay circuit.

Figure 8.8 is a ladder diagram of an MMS wired to control a 3-phase motor through a start-stop station. The auxiliary contacts (Figure 8.7) is connected in parallel to the start button where it latches the coil circuit. The overload contacts are placed in series with the coil so that if any one of them opens the coil is de-energized. The coil shown is 120 V. Coils are usually available at 12, 24, 48, 120, 208, 230, and 460 V.

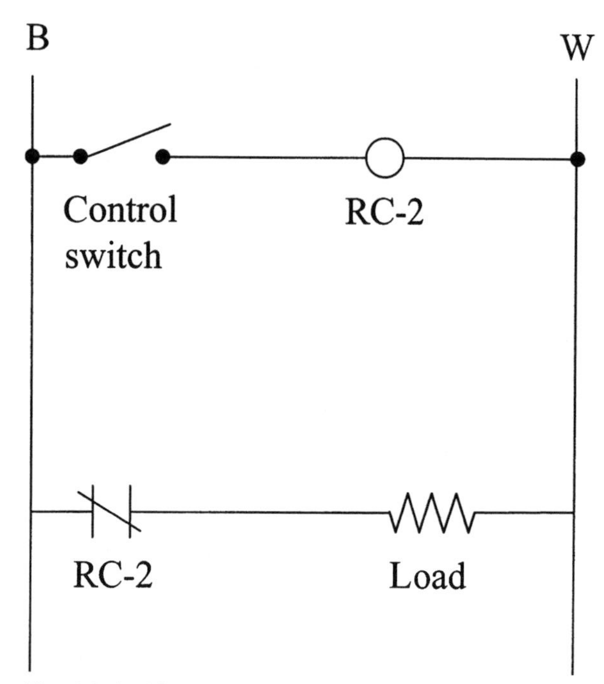

Fig. 8.6. Ladder diagram for Example 8.2.

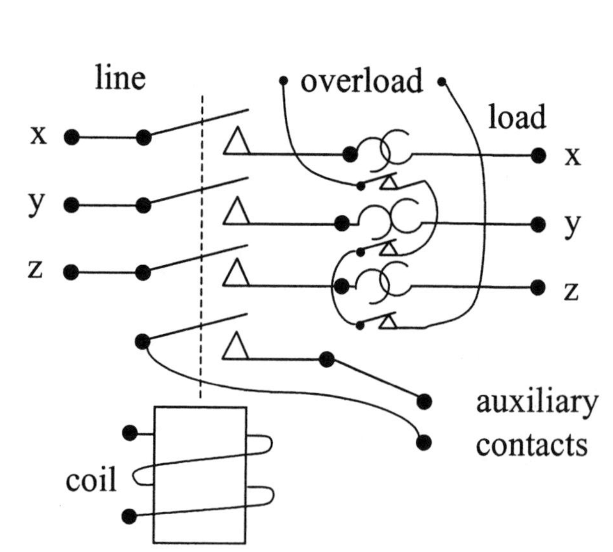

Fig. 8.7. Magnetic motor starter.

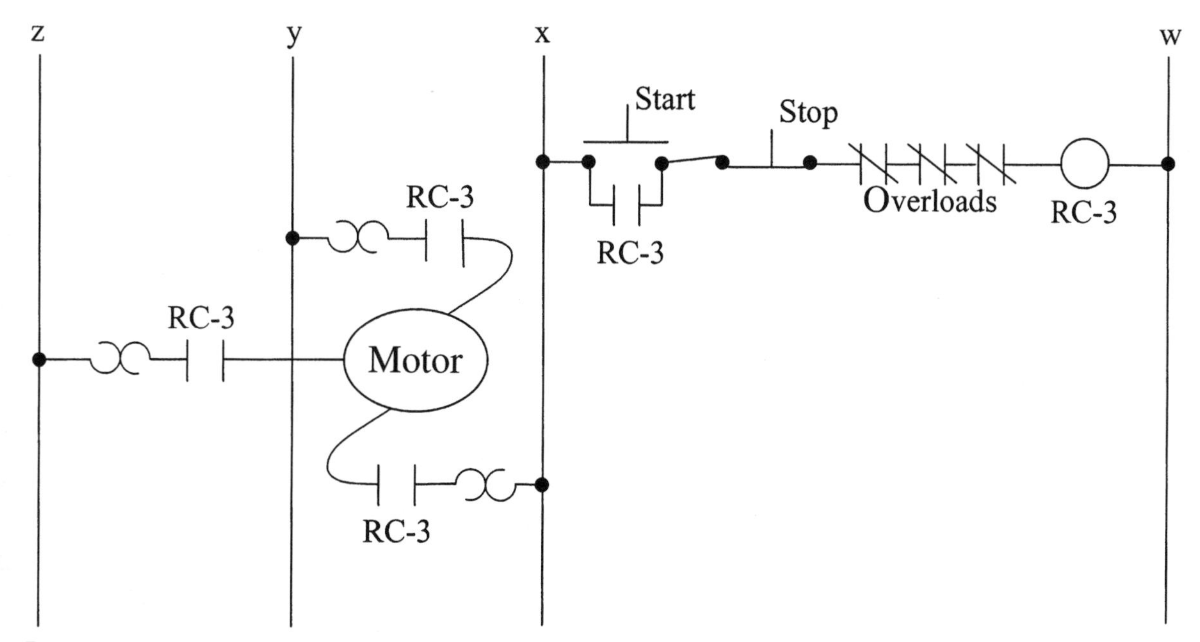

Fig. 8.8. Ladder diagram for magnetic motor starter.

Automatic Controls

Automatic controls are devices which employ a switch that can be opened and closed due to changes in physical conditions such as temperature, relative humidity, pressure, or time. Thermostats, humidistats, pressure switches, and time switches will be discussed in this section.

Snap-Action Switches

Many common automatic control devices employ snap-action switches to turn on and off the electric current being switched (Figure 8.9). With no force on the actuating pin, a closed circuit exists between C (common) and NC. As the pin is depressed, force to depress it increases. When the depression distance exceeds a certain value, a mechanism snaps the pole to its other position. This action closes the circuit from C to NO and opens the circuit from C to NC. The pole's rapid movement (snap) minimizes arcing between contacts. Snap action switches are rated by horsepower or current and voltage capability, and are available in a wide range of sizes and configurations.

Time Switch

A time switch controls an electrical load in response to passage of time. An electromechanical time switch (Figure 8.10) employs a motor-driven disk to change the position of a snap-action switch. When the protruding cam depresses the actuating pin, the desired switching action takes place. The motor runs continuously and turns the disk at a speed appropriate for the application. A time switch controlling lights in

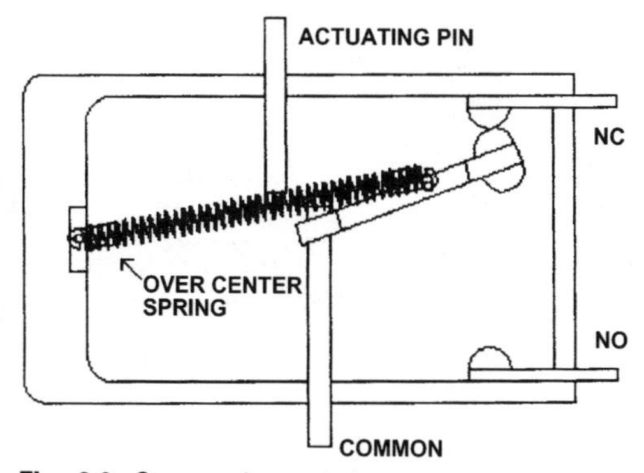

Fig. 8.9. Snap-action switch.

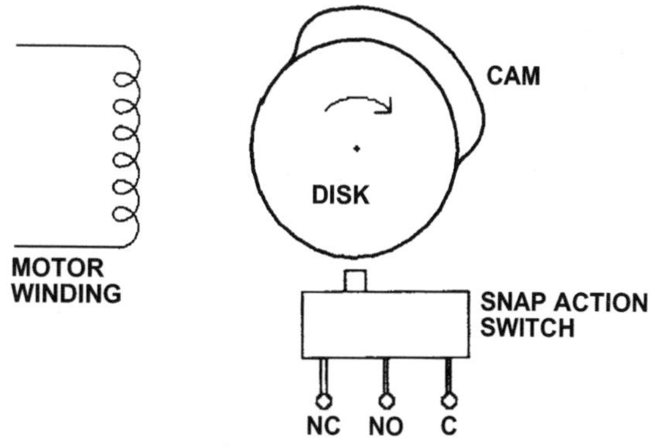

Fig. 8.10. Electromechanical time switch.

a livestock building, for example, typically has a disk which turns at one revolution per 24 h. Other variations include multiple switches and disks on a single shaft, and multiple adjustable cams on one disk.

Thermostat

A thermostat changes the position of a switch in response to a change in temperature of a fluid (air or water, for example). A bimetallic thermostat (Figure 8.11) uses a bimetallic metal strip to actuate a snap-action switch as it bends in response to air temperature changes.

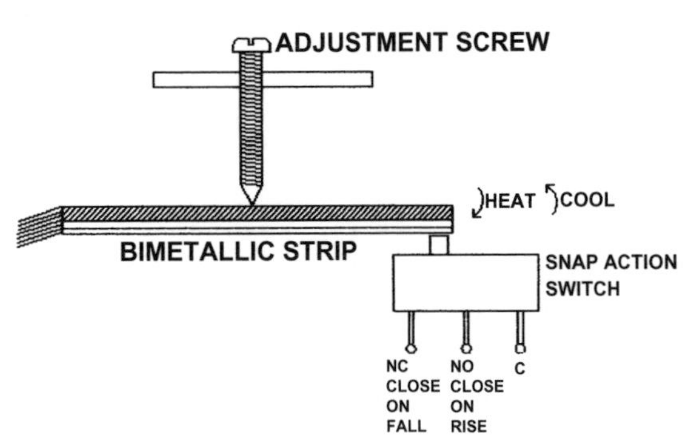

Fig. 8.11. Bimetallic thermostat.

As temperature rises, the actuating pin is depressed and at some temperature, the NO-C circuit closes. Therefore, we can label the NO contact as "close on rise." Likewise, as temperature falls, the NC-C circuit closes at some temperature and we can label the NC contact as "close on fall." The "close on fall" and "common" terminals are used to control a heater, whereas the "close on rise" and "common" terminals are used to control a cooler.

The time response of an environment temperature controlled at a temperature higher than its surroundings by a thermostat can have a shape illustrated by the graph of Figure 8.12. The cut-in temperature is set by the position of the adjustment screw. At the cut-in temperature, the heater turns on. Temperature rises until the shut off is reached and the thermostat turns off the heater. The environment temperature overshoots the shut off by a few degrees, then decreases until cut-in is reached and the heater is turned on. After an overshoot, environment temperature begins to rise and the cycle is repeated. The differential (shut-off–cut in) is typically about 5°F, and is usually not adjustable by the user. Special-purpose thermostats have adjustable differentials.

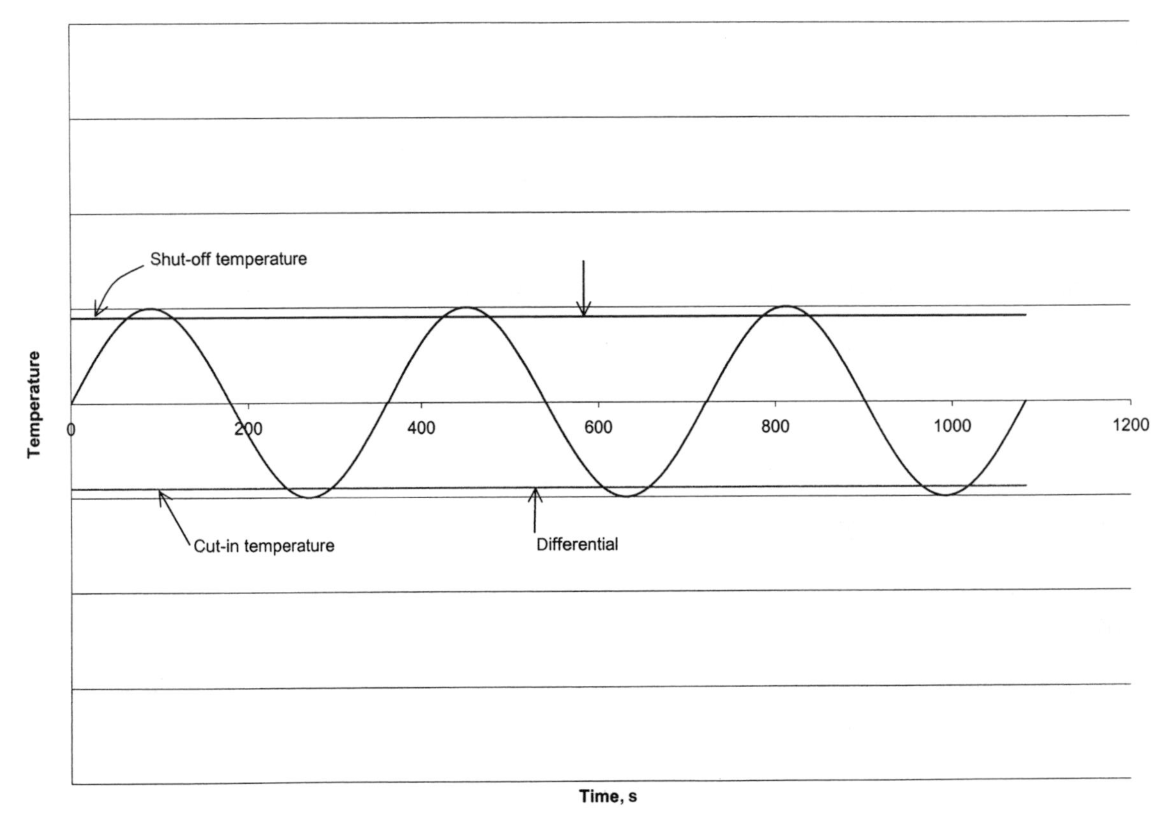

Fig. 8.12. Time response of an environment controlled by a thermostat.

Other types of thermostats may use a fluid-filled tube and bellows as the active element. Electronic thermostats use thermistors or thermocouples to detect temperature. Thermostats are usually reliable devices and retain accuracy for many years.

Humidistats

A humidistat can control relative humidity in an environment by turning on and off a load such as a dehumidifier, ventilation fan, or humidifier. The active element in a humidistat is a strip of hygroscopic material which elongates as air relative humidity rises and shrinks as the air becomes drier (Figure 8.13). The strip is usually plastic, with a thickness of a few mils. Before the advent of plastics, the sensing element was composed of hundreds of strands of human hair. Humidistats tend to be less stable and accurate than thermostats because the elongation-contraction properties of the sensing strip tend to change as the strip picks up contaminants from the air. Since humidistats are rarely designed for use on both humidifiers and dehumidifiers, they are usually fitted with close-on-rise contacts (as shown) or close-on-fall contacts, but not both. Response over time is similar to temperature controlled by a thermostat (Figure 8.12).

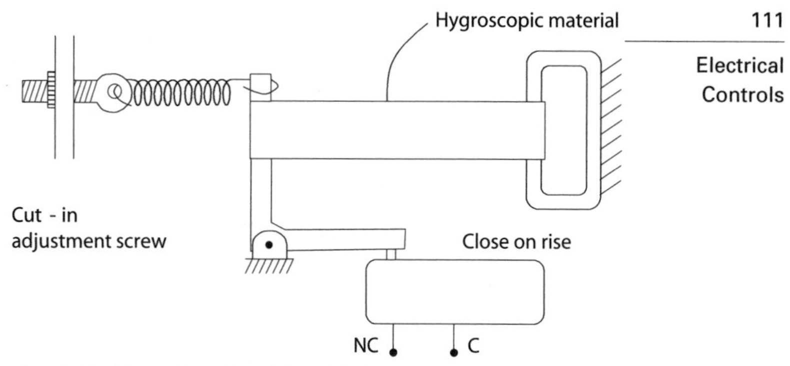

Fig. 8.13. Humidistat for dehumidifier.

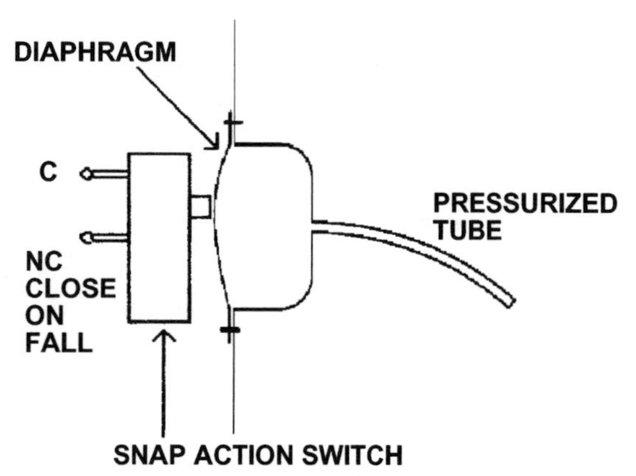

Fig. 8.14. Pressure switch.

Pressure Switches

Pressure switches respond to changes in fluid pressure to control operation of electric motor–driven pumps and compressors. The active element is a diaphragm that deflects due to changes in pressure (Figure 8.14). Switching is done by a snap-action switch. Pressure switches are usually equipped with close-on-fall and common contacts, since the usual application is with pumps and compressors.

Direct Switching vs. Relay Switching

The automatic control devices discussed are usually fitted with snap-action switches capable of directly switching current to heaters or motors up to a specified limit. Electromechanical relays can be used with these devices to control larger loads. When used with a relay, the control devices switch the relay coil current (typically less than 0.5 A). The relay load pole is selected with a capability of switching the load current, whatever it is. A relay can also be used to reverse the operating logic of the device. For example, a dehumidifier humidistat (close on rise) can be used to control a humidifier by use of NC relay contacts.

Problems

8.1. Sketch a circuit in which a lamp is controlled by two switches. Assume the lamp is between the 120-V power source and the switches. Assume wiring is NM 2-wire cable (black, white) and NM 3-wire cable (black, white, red). Show wire colors, and wires coded for hot.

8.2. Sketch an electromechanical relay having a 120-V coil and an SPDT load switch having an NO and an NC contact.

8.3. Sketch a circuit using the relay of problem 8.2 to control 120-V loads A and B by means of an SPST control switch through the relay. Load A is to be off and load B on when the control switch is closed. The opposite operating logic is desired when the control switch is open.

8.4. Draw a ladder diagram of the circuit for problem 8.2.

8.5. Conveyors A, B, and C are all equipped with single-phase 240-V motors, and each has its own start-stop station. Conveyor A directly feeds both B and C. Thus, both B and C must be running before A is started. It should be possible to start B or C independently of the other and in any order. An extra set of contacts is available in each motor starter. All coils are 120 V. Draw ladder diagrams of coil and load circuits.

8.6. A 3-phase motor can be reversed by reversing two of the three conductors to the motor. A push button control is needed in which start button 1 will start the motor in one direction, and start button 2 will start it in the opposite direction. Other information:

- Single stop button is to be used.
- Two magnetic motor starters (called F and R, respectively) are available. Each has both normally open and normally closed auxiliary contacts available.
- If the motor is running, it cannot be reversed without stopping first.
- Draw a ladder diagram of the motor control circuit.

8.7. Conveyor A feeds conveyor B. Both are equipped with 3-phase 120/208-V motors, controlled by magnetic motor starters. Each motor is controlled from a start-stop station. Coils are 208 V. Conveyor B must always be started before and must not be stopped before conveyor A. Draw a ladder diagram of this system.

Chapter 9

Standby Electric Power Generation

Introduction

Occasional line electric power outages are inevitable and must be planned for. On some farms, availability of electric power is essential in order to sustain animal or plant life, or to prevent loss of perishable materials. Examples include electric ventilation systems in animal confinement buildings, electrically controlled heating systems in greenhouses, electrical brooders and incubators in poultry buildings, electric milkers in dairy buildings, and food refrigerators and freezers. Standby electric power generation systems must be available for such installations when line power is not available.

This chapter will describe how generators operate and then will discuss selection and use of standby generators in agriculture.

Principles of Electric Power Generation

A generator is a machine that converts mechanical power to electric power. Figure 9.1 illustrates this conversion process. All electromechanical generators operate by these principles:

- A voltage is produced between the ends of a conductor which is moved through a magnetic field.
- If the ends are connected through an electrically conducting external circuit, the voltage causes a current to flow (through the external circuit) and accomplish useful work.
- The magnitude of the voltage depends on the strength of the magnetic field, the direction of movement of the conductor, and the speed with which the conductor moves.

Components of a Simple AC Generator

In a conventional generator, conductor motion through the magnetic field is in a circular pattern (Figure 9.2a). Notice the parts of the single-loop AC generator. The rotor or armature (rotating part) consists of a loop conductor with its ends each connected to a solid slip ring. The stator (stationary part) provides the magnetic field within which the rotor turns. The stator in Figure 9.2a consists of an iron frame shaped to produce and concentrate a magnetic field when a DC current flows through the field winding. The AC generator must have a source of DC current for the field, unless the field is produced by a permanent magnet. Electrically conducting stationary brushes rub against and carry current from the slip rings as they turn with the rotor. The generator load is connected between the brushes.

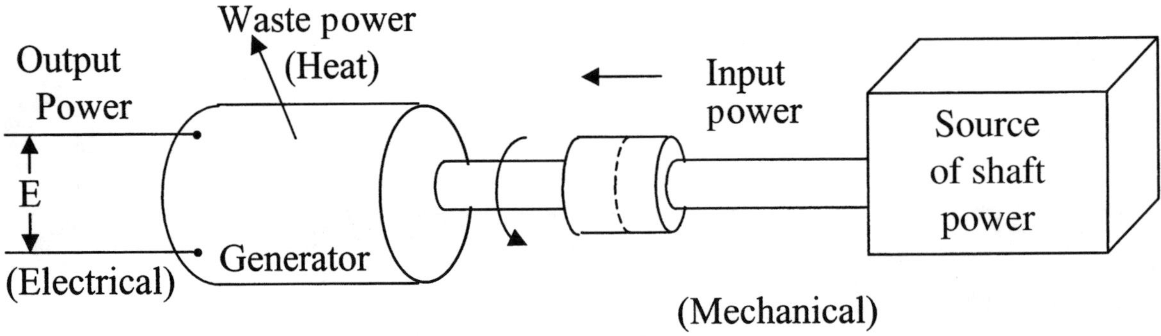

Fig. 9.1. The electric generator functions as a power conversion device.

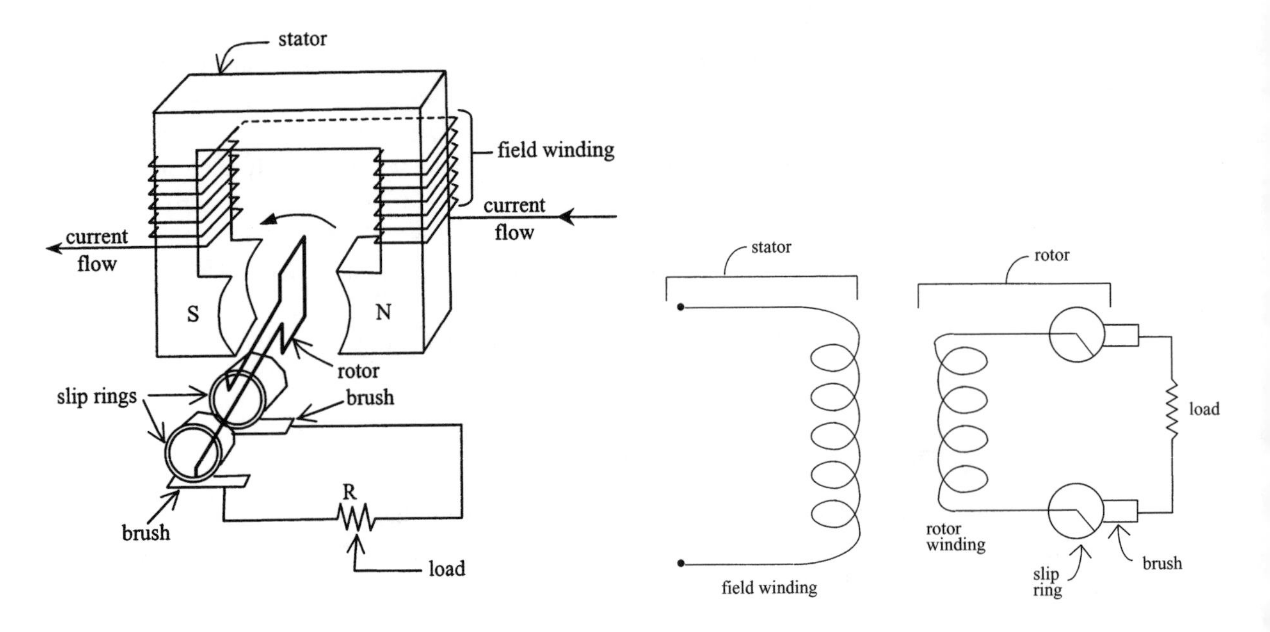

Fig. 9.2. Simple AC generator.

Figure 9.2b shows the circuit of the simple generator schematic in Figure 9.2a. In later sections, different generator types will be described by their circuits.

Generation of an Alternating Voltage

Figure 9.3 illustrates the voltage output of the simple AC generator. As the loop is rotated within the magnetic field, the emf is exhibited as a voltage across the load (shown as a light bulb). Assume that the loop is rotating rapidly. The graph in the figure illustrates the voltage generation effect during this rotation. Notice that at zero degrees the voltage generated is zero. This is because at this instant, the loop is moving parallel to and not cutting any lines of the magnetic field which exist between the ends of the field magnet. At 90° the voltage is at maximum since the loop is at this instant moving perpendicular to the field lines. At 180°, voltage is back to zero. At 270°, the loop sides have reversed their positions from where they were at 90°. Therefore, the voltage across the load has reversed polarity and current through the load has reversed direction. The current through the load will, thus, alternate in direction and is called AC current or power. Rotation through an additional 90° brings the loop back to its original position.

Frequency

Frequency of the generator output voltage is an important parameter of the generator. Frequency is specified in hertz (Hz), defined by Equation 9.1.

$$1 \text{ Hz} = 1 \frac{\text{cycle}}{\text{second}} \tag{9.1}$$

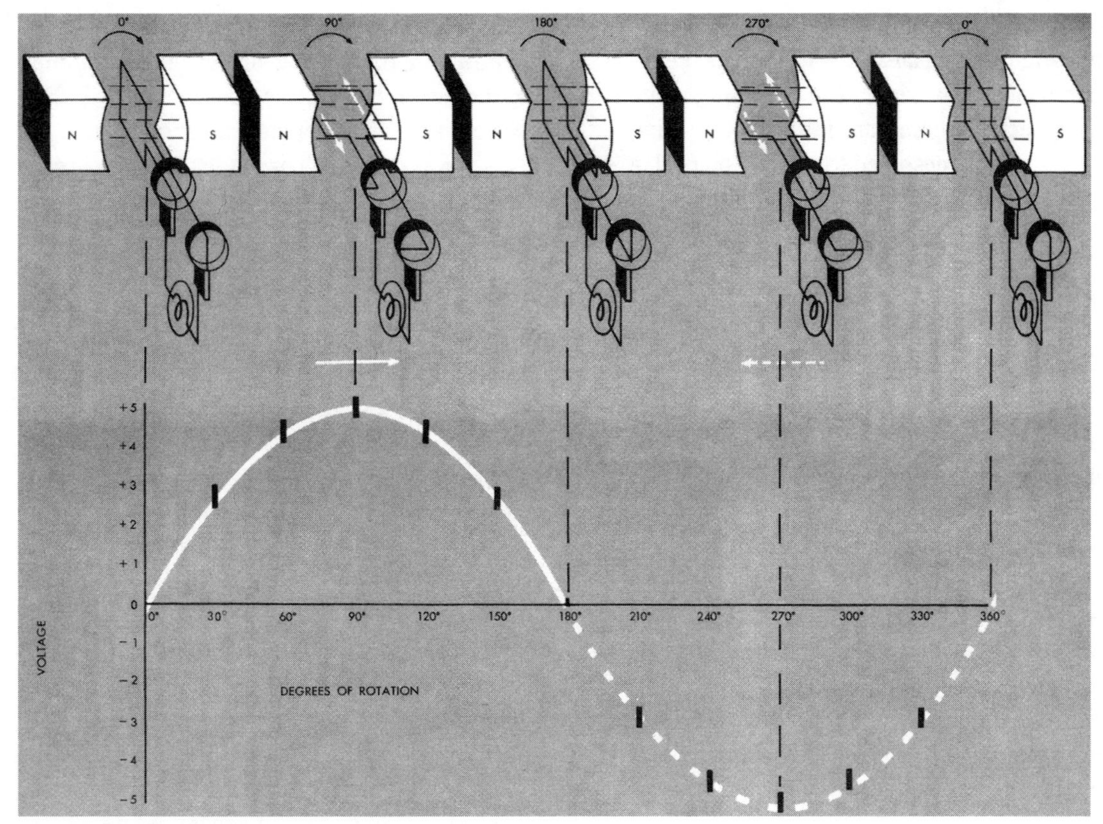

Fig. 9.3. Generation of an AC voltage.

Assuming the rotor loop in Figure 9.3 is turning at, say, 3,600 r/min, we can compute the resulting power frequency. From the waveform we see that one revolution of the rotor is required to complete one voltage cycle:

$$\frac{1\ \text{cycle}}{\cancel{\text{revolution}}} \times \frac{3600\ \cancel{\text{revolutions}}}{\cancel{\text{minute}}} \times \frac{\cancel{\text{minute}}}{60\ \text{seconds}} = \frac{60\ \text{cycles}}{\text{second}} = 60\ \text{Hz} \qquad \textbf{(9.2)}$$

The machine shown in Figure 9.3 is a 2-pole machine. (The field magnet has two poles between which the magnetic field is concentrated.) A 4-pole machine would complete two cycles in one revolution, and its rotor must turn at 1,800 r/min to produce 60 Hz. Equation 9.3 is the generalized relationship among frequency, rotational speed, and number of poles:

$$f = \frac{NP}{120} \qquad \textbf{(9.3)}$$

where f = frequency, Hz
P = number of field poles per phase
N = rotational speed, r/min

Line power frequencies in common use are listed in Table 9.1.

An example problem will illustrate use of Equation 9.3.

Example 9.1

A 4-pole generator has an output frequency of 58 Hz. What is its rotational speed?

$$N = \frac{120(f)}{P} = \frac{(120)(58)}{4} = 1{,}740\ \text{r/min} \qquad \textbf{(9.4)}$$

Power frequency is important to the operation of electrical equipment. The speed of clocks and timing devices driven by synchronous motors is

TABLE 9.1 Line power frequencies

Frequency, Hz	Use
0	All DC
50	European line power
60	USA line power
400	Common aircraft power systems

directly proportional to line frequency. Slight variations in line frequency cause clocks connected to line power to gain or lose time. The operating characteristics of electric motors are influenced by line frequency also. Motors designed to operate on 60-Hz power in the United States usually cannot be used on European 50-Hz systems. This is discussed in chapter 10.

Line frequency is closely regulated by electric power companies, while frequency of power from farm-type standby generators can often vary over several Hz. Manufacturers of engine-driven standby units usually specify that frequency will change no more than 3 Hz from no load to full load and satisfactory equipment performance is generally obtained if frequency is within this range (Seneff and Puckett 1989). The frequency of tractor-driven units can vary more than this because frequency is directly proportional to engine speed. The amount of the frequency variation depends on how careful the operator is in adjusting the tractor or engine governor, and on how precisely the governor is able to maintain a constant engine speed during variations in generator load.

Frequency variations are not usually large enough to cause problems with motor operation. However, motor-driven line-power clocks and timing devices will be inaccurate since their speed is directly proportional to power frequency.

Example 9.2

A 2-pole AC generator is operating at a rotational speed of 3,684 r/min.

a. What is the power frequency of the generator?

$$f = \frac{NP}{120} = \frac{(3684)(2)}{120} = 61.4 \text{ Hz}$$

b. What is the time error for an electric clock powered by the generator for 24 h?

$$\frac{61.4}{60} = \frac{x}{24} \quad x = 24.56 \text{ h}$$

$$\frac{(0.56 \not{h})(60 \text{ min})}{\not{h}} = 33.6 \text{ min}$$

Clock speed is proportional to frequency and the clock will be 33.6 min fast after 24 h.

Field Current

As shown in Figure 9.2, an AC generator must have a source of DC current for the field winding. This current is called the field excitation current and the system to supply this current is called an exciter. Various kinds of exciters can be found on AC generators for standby use.

Transfer of current between a rotating part and a stationary part by means of contact between a stationary brush and a rotating slip ring or commutator is an undesirable procedure. These systems generate sparks, require periodic replacement because of wear, and occasionally fail. Some generators are designed to have these components carry a low current, or even to operate without brushes. Different generator types are described in the next sections.

External Exciter

An external exciter system typically consists of a small compound-wound DC generator mounted on the same shaft as the AC generator, or belt-driven from this shaft (Figure 9.4). In some cases, the exciter, driven with power from storage batteries, is used as a starter motor for the generator engine. The exciter shown is compound wound. That is, part of its field is in series with the rotor and part is in parallel with the rotor. A variable resistance called a rheostat is used in series with the parallel field winding to control exciter output current. Although drawn large in the figure, the exciter is normally much smaller than the generator. With this rotating armature arrangement, the slip rings and brushes must be built to carry full load current, which can be many times as large as the field current.

External Exciter and Rotating Field

Figure 9.5 shows an AC generator with a rotating field and stationary armature. Generator operation is still the same as in Figure 9.4, since this depends on relative motion between the conductor (armature) and the field. This relative motion can be attained with a rotating field and a stationary armature. The advantage of this design is that the slip rings and brushes carry only the field current and not the larger load current. AC generators with rotating fields are often called *alternators.*

Static Exciter Alternators

An alternative to the use of an external exciter is to employ a static (non-moving) exciter. With this

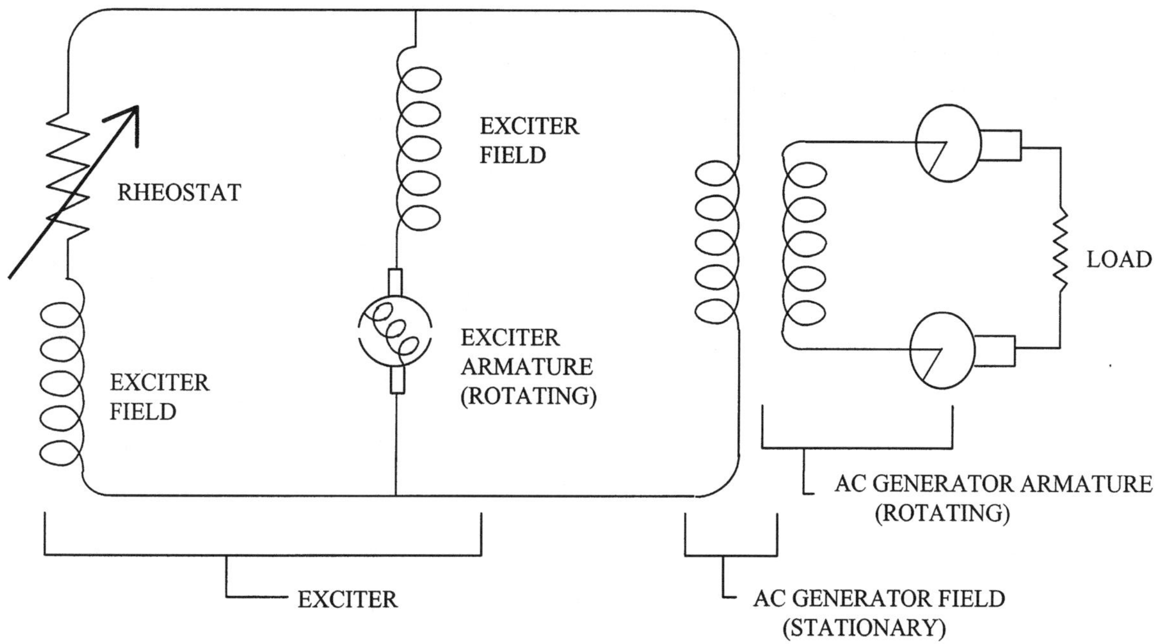

Fig. 9.4. AC generator with external exciter.

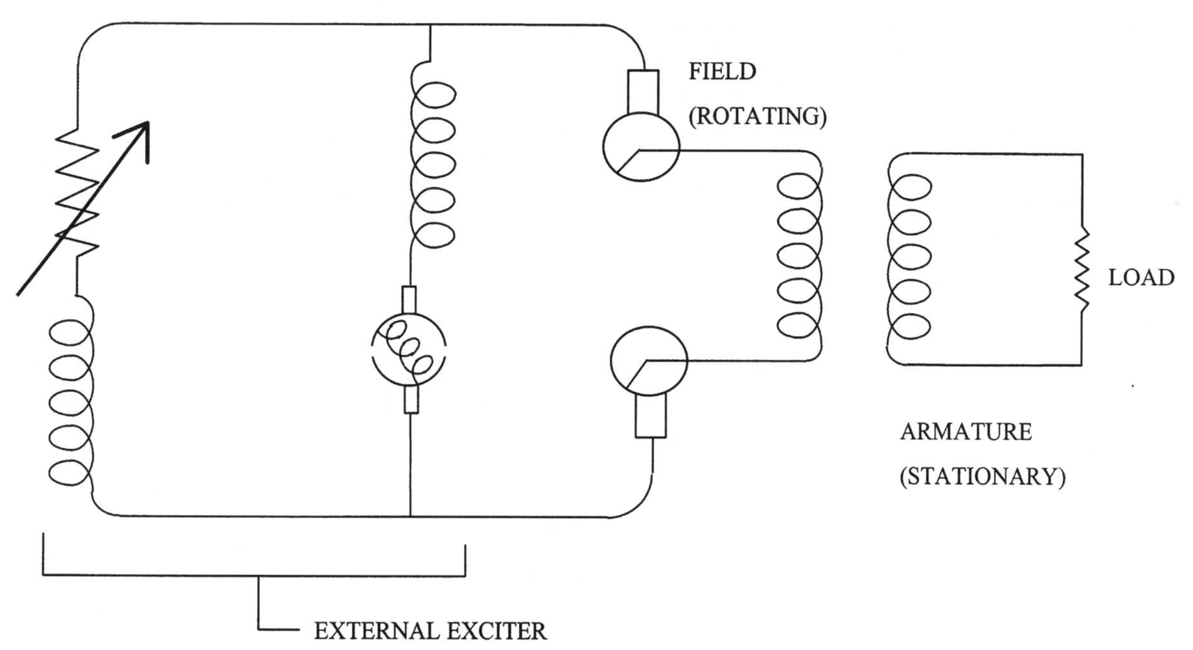

Fig. 9.5. AC generator with external exciter, and rotating field.

arrangement, a portion of the AC alternator output is electronically converted to DC at the proper voltage for the field (Figure 9.6). The rectifier contains transformers, diodes, and other electronic components.

This arrangement eliminates the DC generator, but slip rings and brushes are retained to supply DC power to the rotating field. Automotive alternators are statically excited 3-phase rotating-field AC generators in which output is rectified to 12-V DC for the automotive electrical system.

Brushless Alternators

Some farm-type standby generators operate without brushes. In order to accomplish this, a rotating field–stationary armature design is used. The exciter is a small 3-phase rotating armature–stationary field

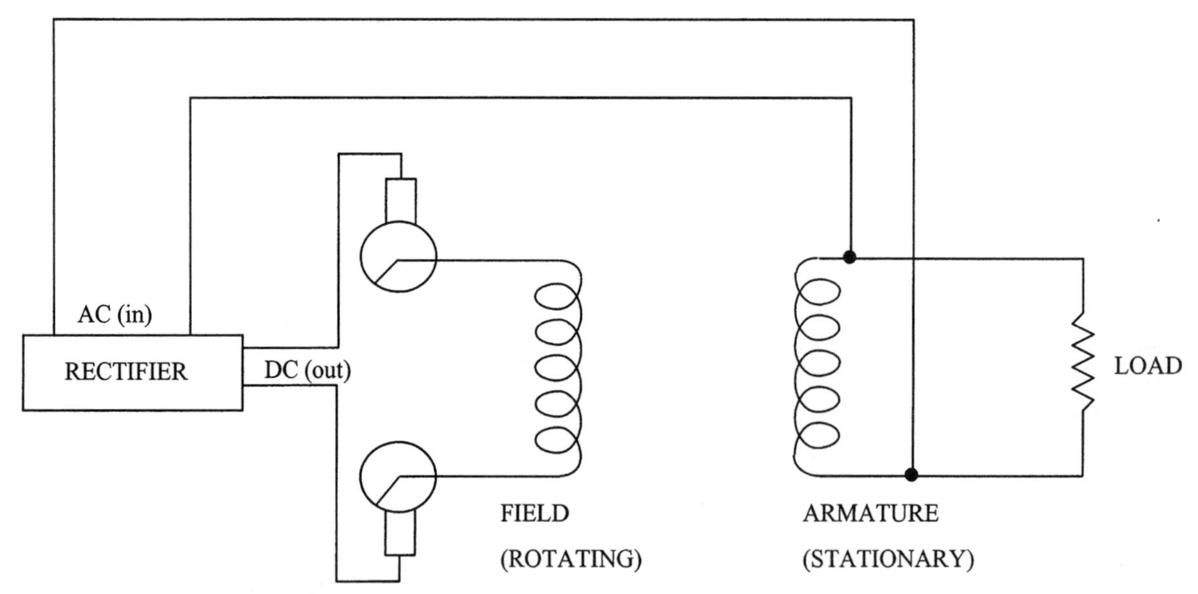

Fig. 9.6. Alternator with static excitation.

generator mounted on the same shaft as the field (Figure 9.7). The exciter field is statically excited from a rectifier receiving AC power from the main exciter armature. The exciter armature, the generator rectifier, and the generator field all rotate. The 3-phase AC power from the exciter armature is rectified on the rotor for use by the rotating field. During start-up, there is sufficient residual field strength in the exciter field so that the exciter can supply field current for the main field at once.

This design, which eliminates all commutators, slip rings, and brushes, became practical to build in the mid 1950s when high-power silicon diodes were developed for use in the rotating full-wave rectifier.

Standby Power Systems

The standby power system consists of these components:

- Power failure detection system
- Power source
- Switches and wiring
- Generator

Power Failure Detection

Power failure detection systems operate when line power goes off, and initiate the process of bringing the standby system into operation. The simplest

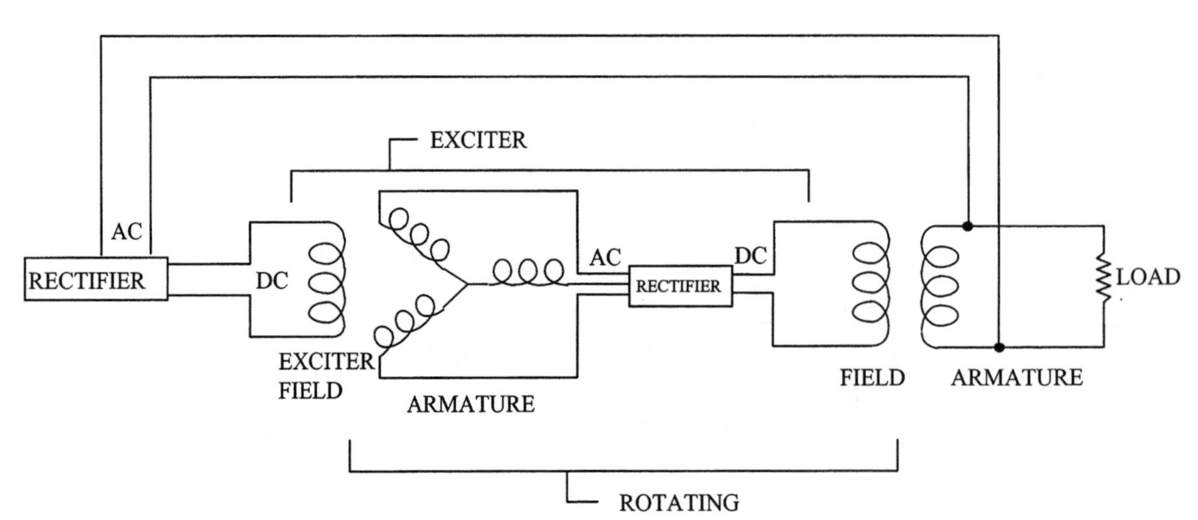

Fig. 9.7. Brushless alternator.

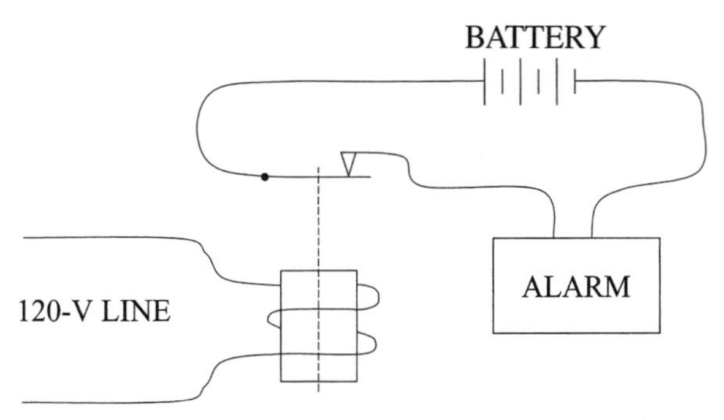

Fig. 9.8. Battery-powered power failure alarm.

approach relies on a person to notice lights and equipment are not operating, and then put the standby system into operation.

Battery-powered failure warning systems can turn on lights, horns, or other alarms (Figure 9.8). The simple alarm in the figure is energized when the normally closed relay contacts close due to the absence of voltage in the 120-V line power circuit. Besides initiating a local alarm, this kind of system can initiate start-up of a standby power system, generate a telephone dialer to call designated persons, or transmit a radio-frequency alarm signal.

Power failure detection may be a component of a more extensive alarm system needed to monitor temperatures, airflow rates, fluid levels, or other parameters. Power failure may be detected through one of these other monitored parameters when loss of power causes it to charge. The power failure detection system needs a preventive maintenance testing program to be performed on a regular basis to ensure the system does not become unworkable due to neglect (ASAE 1997).

Power Sources

Power for a standby generator comes from an internal combustion engine. It may be a direct-connected stationary engine dedicated to the generator, or it may be a tractor engine. In this case, shaft power is supplied through the tractor power take-off (PTO) shaft. Each type system has its advantages.

Advantages of stationary engine generator:

- More convenient to put in operation.
- Can be connected for totally automatic start-up and switching.

Advantages of PTO generator:

- Lower cost (about 50%).
- Easily made portable (on trailer) for use at remote locations.
- Tractor engine is more likely than a stationary engine to be adequately maintained.

Switches and Wiring

Wiring should be done following local codes or the National Electrical Code (NEC 702.6) to provide a safe installation. A double-throw switch must be used to ensure that the standby generator can never be connected to the line. A double-pole–double-throw (DPDT) transfer switch is shown in Figure 9.9. This type of switch (with suitable amperage rating) is used for a single-phase 240-V system. The switch is located between the watthour meter and the service disconnect (main fuse box) for the farm. Notice that the white (neutral) is not switched. With the handle in the up position, black and red load conductors are connected to the black and red line conductors respectively. With the handle in the down position, the load conductors are disconnected from the line and connected to the red and black conductors from the standby generator. Switch design makes it impossible to connect the load to line power and standby generator power simultaneously. Single-phase 120-V and 3-phase 4-wire systems need to use SPDT and TPDT transfer switches, respectively.

Although the transfer switch is an expensive item, it is necessary for safety, and is also convenient. Cheaper systems of plugs and receptacles, or wire connectors, can allow the line and generator to be

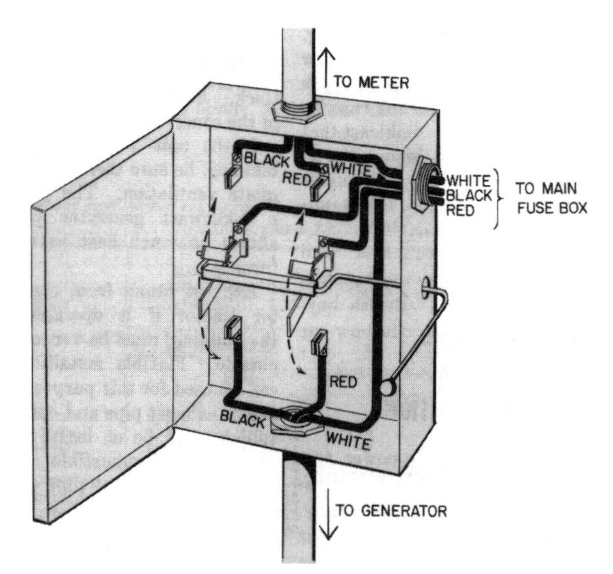

Fig. 9.9. Double-pole–double-throw transfer switch (Puckett 1980).

connected. Such arrangements do not meet NEC rules, are unsafe, and should never be used. Such connections endanger workers repairing a line energized by a standby operator but believed to be dead because line power is off. The standby generator can also be damaged when line power is restored.

Generators

The most common generator types used in standby systems are the brushless, the statically excited rotating armature, and the statically excited rotating field. The generators are usually rated in watts or kilowatts. Two ratings are usually listed. A continuous rating is listed along with a second higher rating, which may be called a "surge," "intermittent," "overload," "standby," "maximum," or "inrush" rating. For example, the Katolight D85FJJ4 has this rating:

Continuous (prime):	70 kW
Standby (inrush):	85 kW

Unfortunately, there is no standardization among manufacturers as to precisely what the higher rating means. In some instances, the manufacturer's literature defines it.

In this chapter, we will assume the generator can supply a continuous output for as long as standby power is needed and can supply its higher (we will call it intermittent) capacity long enough to start all connected loads. Precise definitions of these ratings should be obtained from the dealer or manufacturer before purchase.

Generator Sizing

A standby generator should be carefully sized so that it is capable of reliably starting and operating necessary loads at a minimum investment cost. Appendices A and B list wattages of common loads. Table 9.2 lists starting and running wattages of different sizes of single-phase motors.

TABLE 9.2 Wattage requirements of single-phase motors (Puckett 1980)

Motor Horsepower	Starting Watts	Running Watts (Under Load)
1/6	1,000	215
1/4	1,200	300
1/3	1,600	400
1/2	2,300	575
3/4	3,345	835
1	4,000	1,000
1.5	6,000	1,500
2	8,000	2,000
3	12,000	3,000
5	18,000	4,500
7.5	28,000	7,000
10	36,000	9,000

Starting induction motors is the most difficult factor related to standby generator selection. Induction motors draw up to four times the running watts during starting. For example, look at the power requirements for a 1-hp motor in Table 9.2. The motor draws about 1,000 W while running, but can draw up to 4,000 W while starting. Wattage of loads other than motors can be assumed constant during starting and during normal operation.

Generator size specifications are determined using either the total-load system or the critical-load system. The total-load system can start and run all loads on the system but requires no management of loads during operation. It must be used for sizing automatic standby systems. The critical-load system requires load management since some loads cannot be used. Loads used must be started in a definite sequence. The sizing procedures recommended by Campbell (1965) are described here.

Total-load sizing procedure:

1. Itemize all the connected loads indicating the wattage, horsepower, or amperes from the nameplates of the equipment.
2. For motors, determine from Table 9.2 both the starting and running watts.
3. Determine the watts required for other electrical equipment (Appendix B).
4. Allow wattage for equipment to be added in the future.
5. The total starting watts of all motors plus other equipment watts is the minimum intermittent capacity required for the generator. The total running watts of all motors plus other equipment watts is the minimum continuous rating of the generator.

Critical-load sizing procedure:

When a standby generator is sized by the critical-load method, special operating procedures must be used. In the event of a power failure, all electrical equipment on the farm must be switched off. Then the generator is placed in operation and the critical equipment is started. Ideally, motors are started first, one at a time and in order of decreasing hp.

1. Itemize the critical load indicating wattage, horsepower, or amperes from the nameplates of the equipment.

2. For critical motors, determine from Table 9.2 the starting and running watts.
3. Determine the load for other critical electrical equipment.
4. Add the loads determined above (Step 2 + Step 3).
5. Add the starting watts of the largest motor (No. 1) and Step 3.
6. Add the running watts of motor No. 1 and Step 3 and the starting watts of motor No. 2.
7. Add the running watts of No. 1, the running watts of No. 2, Step 3, and the starting watts of No. 3.
8. Add the running watts of Nos. 1, 2, and 3, Step 3, and the starting watts of No. 4.
9. Allow for future needs.
10. Continue calculating until all motors to be operated at the same time have been considered. You will need a generator with an intermittent capacity equal to the maximum wattage demand at any point in the calculations.
11. Total the running watts of all critical motors and other critical electrical equipment watts. This will be the minimum continuous output rating of the generator required.
12. Check the power available for other non-critical electrical equipment. Subtract the total found in Step 10 from the continuous power output capacity of the generator selected and the remaining watts are available for other non-critical electrical equipment.

An example problem will illustrate the procedure.

Example 9.3

Given: A dairy barn with the following equipment:

Ventilation fan	0.5 hp
Ventilation fan	0.5 hp
Milker	1.5 hp
Milk cooler	3 hp
Barn cleaner	5 hp
Feed conveyor	5 hp
Silo unloader	7.5 hp
Water heater	2 kW
Lighting	1 kW

Find: Generator size by (a) total-load method and (b) critical-load method. Assume the future needs are 10% of present needs.

a. Total-load method

Load		Intermittent Capacity	Minimum Continuous Capacity
Ventilation fan	0.5 hp	2.3 kW	.575 kW
Ventilation fan	0.5 hp	2.3	.575
Milker	1.5 hp	5.0	1.5
Milk cooler	3 hp	11.0	3.0
Barn cleaner	5 hp	15.0	4.5
Feed conveyor	5 hp	15.0	4.5
Silo unloader	7.5 hp	21.0	7.0
Water heater	2 kW	2.0	2.0
Lighting	1 kW	1.0	1.0
		74.6 kW	24.65 kW
+ future needs		7.5	2.46
		81.1 kW	27.11 kW
		or	or
		81 kW	27 kW

For this example, the critical loads include one fan, the milk cooler, water heater, 0.5 kW of lights, the milker, feed conveyer, and silo unloader. The starting order of these loads will influence the intermittent rating required for the generator. The smallest generator size will be obtained if the motors are started first, one at a time, with the largest first. Then other loads like heaters and lights are started last. Often management considerations will make this ideal starting order impossible.

b. Critical-load method

Critical loads are: one ventilation fan
milk cooler
water heater
0.5 kW of lighting
milker
feed conveyor
silo unloader

Assume that critical lighting must be on first and that the water heater cycles on and off automatically. Assume motors can be started in any sequence.

Critical lighting		
+ water heater	0.5 + 2	= 2.5 kW
+ silo unloader	21.0 + 2.5	= 23.5
+ feed conveyor	15.0 + 7.0 + 2.5	= 24.5
+ milk cooler	11.0 + 4.5 + 7.0 + 2.5	= 25.0
+ milker	5.0 + 3.0 + 4.5 + 7.0 + 2.5	= 22.0
+ ventilation fan	2.3 + 1.5 + 3.0 + 4.5 + 7.0 + 2.5	= 20.8
Continuous load	0.575 + 1.5 + 3.0 + 3.4 + 7.0 + 2.5	= 19.1 kW

With this starting sequence, the maximum intermittent load is 25.0 kW, and it occurs when the milker is turned on.

Intermittent capacity = 25.0 + future needs
= 25.0 + 2.5 = 27.5 kW

Minimum continuous capacity required
= 19.1 + future needs
= 19.1 + 1.91 = 21 kW

Notice that the intermittent and continuous capacities required from the generator are substantially reduced when the critical load method is used.

Example 9.3

Modified total-load procedure (Cotthoff 2002)

A 125,000-bird laying complex contains these single-phase loads, which must be picked up automatically by a standby generator in the event of a power outage:

Load		Intermittent Capacity, W	Continuous Capacity, W
50 1-hp ventilation fans (controlled so 4 fans come on every 2 s)	4(4000)	16,000	50,000
48 ⅓-hp feed motors (assume 25% will need to start and will be on)	(.25)(48)(1600) =	19,200	4,800
400 13-W fluorescent lamps		5,200	5,200
3-hp well pump		12,000	3,000
7 ½-hp egg collection motors		16,100	4,025
12 ¾-hp manure handling motors		40,140	10,020
Other incidental loads		5,000	5,000
	Totals	113,640	82,045

The generator will need an intermittent capacity of 114 kW and a continuous capacity of 82 kW.

Generator Installation

Portable PTO-driven standby generators which are trailer-mounted do not require construction of special facilities. However, the generators should be stored inside when not in use. PTO-driven units which are to be permanently mounted at the meter pole should be anchored on a concrete pad and a shelter should be constructed to protect them from the weather. The shelter must provide adequate ventilation so the generator does not overheat.

Stationary engine units have more detailed installation requirements. Following is a list of points for consideration before installation of a stationary engine:

- clearance for service accessibility
- vibration damping pads
- air inlets to prevent hot air recirculation
- the engine exhaust system
- fuel storage

Manufacturers generally provide detailed instructions about the installation of their stationary engine units and these should be carefully followed.

System Implementation

Power suppliers, insurance companies, and local codes may have rules about standby generator installations. All these sources should be consulted before a system is implemented.

References

ASAE. 1997. ASAE standards 1997. ASAE S417.1. Specifications for alarm systems used in agricultural structures. ASAE, St. Joseph, MI.

Campbell, L. E. 1965. Standby electric power equipment for the farm. Leaflet No. 480. USDA, Washington, DC.

Cotthoff, T. 2002. Personal communication. Katolight Corporation, Mankato, MN. <<*www.katolight.com*>>

Puckett, H. B. 1980. Standby electric power equipment for the farm and home. Farmers' Bulletin Number 2273. USDA, Washington, DC.

Seneff, W., and Hoyle Puckett. 1989. Electric motors: Selection, use and management. In: Electrical energy in agriculture. K. L. Mc Fate, ed. National Food and Energy Council, Columbia, MO.

Problems

9.1. Compute the cost per kWh of electrical energy obtained from a size D cell. Assume the cell costs $1 and supplies 2 A at an average voltage of 1.5 V for 6 h.

9.2. Compute (in ¢ per kWh) the fuel cost for producing electrical energy with a tractor-driven alternator. Assume the alternator is 50% efficient, the tractor fuel efficiency is 10-hp h/gal, and gasoline costs $1.80/gal. Recall 1 hp = 746 W.

9.3. While a 4-pole standby generator is in operation, a motor-driven clock gains 2 min during a 4-h period. What is the line frequency and the rotational speed of the generator?

9.4. A standby generator is being selected for a farm containing this single-phase equipment:

House: 12 kW of resistance load (lights, heaters)
0.33-hp furnace
0.33-hp freezer
0.25-hp refrigerator
0.75-hp water pump

Farmstead: 10-hp silo unloader
1.5-hp feed auger
2-hp corn auger
2-hp hammer mill
400-W lighting
3-hp grain aeration fan

Critical loads include the furnace, water pump, silo unloader, feed auger, corn auger, mill, and 300 W of lighting. All of these loads must be able to operate at once. Starting can be in any sequence, except that the lights must come on first, the feed auger must come on before the silo unloader, and the mill must come on before the corn auger.

Determine the necessary continuous and intermittent generator capacity necessary using the total load and the critical load procedures.

9.5. A standby generator system is to be designed to serve these critical loads:

1000-W lighting
one 3-hp motor
one 5-hp motor
two ⅓-hp motors
one 10-kW heater

The loads are all manually controlled. The lights must be turned on first.

(a) Compute the minimum intermittent and continuous rating for the generator assuming that after the lights are on, the operator will start the other loads in an order which will minimize starting watts.

(b) Compute the minimum intermittent and continuous rating for the generator assuming that the lights are turned on first and that the heater is automatically switched and, thus, may be on at any time. Other loads might be started one by one by the operator, but in any order.

(c) Compute the minimum intermittent and continuous rating for the generator, assuming everything can be turned on simultaneously.

9.6. A standby power generator having a capacity of 50 kW (intermittent) and 35 kW (continuous) is to be used for a drying system consisting of 1 kW of lights and several 5-hp motors running fans. The lights must be turned on first. Then the motors are turned on one at a time. How many motors can the generator handle, and what is the intermittent and continuous power needed for this number of motors? Do not plan for future needs.

Chapter 10

Electric Motors

Introduction

An electric motor is a machine that changes electrical power to shaft power (Figure 10.1). Output power comes through a shaft which turns against the resisting torque of the load. Like all machines, electric motors are less than 100% efficient; that is, their useful output power is less than the input of electrical power. The difference between input power and output power is waste power, given off as heat.

The output power of electric motors is generally 60% to 95% of the input power, that is, motors are generally 60% to 95% efficient. Large motors are usually designed for higher efficiency.

Compare electric motor function to electric generator function (Figure 10.2). Compared to other sources of shaft power (such as internal combustion engines, hand cranks, pedal shafts, water turbines, wind turbines, steam turbines, gas turbines, and treadmills), the electric motor is relatively efficient, inexpensive, quiet, clean, small, long lived, low maintenance, easy to control, non-polluting, and safe. The electric motor is relatively difficult to make portable due to problems associated with storing electrical energy.

Application of the electric motor has eliminated a great deal of human drudgery (and also many jobs) in applications where the motor has replaced a person turning a hand crank or pedaling. An electric motor can, for an energy cost of about 10 ¢, do as much work as a strong person working a full day, assuming output of the person at 112 W (0.15 hp) for 8 h, 8 ¢/kWh electrical energy cost, and motor efficiency of 75%.

This chapter deals with how motors work, their characteristics, and how to apply motors.

Classification of Motors

Figure 10.3 is a classification of electric motor types in common use. The first subclassification is into AC and DC motors. This chapter will concentrate on AC motors. Because most electrical power is used as AC, AC motors are widely used. We will discuss in detail several of the types of single-phase motors listed in Figure 10.3, and also the 3-phase squirrel cage type.

Induction Motor Operation (An Intuitive Approach)

Conduction motors. During operation, electrical energy is transferred from the stationary part (stator) of a DC motor (and also in one type of AC motor) to the rotating part (rotor) by conduction through a solid path. The interface between the rotating part and the stationary part is the point where a moving commutator rubs against the stationary brush. This type motor can be called a *conduction* motor since electric current is conducted along a solid path.

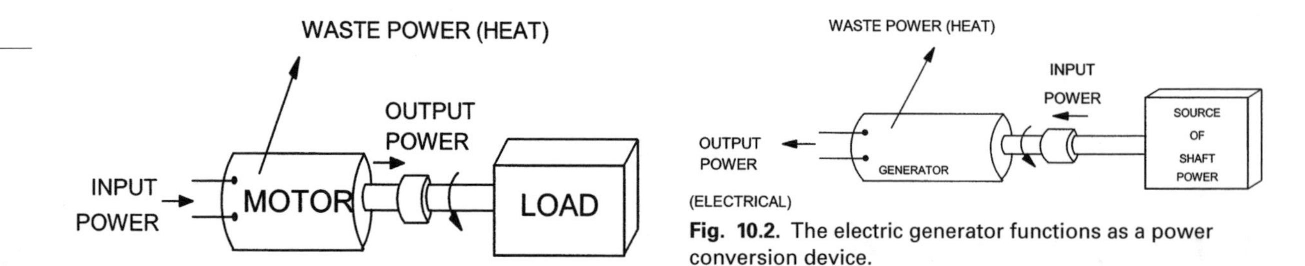

Fig. 10.1. The electric motor functions as a power conversion device.

Fig. 10.2. The electric generator functions as a power conversion device.

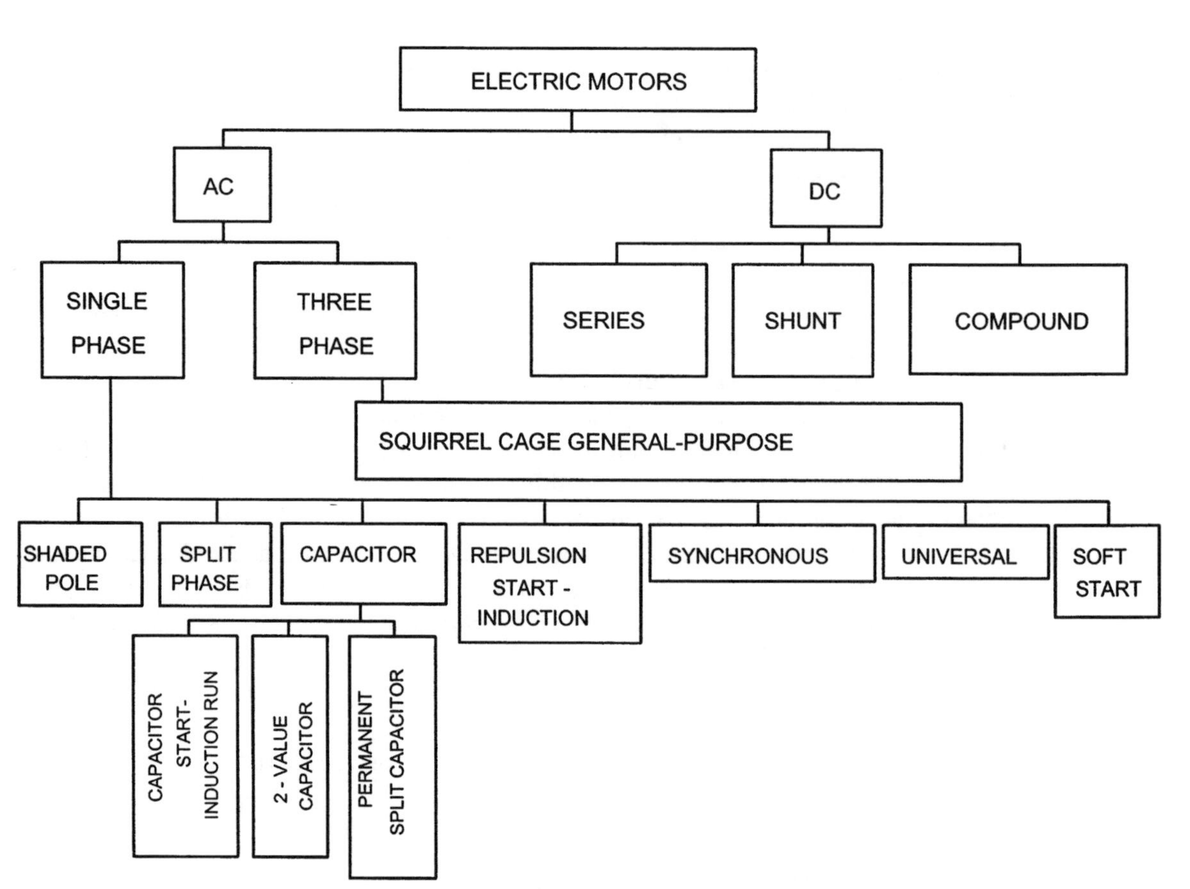

Fig. 10.3. Electric motor classification.

Induction motors. In most AC motors, electrical energy is transferred from stator to rotor inductively (like the secondary winding of a transformer receives energy from the primary). In these so-called *induction* motors, there is no solid electrical path (conductor) from stator to rotor. Magnetic energy, produced from electricity in the stator, is transferred across the air gap by induction. An AC induction motor is a very simple, versatile rotating electrical machine and is used in many applications.

Rotating field. Some form of rotating field is necessary for operation of every induction motor, and a clear notion of the concept of rotating field is necessary for understanding motor operation. The concept is easiest to see by an analysis of a 3-phase induction motor. Figure 10.4 represents the circuitry within a 2-pole, 3-phase stator. Lines crossing without a dot signify no connection. Poles will be discussed a little later.

Let's first point out some features of this circuit. Notice that there are three terminals A, B, and C (as all 3-phase loads have). Notice that the windings could be drawn in a wye configuration as in Figure 10.5 and that the windings for each phase are split in half. This signifies that half of the windings of that

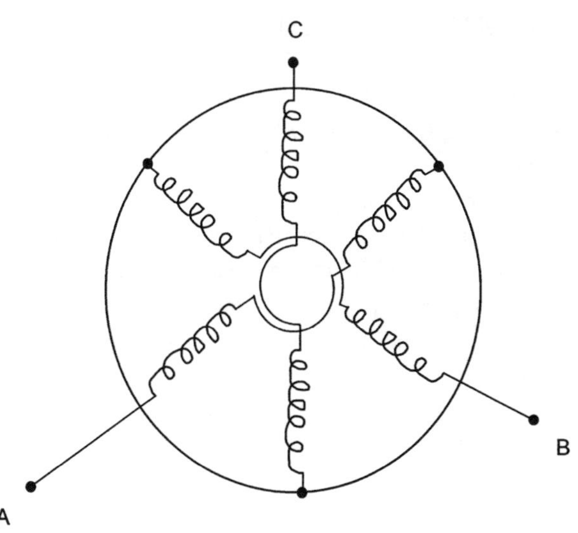

Fig. 10.4. A 2-pole, 3-phase motor stator. (The circle is a solid conductor.)

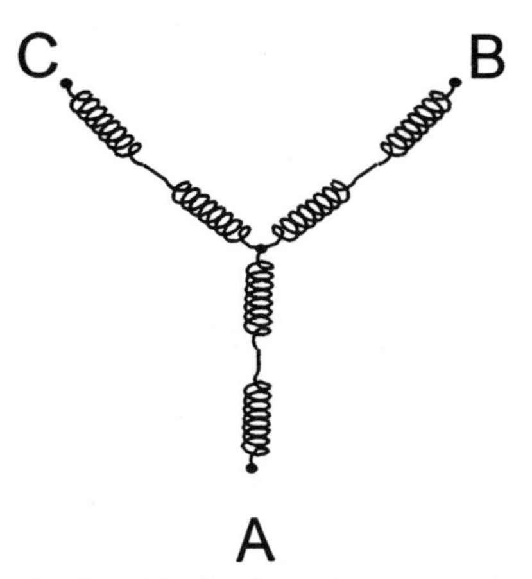

Fig. 10.5. The 2-pole, 3-phase stator circuit drawn in a wye load configuration.

phase are wound around one stator pole, and half are wound around another pole across the center from the first. This stator has 2 poles (the least it can have) for each of its 3 phases. Motors can have 2, 4, 6, 8, or more poles per phase.

When this stator is connected to 3-phase lines, current flows through all the windings. Because current in each of the lines is 120° out of phase with current in the other two, a rotating field is established within the stator. Analysis of the field within the stator at successive times will establish the direction of rotation and aid in understanding the rotating field concept.

Figure 10.6 shows a graph of current in each phase versus time, and the orientation of the field at six successive times. Look at the graph first. Notice that the current lines go through zero on the way up in this sequence: C, A, B. We then say that the phase sequence is CAB (or ABC or BCA).

At time (1), current in phase B is 10 A in a + direction (taken in this case to mean away from the center of the wye and toward the generator). Currents in A and C are 5-A negative (toward wye center). Assuming that currents toward the center of the stator establish north poles and currents away from the center of the stator establish south poles, we see that the three poles on the top and left are north, and three poles on the bottom and right are south. Also, the

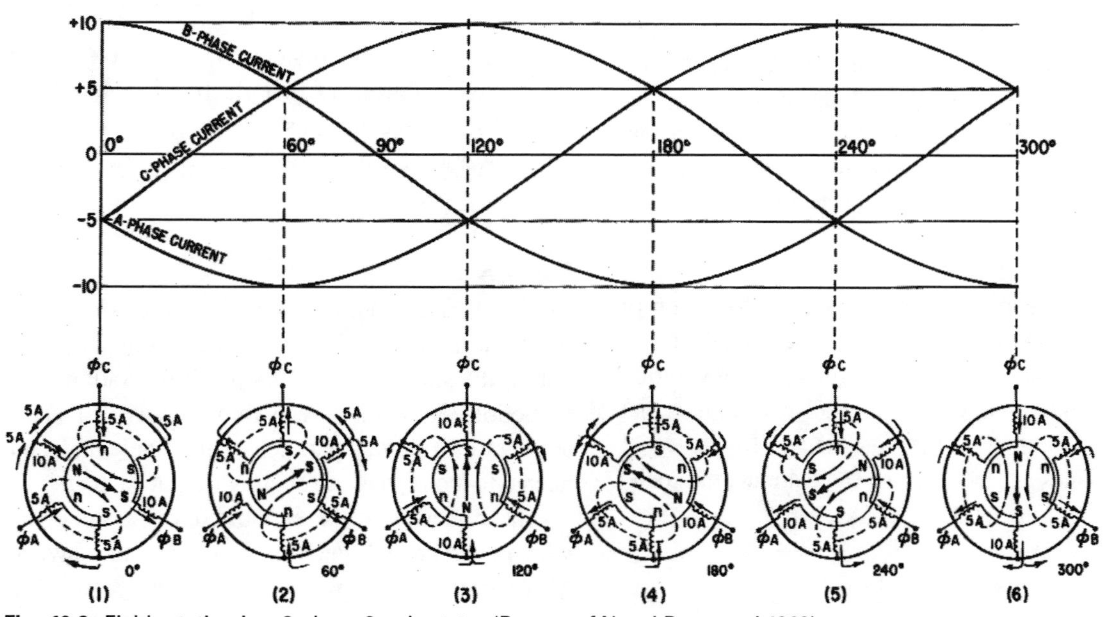

Fig. 10.6. Field rotation in a 3-phase 2-pole stator (Bureau of Naval Personnel 1969).

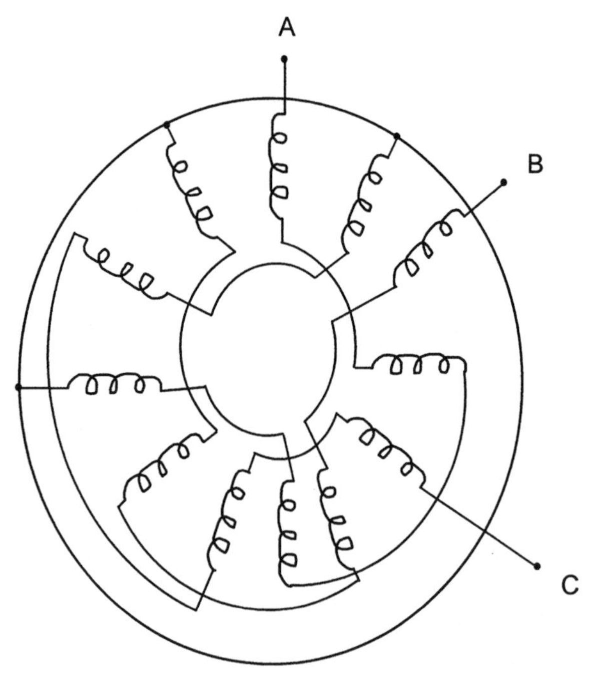

Fig. 10.7. A 3-phase, 4-pole stator circuit. (The circle is a solid conductor.)

windings carrying 10 A have stronger fields (N and S) than the windings carrying 5 A (n and s). A north-to-south arrow signifying field orientation points down and to the right.

If the same procedure is used at time (2), the arrow is seen to have rotated $^1/_6$ revolution (60°) counterclockwise. Analysis at successive times indicates that the field within this non-moving stator is rotating in a counterclockwise direction.

If connections of any two phases (wires) to the stator are reversed, the arrow (and hence the field) will reverse its direction of rotation. Problem 10.1 is an exercise in proving this. In like manner, switching any two of the leads to a 3-phase squirrel cage motor causes reversal of the rotating field direction and hence reversal of rotor direction. (Be sure you understand the preceding explanation before you go further into the chapter.)

Synchronous Speed

Now we are able to compute speed of rotation of the rotating field in Figure 10.6. Notice that the rotating field and the arrow rotate through one revolution in one complete power cycle (360°). Assuming that the power supplied to the stator is at a frequency of 60 Hz (cycles per second), we can compute the speed of rotation of the field:

$$N_s = \frac{1\ \text{r}}{\cancel{\text{cycle}}}\ \frac{60\ \cancel{\text{cycles}}}{\cancel{\text{s}}}\ \frac{60\ \cancel{\text{s}}}{\text{min}} = 3{,}600\ \text{r/min} \quad \textbf{(10.1)}$$

The field is thus turning at 3,600 r/min. This is called the synchronous speed of the motor and is denoted as N_S. All 2-pole induction motors have a synchronous speed of 3,600 r/min (when connected to a 60-Hz supply).

A 4-pole (3-phase) stator has windings arranged as in Figure 10.7. Notice that if you start at a terminal (A, B, or C) and follow the lead until you reach the solid ring (or wye center), you go through 4 windings. Hence the motor is said to have 4 poles (per phase).

If a rotating field analysis is done here, the arrow will advance only one-half revolution in one power cycle. Thus:

$$\frac{0.5\ \text{r}}{\cancel{\text{cycle}}}\ \frac{60\ \cancel{\text{cycles}}}{\cancel{\text{s}}}\ \frac{60\ \cancel{\text{s}}}{\text{min}} = 1{,}800\ \text{r/min} \quad \textbf{(10.2)}$$

$$N_s = \frac{2(60)\ f}{P}$$

or,

$$N_s = \frac{120\ f}{P} \quad \textbf{(10.3)}$$

where N_s = synchronous speed (speed of field rotation), r/min
f = power frequency, Hz
P = number of poles per phase

Equation 10.3 shows that the synchronous speed is constant for a given motor and given power frequency.

Most induction motors employ the 4-pole design. Successively fewer are manufactured in 2-pole, 6-pole, 8-pole, and designs with higher numbers of poles. A motor cannot be built with fewer than 2 poles.

Torque

In order for a motor to supply a torque or twisting force through a shaft to its load, a rotor is placed within the stator. Figure 10.8 is an illustration of an induction motor rotor. The rotor consists of heavy conductors running parallel to the shaft and connected to a common conducting ring at each end.

The conductors are supported on a steel core which is laminated (built up from many thin layers). This construction provides desirable magnetic characteristics in the rotor. In rotor designs where the core

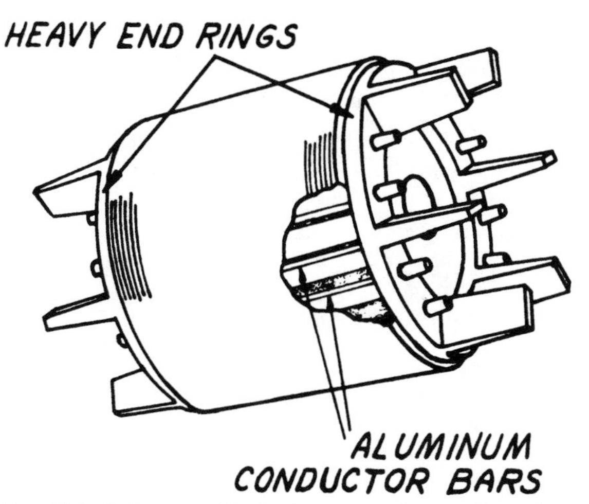

Fig. 10.8. Cut-away of induction motor "squirrel cage" rotor.

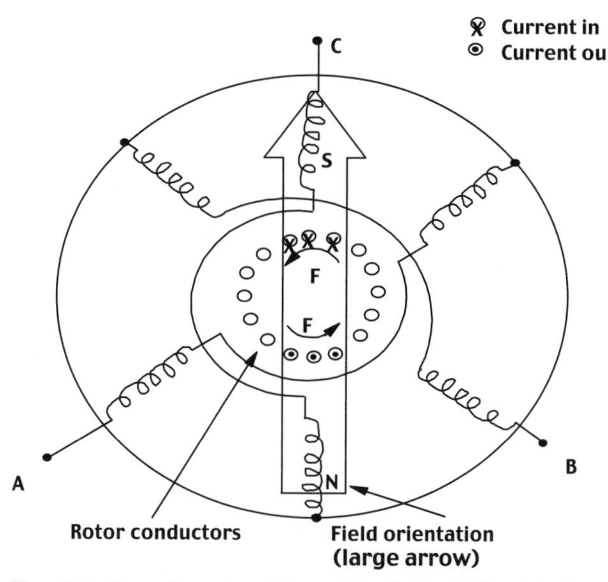

Fig. 10.9. Rotor in stator with rotating field position (3) of Figure 10.6.

does not fill the volume between the shaft and the conductors, the rotor resembles a pet squirrel's exercise wheel. For this reason, an induction motor rotor is sometimes called a *cage* or *squirrel cage* rotor.

The rotor is not electrically connected to the stator. However, the air gap between the inside surface of the stator and the outer surface of the rotor is often less than 1 mm, and the rotating field of the stator induces a current to flow in the rotor conductors if there is any relative motion between the rotating field (which always rotates at synchronous speed) and the rotor conductors. This relative motion is called *slip*. Percent slip is defined in Equation 10.4. This is illustrated in Figure 10.9.

$$S = \frac{N_s - N}{N_s}(100) \qquad \textbf{(10.4)}$$

where S $=$ slip, %
N_s = synchronous speed, r/min
N = rotor speed, r/min

Current flowing in the direction shown will result in a force to the left on the conductors in the top portion of the rotor and a force to the right in the bottom portion of the rotor. The forces will cause a counterclockwise torque on the rotor shaft.

The rotor will accelerate if this torque exceeds the opposing torque of the load. The magnitude of the torque depends on the magnitude of the slip since a greater relative motion between field and conductor will result in a greater induced rotor current. A 4-pole induction motor (N_s = 1,800 r/min) will turn at about 1,795 r/min with no load connected to the shaft. This means that 5 r/min of slip (S = 0.3%) is required for the motor to develop enough torque to overcome bearing friction and air turbulence around the rotor. When loaded to nameplate hp (full load), induction motors operate at about 4.2% slip. This means that the load of a 4-pole induction motor will decrease to about 1,725 r/min (95.8% of 1800) if the motor is loaded to its nameplate hp. Table 10.1 shows speeds for various induction motors operated at 60 Hz (U.S. standard) or 50 Hz (European standard).

TABLE 10.1 Induction motor speeds

Number of Poles	60-Hz Power		50-Hz Power	
	Synchronous Speed N_s, r/min	Full-Load Speed N_{FL}, r/min	Synchronous Speed N_s, r/min	Full-Load Speed N_{FL}, r/min
2	3,600	3,450	3,000	2,875
4	1,800	1,725	1,500	1,438
6	1,200	1,150	1,000	958
8	900	863	750	719
n	7,200/n	0.958(7200/n)	6,000/n	0.958(6,000/n)

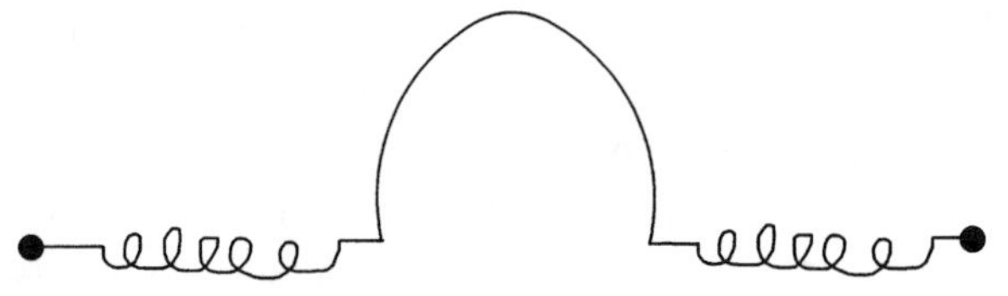

Fig. 10.10. Two-pole, single-phase stator.

Single-Phase Induction Motors

The preceding discussion of motor operation dealt with a 3-phase induction motor since the principles are easiest to understand with this type of motor. Figure 10.10 is the circuit of a 2-pole, single-phase stator. Note that it has a total of 2 poles (instead of 6 total poles in a 2-pole, 3-phase stator). When an AC voltage is applied across the terminals, current flows through the windings in one of two directions. If an attempt is made to map field orientation, as was done with the 3-phase stator, one finds that the field orientation arrow is either straight up or straight down. There are no intermediate positions and hence the field is pulsating but not rotating. If a squirrel cage rotor is placed in this stator, it will not start turning by itself because the pulsating field produces no torque on the rotor. However, if it is given a start (in either direction) by some other means, it will quickly accelerate up to running speed. (This can be demonstrated by opening the starting winding circuit of a motor such as a split-phase type and then applying rated voltage to the terminals. A 60-Hz hum will be heard, but the rotor will not begin to turn. Twirling the shaft by hand will start the motor in either direction. Since the windings can overheat quickly, don't leave the power on with the rotor stationary for more than 2 or 3 sec.)

Since single-phase induction motors have no inherent rotating field, they must have an additional starting system to provide a torque for starting. The single-phase motor types vary in the starting system employed. These different types will be discussed later in the motor types section.

Motor Characteristics

Torque-Speed

The torque-speed characteristics of motors are commonly expressed using graphs of the type shown in Figure 10.11. The graph shows typical characteristics of a 3-phase, general-purpose type (NEMA Design B) motor. This is the most common type of 3-phase motor. The term, "NEMA Design B" will be explained later. Because both axes are percentages, the curve is applicable to any size of motor of this type. Motor hp output is a function of speed and torque and can be computed using the equation:

$$HP = \frac{2\pi TN}{33{,}000} \qquad \textbf{(10.5)}$$

where: HP = power output, hp
T = torque, lb·ft
N = rotational speed, r/min

With the rotor locked (restrained from turning) $\% N_s = 0$ and the graph shows that the motor develops 210% of full-load torque. This is the maximum starting torque available from the motor and is called the locked rotor torque, T_{LR}. If T_{LR} exceeds the torque required to start the load, the motor will accelerate the load up to operating speed. The rotor will operate at the full-load operating point if the load requires nameplate hp from the motor. Notice that this is at a speed of about 96% of N_s ($S = 4\%$). If the motor is not connected to a load, the no load rotor speed, N_{NL}, will be about 99.5% of N_s. Notice that the speed decrease is approximately linear with increasing torque between $N = N_s$ and $N = 0.95\ N_s$.

If the motor is loaded in excess of its nameplate hp, speed will decrease. With a decrease in speed, the motor is able to produce an increase in torque up to a certain point called the breakdown torque, T_{BD}. In Figure 10.11, $T_{BD} = 270\%$ of T_{FL}. The breakdown torque is the last point on the torque-speed curve where a decrease in speed will result in an increase in torque. If motor speed is pulled down below the breakdown torque speed, the motor torque will decrease abruptly and the motor may stall, depending on how the load behaves when speed decreases. An example problem will help explain information presented on the torque-speed graph.

Example 10.1

Compute the full-load speed, full-load torque, locked rotor torque, no-load speed, and breakdown torque for a 1-hp, 3-phase, 4-pole, general-purpose motor operated on 60-Hz power. From Equation 10.3:

$$N_s = \frac{(120)(60)}{4} = 1{,}800 \text{ r/min}$$

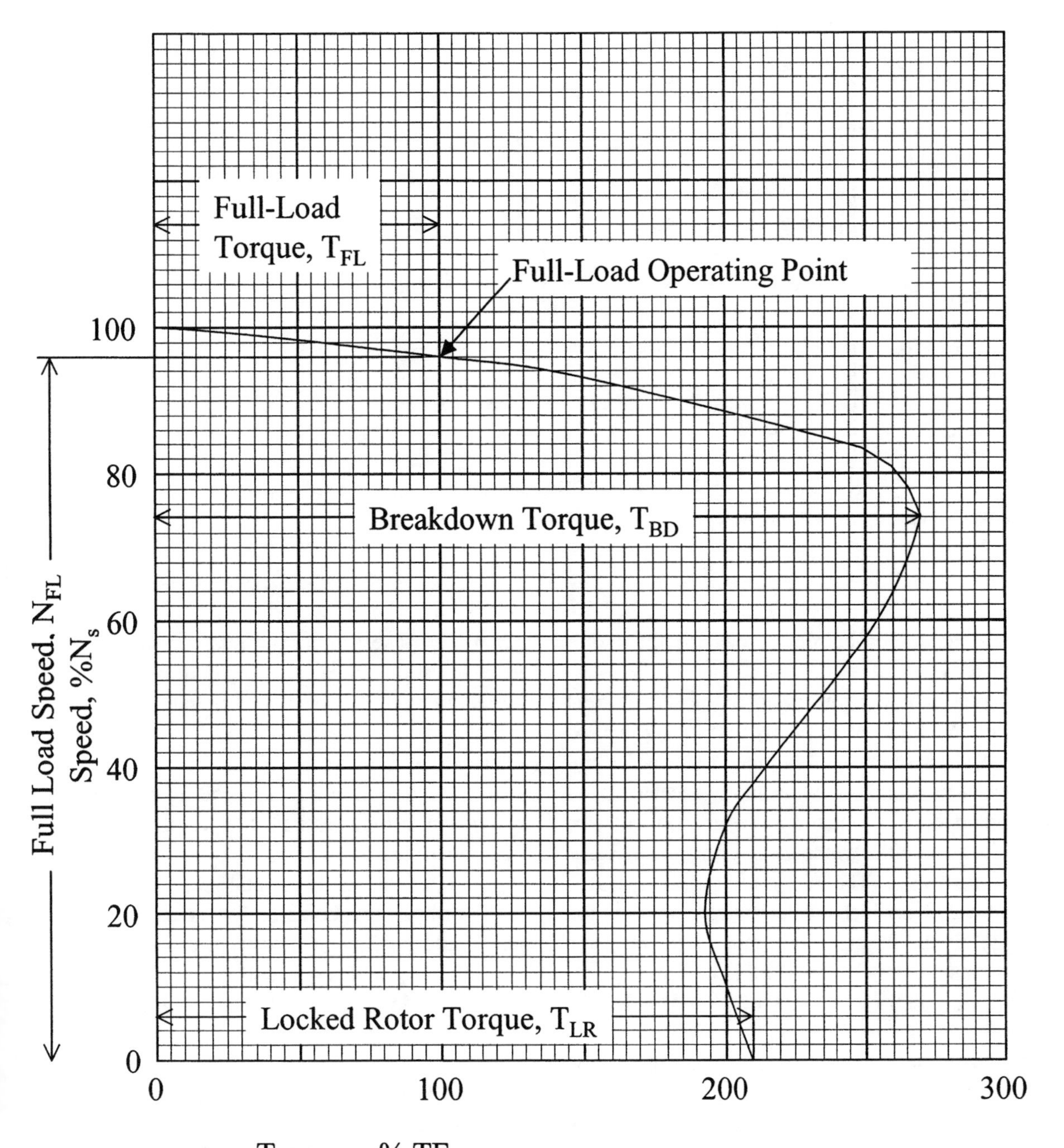

Fig. 10.11. Torque-speed characteristics of a 3-phase, general-purpose motor.

From Figure 10.11, full-load speed (N_{FL}) is about 96% (actually 95.83%) of N_s:

$$N_{FL} = (0.9583)(1{,}800 \text{ r/min}) = 1{,}725 \text{ r/min}$$

Note that these calculated speeds match those for a 4-pole motor in Table 10.1. Full-load torque can be computed by solving Equation 10.5 for T and substituting in full-load (nameplate) parameter values:

$$T_{FL} = \frac{(\text{HP})(33{,}000)}{2\pi \, N_{FL}} = \frac{(1)(33{,}000)}{2\pi(1{,}725)} = 3.0 \text{ lb}\cdot\text{ft}$$

From Figure 10.11, the locked-rotor torque is read as 210% of T_{FL}:

$$T_{LR} = 2.10(3.0) = 6.3 \text{ lb}\cdot\text{ft}$$

Breakdown torque is read as 270% of T_{FL}:

$$T_{BD} = 2.70(3.0) = 8.1 \text{ lb}\cdot\text{ft}$$

Starting Current

Most electric motors draw high currents while they are accelerating up to speed. Figure 10.12 illustrates this characteristic, with current expressed as % of I_{FL},

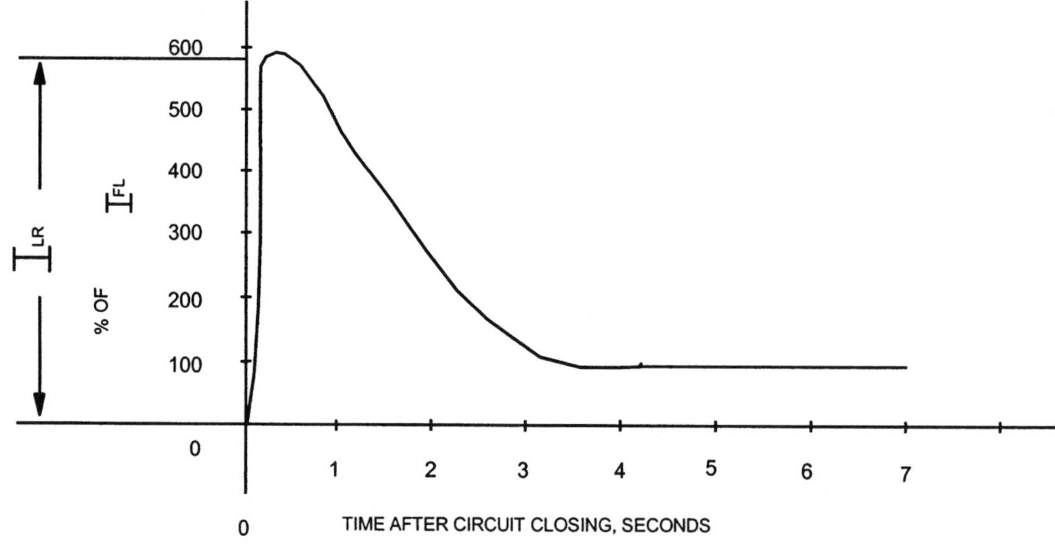

Fig. 10.12. Example of starting current for a loaded induction motor.

the full-load motor current shown on the motor nameplate.

As soon as the motor circuit is closed, current jumps up to a high value (nearly 600% of I_{FL} in this example) because current is limited only by the impedance of the windings. As the motor accelerates, an electromotive force which opposes the flow of current is induced in the windings and current decreases to a constant value. The locked rotor current (I_{LR}) is the peak value to which the current can climb during starting. It is also the current the motor will draw (and will continue to draw) if the rotor is locked so that it cannot turn. I_{LR} varies with motor type and tends to be higher for cheaper motor types. The code letter on the motor nameplate provides a method of estimating the locked rotor current for the motor. This will be discussed more in the section on nameplate information.

The time required for the current to decrease from I_{LR} to a steady value depends on the type of load connected to the motor. High-inertia loads like heavy fan rotors require long acceleration times.

Motor starting current becomes increasingly troublesome as motor size increases. In order to limit starting current, some motor starters supply a reduced voltage to the motor as it starts.

The starting currents of motors are very troublesome. They cause line voltage drops which may visibly dim lights. They tend to trip (open) overcurrent devices even though the high current does not continue long enough to damage components protected by the overcurrent devices. Such a trip is called a *nuisance trip*. These high currents can also overload generating equipment—especially if the motor uses a large portion of the generator's output.

Operating Characteristics

Typical induction motor operating characteristics are shown in Figure 10.13. Shapes of curves would be similar regardless of horsepower rating or motor type. Notice that the horizontal graph axes extend to 200% of hp_{FL}. It is common for induction motors to have the capability of delivering twice their nameplate hp. Motor damage will occur if the motor is overloaded long enough for its windings to overheat.

- *Speed.* Rotor speed decreases almost linearly as load increases. Speed changes only about 4.2% between no load and full load. This is a slight decrease and in comparison to some shaft-power sources, the induction motor can be thought of as a constant speed device.
- *Watts.* Power use increases approximately linearly as power output goes up.
- *Current.* Motor current changes little between no load and full load, but rises more rapidly under overload conditions.
- *Efficiency.* Motor efficiency increases up to full-load conditions and then stays about constant or decreases for overload conditions.
- *Power factor.* The power factor increases (improves) as load increases.

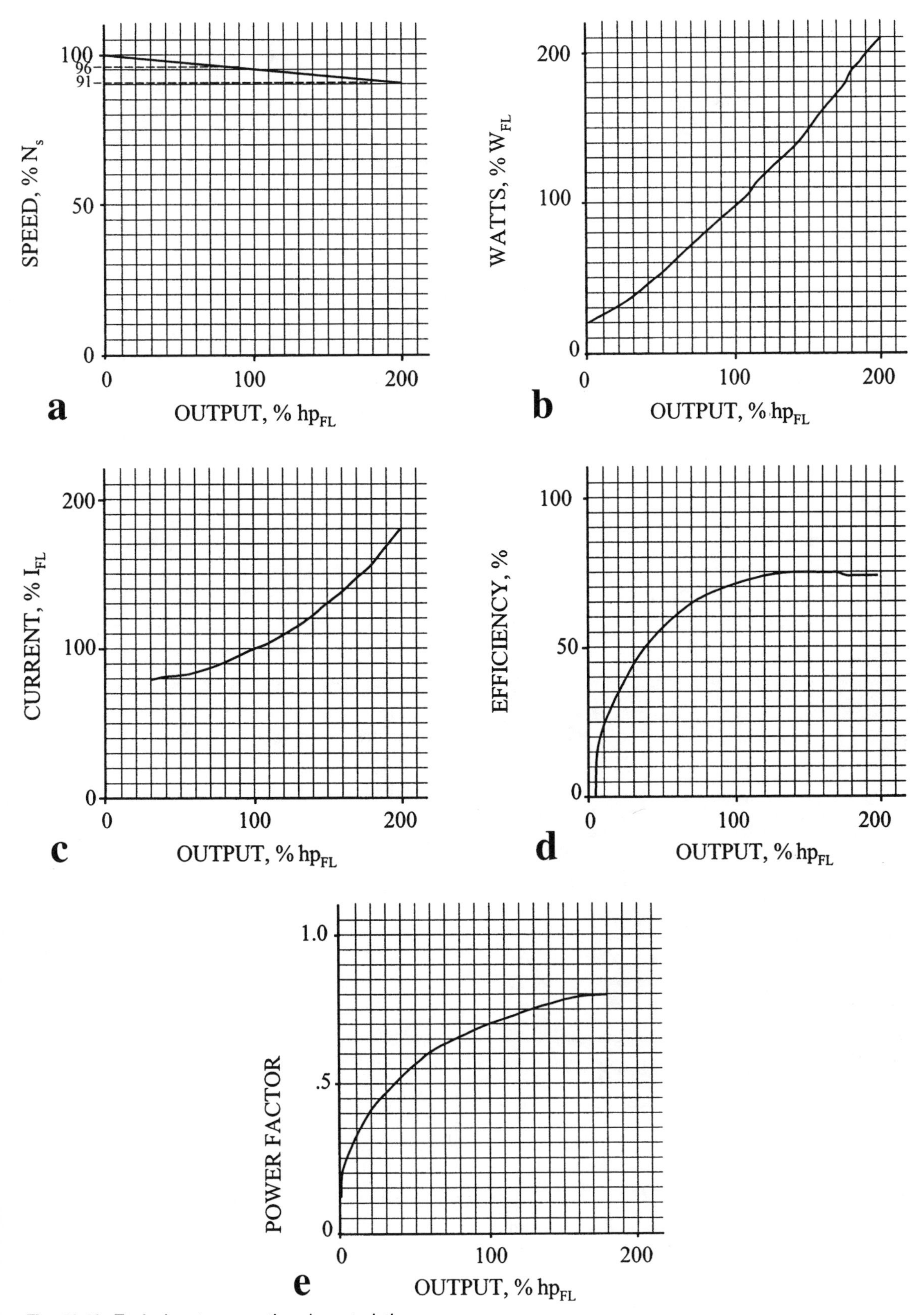

Fig. 10.13. Typical motor operating characteristics.

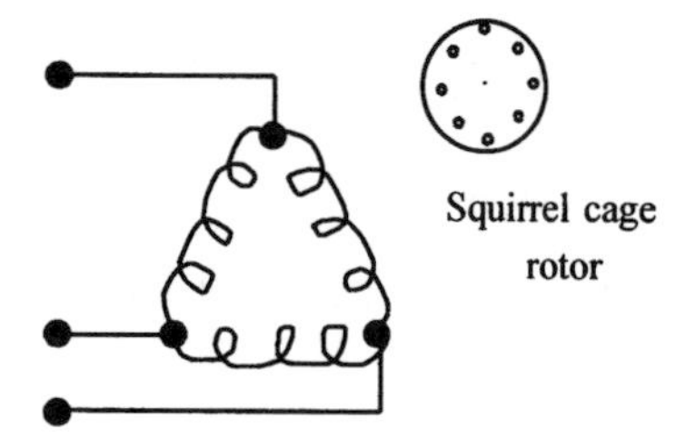

a. Circuit with delta-connected stator windings

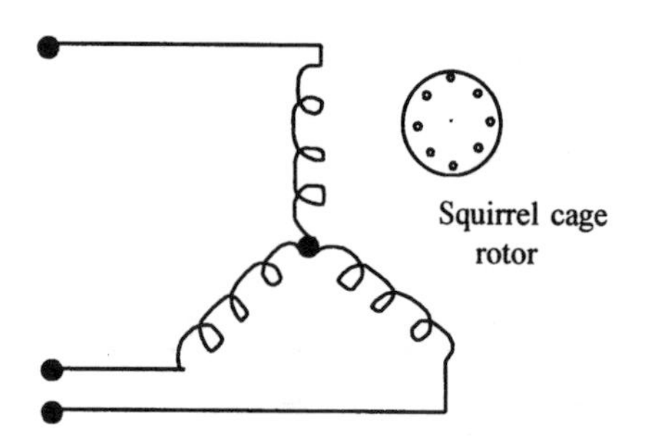

b. Circuit with wye-connected stator windings

c. Typical model 3-phase squirrel cage motor

Fig. 10.14. 3-phase squirrel cage motor.

Types of AC Electric Motors

In this section, some AC electric motor types listed in Figure 10.3 will be discussed. Table 10.2 summarizes their characteristics.

Intended Use

In discussing motor types, reference will be made to the purpose for which a motor is made:

General-purpose motors are the type you buy off the shelf for many varied applications. The motor driving a grain auger in a grain handling system is a general-purpose motor. Pertinent physical and electrical features are standardized among brands (more on this later). These motors are made to be applicable for a wide variety of uses.

Definite-purpose motors are designed for specific applications and are also standardized among brands for this application. Submersible pump motors are definite purpose motors. Their use for another purpose (such as running a shop grinder) would be difficult to accomplish due to their physical features, and might be detrimental to the motors due to their electrical characteristics. Because of high volume of production, definite purpose motors tend to be lower in cost than general-purpose motors of comparable characteristics.

Special-purpose motors are built for a specific use. Standardization among manufacturers is limited. The motor in an electric drill is a special-purpose motor.

Load Starting Ability and Starting Current

Load starting ability and starting current are important characteristics of an electric motor. It is desirable for motors to have high load starting ability and low starting currents. In Table 10.2, typical load starting ability and starting current are listed for several motor types. Ratings are defined in terms of locked rotor torque (T_{LR}) and full-load torque (T_{FL}), and locked rotor current (I_{LR}) and full-load current (I_{FL}).

Three-Phase Squirrel Cage

Operation of the 3-phase squirrel cage motor was discussed in an earlier section. Figure 10.14 shows other features of it. The motor stator can be wired in either a delta or a wye configuration. Construction is simple because no separate starting mechanism is required. Starting current and starting torque are both medium. Its torque-speed characteristics were shown in Figure 10.11.

Rotational direction can be reversed by interchanging any pair of input leads. Doing this reverses direction of the rotating field. (To prove that this occurs, work problem 10.1.) If a pair of input leads is reversed while the motor is running, it will stop and then start in the opposite direction. However, reversing on-the-go may damage the motor, or the connected load, and cause abnormal currents in the motor circuit.

The 3-phase general purpose motor has many features to recommend it:

TABLE 10.2 AC motor types and their characteristics

Motor Type	Common hp Range	Typical Load Starting Ability[1]	Typical Starting Current[2]	Distinguishing Physical Features	Rotational Direction Reversal	Other Characteristics	Common in General-Purpose Designs?	Typical Uses
3-Phase squirrel cage (Design B)	$^1/_{10}$–5,000	Medium	Medium	Squirrel cage rotor 3 lead wires	Interchange any two input leads	Simple construction	Yes	Everything except where very high starting torque is required
Shaded pole	$\leq {}^1/_2$	Low	Medium	Squirrel cage rotor Single winding Shades on each pole	Not reversible	Simple construction High slip	No	Small fans
Split-phase	$^1/_{20}$–$^3/_4$	Medium	High	Squirrel cage rotor Two field windings Centrifugal switch	Interchange starting winding leads with respect to running winding leads	Lowest cost of single-phase motors for general-purpose use	Yes	Fans, centrifugal pumps, stationary shop tools, clothes washers and driers
Capacitor start-induction run	$^1/_8$–10	Medium to high	Medium	Squirrel cage rotor Protruding capacitor enclosure 2 field windings Centrifugal switch	Interchange starting winding leads with respect to running winding leads	The common "farm duty" motor	Yes	Grain augers, compressors, water pumps
Permanent split capacitor	$^1/_{50}$–1	Low	Low	Squirrel cage rotor Protruding capacitor enclosure Two field windings Short length	Interchange auxiliary winding leads with respect to main winding leads	Adaptable to speed control	Yes	Direct-drive fans and blowers
Two-value capacitor	2–20	High	Medium	Squirrel cage rotor Capacitor-enclosure box adjacent to motor Two field windings Centrifugal switch	Interchange auxiliary winding leads with respect to main winding leads		Yes	Grain augers, compressors, silo unloaders, barn cleaners, dryer fans
Repulsion-start induction run	$^1/_6$–10	High	Low	Wound rotor One field winding Brushes and commutator	Shift brush holder	Expensive to build	Yes	Silo unloaders, barn cleaners, water pumps
Soft-start	15–50	Low	Low	Squirrel cage rotor Capacitor and contactor box adjacent to motor	Interchange auxiliary winding leads with respect to main winding leads		Yes	Forage blowers, crop dryer fans, irrigation pumps, saws
Universal	$^1/_{50}$–$^1/_2$	High		Wound rotor brushes, and commutator Field and rotor in series		No synchronous speed	No	Power tools, vacuum cleaners

[1]Load starting ability: Low ($T_{LR} \leq 150\%\ T_{FL}$)
Medium ($150\%\ T_{FL} < T_{LR} \leq 350\%\ T_{FL}$)
High ($T_{LR} > 350\%\ T_{FL}$)

[2]Starting current: Low ($I_{LR} \leq 300\%\ I_{FL}$)
Medium ($300\%\ I_{FL} < I_{LR} \leq 500\%\ I_{FL}$)
High ($I_{LR} > 500\%\ I_{FL}$)

- It is very simple and reliable.
- It is available in a very wide range of horsepower ratings.
- It has inherently low vibration because the 3-phase line supplies constant power and hence constant (not pulsating) torque.
- It is manufactured in large numbers.
- It has a comparatively low original cost.

In some instances, especially in applications below about $^{1}/_{3}$ hp, wiring and switching equipment for a 3-phase motor will be more complex and more expensive than that required for a single-phase motor. This may result in the single-phase system being the best choice.

The common general-purpose (Design B) 3-phase motor may have inadequate starting torque for some hard-to-start loads. In these cases, a different type of 3-phase motor can be selected. See the discussion of NEMA Design in the section on motor nameplate information. Because of its many advantages, the 3-phase motor is usually the best motor choice if 3-phase power is available.

Phase Converters

A phase converter is a device which creates a 3-phase service using a single-phase, 240-V input. Phase converters make it possible to operate 3-phase motors in rural areas where 3-phase service is not available, and the cost of providing it is prohibitive. This use of 3-phase motors is attractive when:

- Motors larger than 10 hp are required. Rural lines usually cannot support single-phase motors larger than 10 hp because of their excessive starting currents. The lower starting currents of 3-phase motors allow larger hp sizes.
- Three-phase motors are the only choice for the application. Some machines are only available with 3-phase motors.

Three-phase power provided by a phase converter is not equal in quality to 3-phase power supplied by a 3-phase distribution transformer. Motors running on phase-converter power sometimes must be derated to a maximum horsepower value less than their nameplate value.

Motors that work best on phase converters drive loads which have low starting torques and which do not have abrupt fluctuations. Examples include crop drying fans and irrigation pumps. Some phase converters are designed for specific applications. Although phase converters allow use of 3-phase motors, which are cheaper to purchase than comparable size single-phase motors, the price of the 3-phase motor plus its phase converter will exceed the price of the single-phase motor.

Shaded-Pole

In the section on electric motor operation, the concept of a rotating field was discussed. The 3-phase motor, having a rotating field by simply being connected to the 3-phase system, was simplest. All single-phase induction motors require an additional system to provide a rotating field effect for starting. These systems are all imperfect and rarely reach the symmetry, simplicity, and smoothness of the 3-phase system. All single-phase induction motors supply a pulsating torque, and motor vibration is inherent. We will now discuss these motors, beginning with the simplest, the shaded-pole type.

The least expensive approach to providing a rotating field (and thus starting torque) for a single-phase motor is to "shade" a portion of each pole face of the motor by encircling it with a conduction ring made of a material such as copper. Figure 10.15a illustrates this. This shade is, in effect, a shorted field winding (Figure 10.15b). It causes the magnetic flux under the shaded portion of the pole to lag (in time) the flux in the remainder of the pole. A rotating field effect results and torque on the squirrel cage rotor is produced. Its characteristics include a higher full-load slip than most induction motors (see Figure 10.15c) and a full-load efficiency of only 30% to 40%. Because the shade position determines direction of rotation, the shaded pole is not reversible.

The shaded-pole motor is not common as a general-purpose motor. Definite and special-purpose designs rated less than $^{1}/_{10}$ hp for small fans are where it is used most.

Split-Phase

The split-phase motor is the cheapest single-phase type available in general-purpose designs. In order to achieve a rotating field effect for starting, a second field winding is added to the stator (Figure 10.16a). This parallel auxiliary or starting winding has a higher resistance and lower reactance than the main winding. As a result, there is a phase difference between the currents in the two windings, with the starting winding current (I_S) leading the running winding current (I_R) (Figure 10.16b). The phase supplying current to the motor has been split in two, hence the

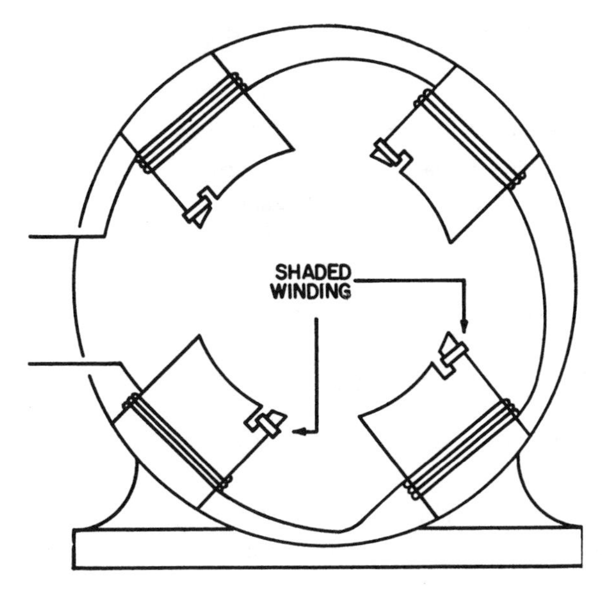

a.

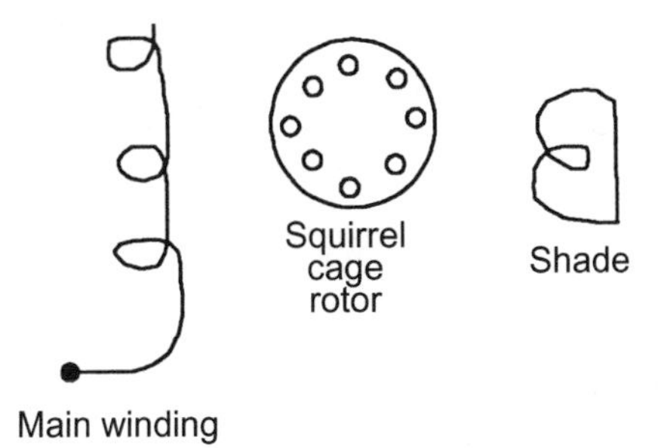

b.

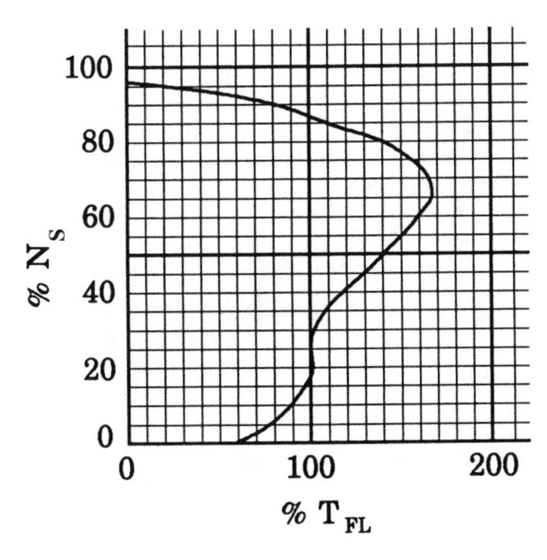

c. Typical torque-speed characteristics of a 4-pole motor

Fig. 10.15. Shaded-pole motor.

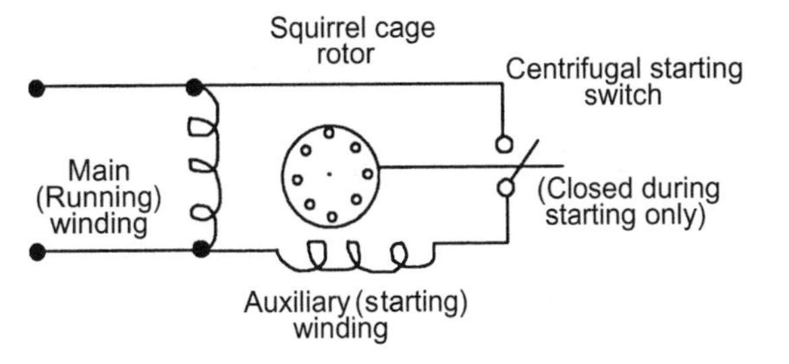

a. Circuit schematic

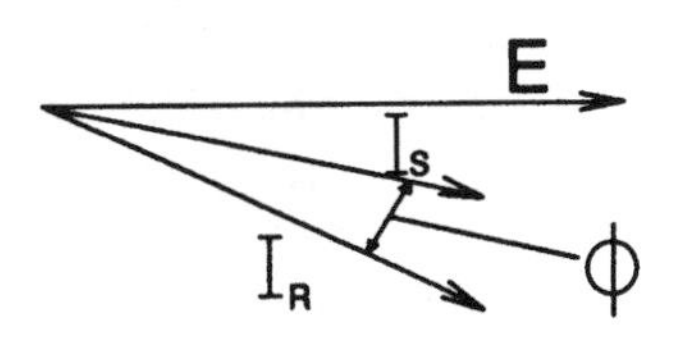

b. Phase relationship

c. Internal centrifugal mechanism

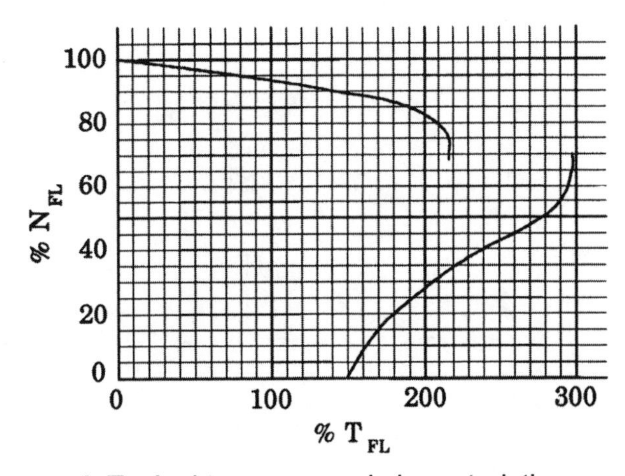

d. Typical torque-speed characteristics

Fig. 10.16. Split-phase motor.

name, *split phase*. This phase difference (angle ϕ in Figure 10.16b) causes a rotating field effect which produces a torque on the squirrel cage rotor although this angle is far less than the 120° existing in the 3-phase stator.

When the motor attains 60% to 80% of N_s, the starting switch opens and the starting winding is taken out of the circuit. The running winding alone is in use after starting. As the starting winding is switched out of the circuit, there is an abrupt drop in torque. This causes the discontinuity visible in the torque-speed curve (Figure 10.16d). This discontinuity is exhibited by all single-phase motors employing an auxiliary winding used only during starting.

The switch in the starting winding circuit of the split-phase motor (and in many single-phase motors employing an auxiliary winding used only during starting) usually consists of open contacts actuated through a rotating bushing by a centrifugal mechanism on the rotor (see Figure 10.16c). If this switch is not closed, the motor will not start because only the running winding is energized. If this switch does not open after the motor starts, the motor will operate with little immediate indication of a problem, but the winding will overheat and may destroy its insulation and short out within a few minutes. Failure of this starting switch is a common cause of single-phase motor failure.

The operating characteristics of the split-phase motor are adequate for many easy-to-start applications. This, together with its low cost, make it a popular choice for fans, centrifugal pumps, and stationary shop tools.

Reversing Direction of Rotation

The split-phase motor is reversed by interchanging starting winding leads with respect to running winding leads. Figure 10.17 illustrates this procedure. When starting winding leads are reversed with respect to main winding leads, the motor will start and run in the opposite direction of rotation. Note these important points in regard to this procedure:

- This procedure will work on all single-phase motors having a starting winding.
- The motor manufacturer normally brings winding ends into the terminal box and provides instructions somewhere on the motor for reversal of direction with winding ends identified by color or number.
- Some motors are manufactured without the necessary winding ends accessible and are, thus, not reversible.
- If this connection change is done while the motor is running, the motor will not reverse until it has been shut off and re-started.

Capacitor Motors

When a capacitor is placed in series with a load such as a motor winding, its current phasor shifts counterclockwise. If the capacitor is placed in series with the starting winding, the phase angle between the starting winding current and the running winding

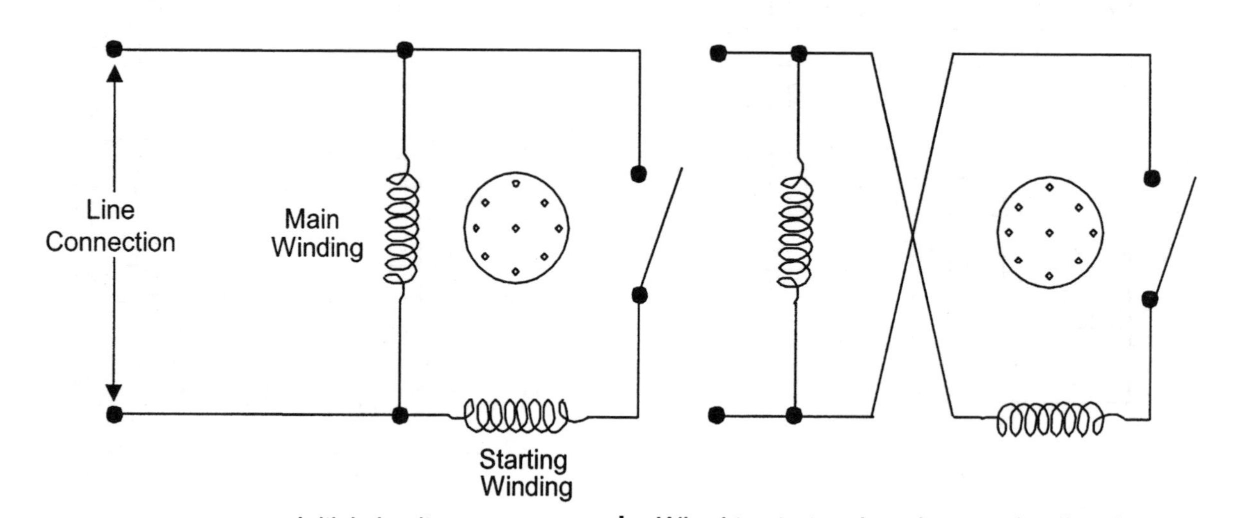

Fig. 10.17. Reversing direction of a single-phase motor equipped with starting windings (split-phase motor circuit illustrated).

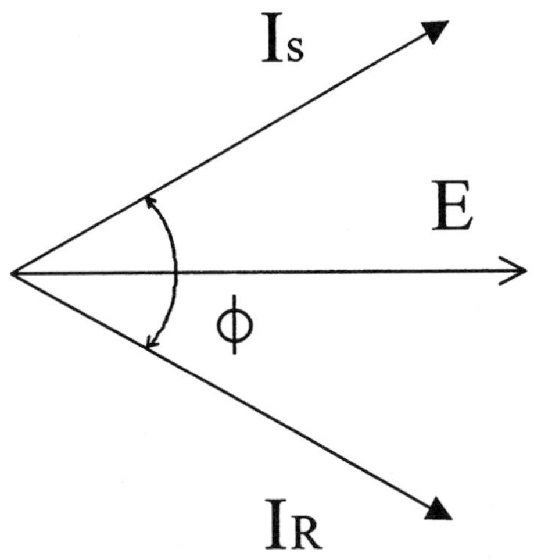

Fig. 10.18. Typical phase relationship for a capacitor-type motor.

current is increased. If a capacitor of appropriate size is used, current in this winding will lead the voltage by a few degrees and may lead the run winding current by nearly 90° (see Figure 10.18).

As in the split-phase motor, the phase difference between currents in the two windings causes a rotating field effect which results in a torque on the rotor. The larger phase difference can result in a relatively higher starting torque compared to the split-phase motor.

A capacitor motor splits the single phase as does the split-phase motor; however, it is not called a split-phase motor. The three common types of capacitor motors are the capacitor start-induction run, the permanent split capacitor, and the 2-value capacitor. Capacitor motors are the most common type of single-phase, general-purpose motor in use on farms.

Capacitor Start-Induction Run

The circuit of a capacitor start-induction run motor is shown in Figure 10.19a. The auxiliary winding and the capacitor in series with it are connected across the line during starting. Both are de-energized when the starting switch opens at 60% to 80% of N_s. Typical torque-speed characteristics are shown in Figure 10.19b. Starting characteristics are seen to be better than those of the split-phase motor. This makes the capacitor start motor suitable for hard-to-start loads such as grain augers and compressors. The capacitor start-induction run motor usually has a capacitor enclosure protruding from the side or top (see Figure 10.19c).

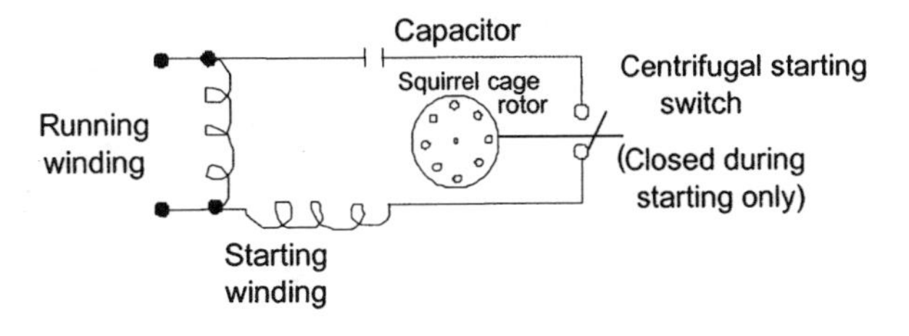

a. Schematic circuit

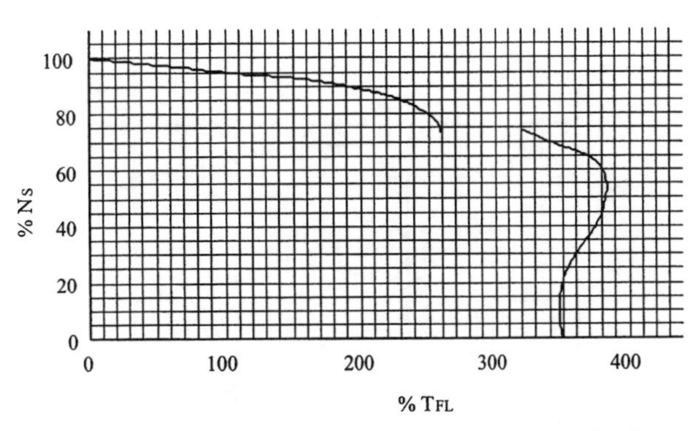

b. Typical torque-speed characteristics

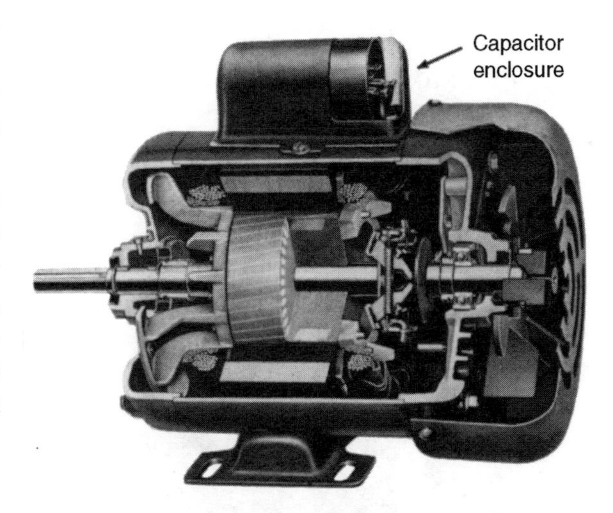

c. Capacitor start-induction run motor

Fig. 10.19. Capacitor start-induction run motor.

Permanent Split Capacitor

The permanent split capacitor motor is similar in circuitry to the capacitor start motor, except it has no starting switch (see Figure 10.20a). The single phase is split using an auxiliary winding and capacitor which both remain energized after starting.

Since the auxiliary winding circuit is designed for continuous use, starting torque is poorer than that of the capacitor start motor and there is no discontinuity

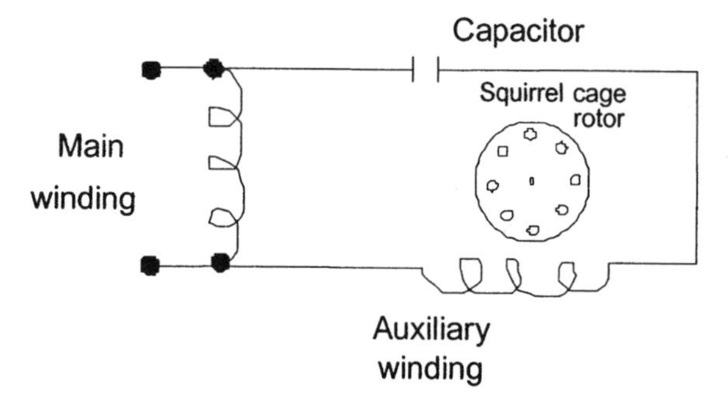

a. Schematic circuit

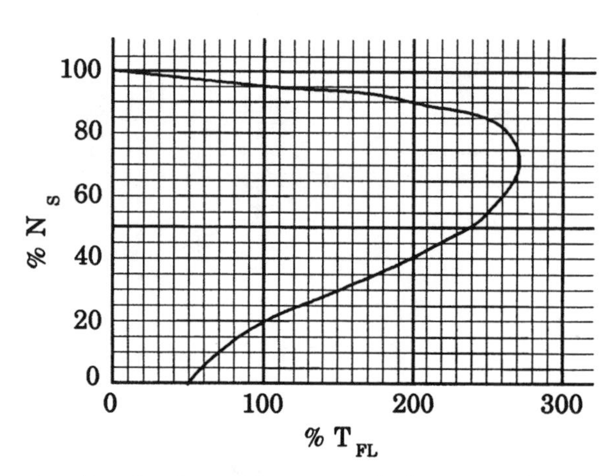

b. Typical torque-speed characteristics

c. Permanent split capacitor motor

Fig. 10.20. Permanent split capacitor motor.

in the torque-speed curve where the starting switch opens (see Figure 10.20b). Lacking this discontinuity, the motor can be used with variable speed controllers and operated at speeds down to near the breakdown torque speed (about 75% of N_s).

The permanent split capacitor motor often has its capacitor mounted remote from the motor, and sometimes the capacitor must be purchased separately. The motor is often physically shorter than other capacitor motors because there is no starting switch (see Figure 10.20c). The running or permanent capacitor counteracts the inductive characteristics of the field windings and improves (increases) the lagging power factor during running, as compared to the split phase or capacitor start designs.

The permanent split capacitor motor is most commonly used for direct-drive fans and blowers. These loads do not require a high starting torque and often employ speed control.

Two-Value Capacitor

The two-value capacitor (or capacitor start-capacitor run) motor employs a main and auxiliary winding during both starting and running as does the permanent split capacitor type. However, a second capacitor is switched in during starting by a centrifugal switch (see Figure 10.21a). The second capacitor increases starting torque by increasing the phase angle between the currents in the two windings. Typical torque-speed characteristics are shown in Figure 10.21b. The motor may have two capacitor enclosures (Figure 10.21c).

This motor type finds use for difficult-to-start loads in the 2- to 10-hp range, and for all types of loads in the 10- to 20-hp range where other single-phase motor types are not available.

Repulsion-Start Induction Motors

Repulsion-start induction motors employ a single field winding and a wound rotor with a commutator and brushes. The brushes have a conductor extending between them but are not connected to any external circuit. A closed loop of conductor exists at all times in the rotor.

During starting, the stator winding is energized and a current is induced in the loop described above. After starting, the motor operates as an induction motor.

Repulsion-start induction motors have the best torque-speed characteristics of any single-phase motor. This, along with a very low starting current, make it a natural choice for loads such as silo unloaders, barn cleaners, or reciprocating water pumps, which require very large starting torques. The wound rotor construction is, however, very expensive to build. Consequently, the cheaper two-value capacitor motor

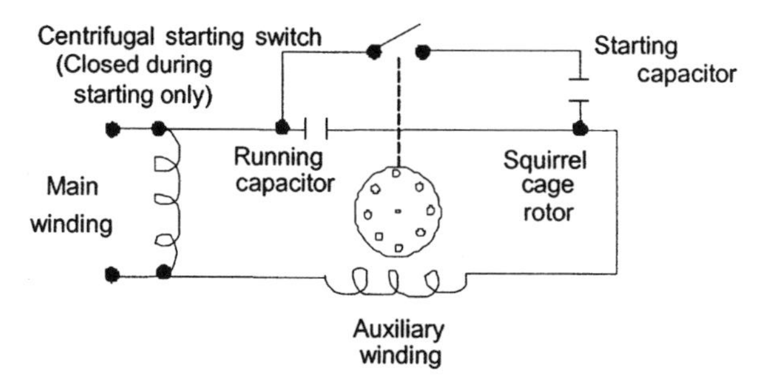

a. Schematic circuit

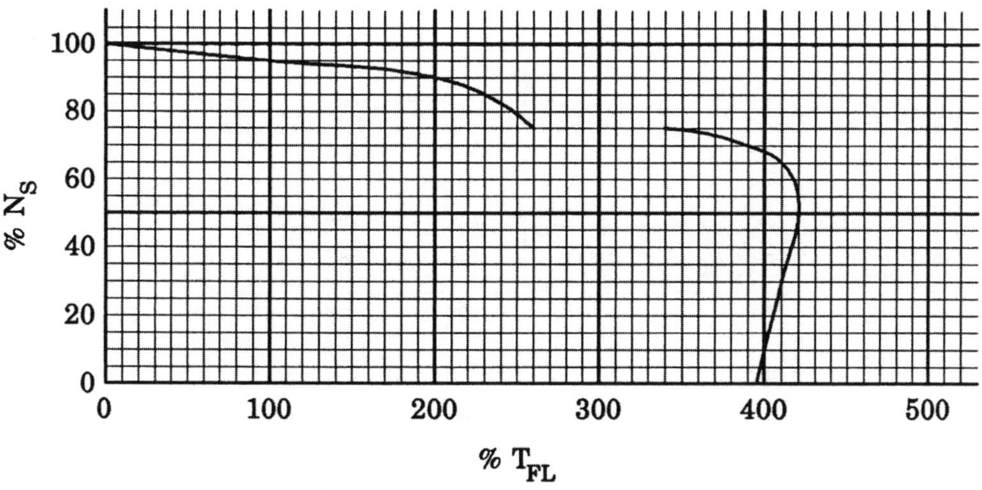

b. Typical torque-speed characteristics

c. Two-value capacitor motor

Fig. 10.21. Two-value capacitor motor.

is usually chosen instead and repulsion motors are seldom available.

Soft-Start Motor

The high starting current of single-phase motors larger than 10 hp may prevent their use on rural distribution lines (390 A is a typical starting current for a 230-V, 15-hp single-phase motor). For farm applications in this hp range which have a low starting torque requirement, the soft-start motor may be applicable. Typical applications include forage blowers, crop dryer fans, irrigation pumps, and saws.

Starting current for the soft-start motor is kept very low by a switching mechanism which connects the halves of the main winding in series for more impedance during starting (Figure 10.22a). After starting, the mechanism changes connections so that the main winding halves are connected in parallel for running. The result is a very low starting current (Figure 10.22c), but also a very low starting torque (Figure 10.22b). In appearance, the soft-start motor resembles the two-value capacitor type. However, the adjacent control box is larger to accommodate the extra switching gear (Figure 10.22d).

Synchronous Motor

A synchronous motor is so named because it operates at synchronous speed. The most common applications in fractional hp sizes are for clocks and timing

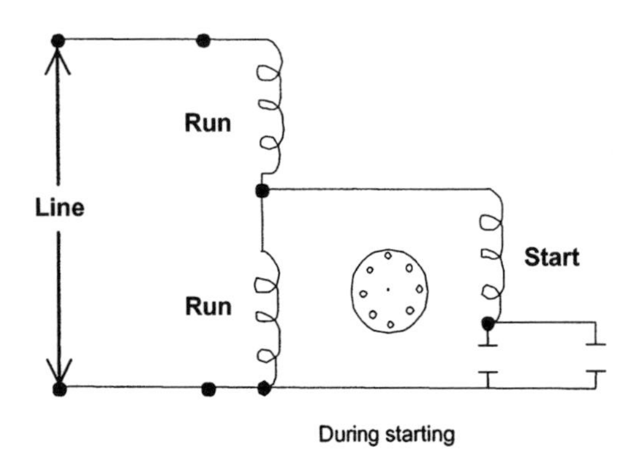

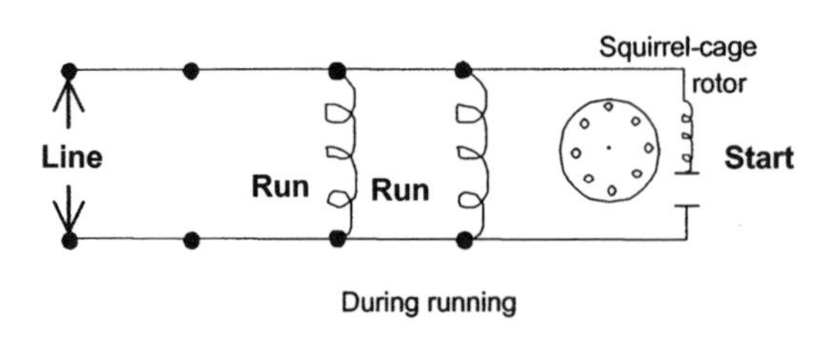

a. Schematic circuit

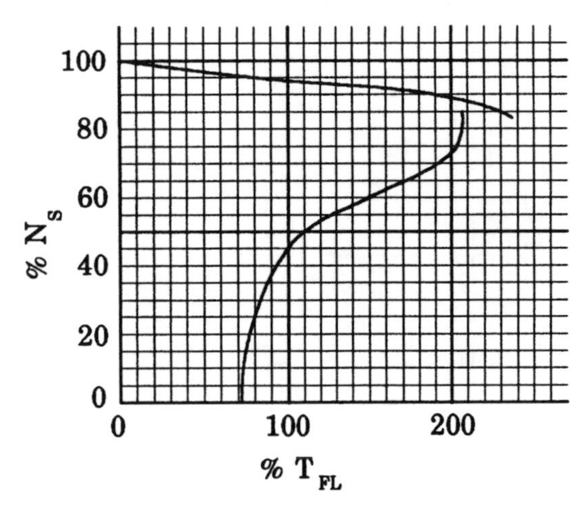

b. Typical torque-speed characteristics

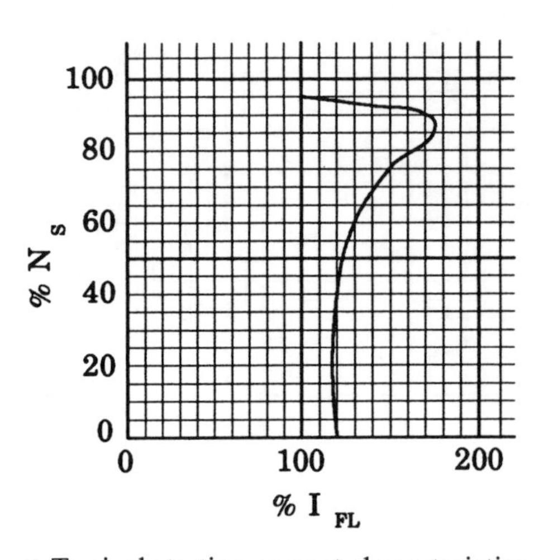

c. Typical starting current characteristics

d. Soft-start motor

Fig. 10.22. Soft-start motor.

switches where extreme constant speed is necessary (Figure 10.23).

The principle of operation is based on having a rotor with poles which are permanently defined. Then, at synchronous speed, the rotor poles are always in the correct position so that a force is exerted on them by the stator field. During starting, the rotor is brought up to near synchronous speed by some other system, and then the rotor and field lock in at synchronous speed and remain so throughout the motor's load range.

Universal Motor

The universal motor is so named because it can operate on either AC or DC power. Its circuit is that of a series-wound DC motor (Figure 10.24a). The circuit shows a field winding in series with a wound rotor having commutator and brushes.

To understand how it works, consider the circuit connected to a DC source. The rotor is constructed so that the rotor loop carrying current at any instant is in the correct position to interact with the field of the main winding, resulting in a torque on the rotor. Since the rotor and field winding are in series, field and rotor current are always equal and in the same direction, so when connected to AC, the principle of

Fig. 10.23. Synchronous timing motor. (Hansen Corp., Princeton, NJ; reprinted with permission.)

operation remains the same as when connected to DC.

The universal motor has many characteristics distinctively different from other motors discussed so far:

- It is a conduction (not induction) motor, since electric power is transferred to the rotor by a solid conducting path.
- It has no synchronous speed. Speed is determined by the load (Figure 10.24b).
- It can be designed for operating speeds up to 15,000 r/min.

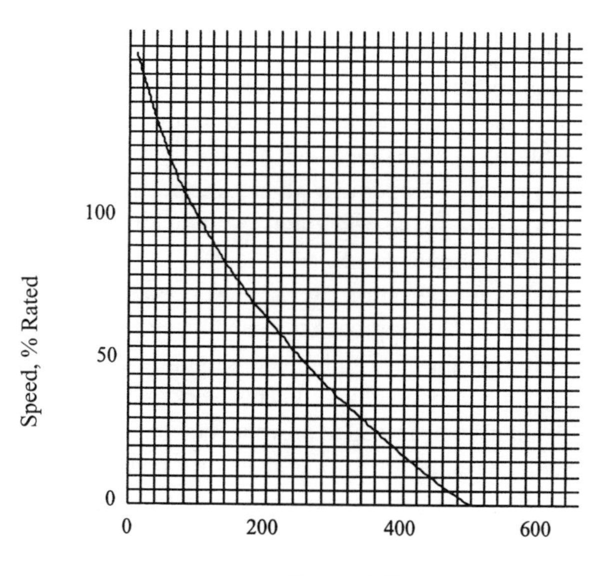

b. Typical torque-speed characteristics

c. Universal motor in a drill

Field →
Commutator
Brush

a. Schematic circuit

Fig. 10.24. Universal motor.

- Power-to-weight ratio is very high. A universal motor is very powerful for its size and weight.
- Motor life is relatively short because of brush and commutator wear.
- Starting torque is very high (Figure 10.24b).

The features of the universal motor make it attractive for applications like hand tools, vacuum cleaners, and home appliance motors where high speed and light weight are required, and where use is intermittent (Figure 10.24c). Speed control of universal motors is common and can be accomplished by varying applied voltage.

Varying Motor Speed

An induction motor operates at some speed between its synchronous speed and its full-load speed (Table 10.1). Many loads can be directly connected to the motor and operated at motor speed. This approach is desirable because it's simple, low-cost, and doesn't consume power.

In other applications, loads must operate at different speeds. One approach to changing speed is to use a belt or chain drive and to choose pulley sizes to attain the desired load speed. This works well, but the drive must be guarded to prevent personal injury, and it will consume 7 to 10% of the motor power.

Induction motor speed can be varied electrically. However, simply varying voltage to the motor is seldom satisfactory. Adjustable frequency drives rectify the AC input to DC and then invert this to 3-phase AC to drive a 3-phase induction motor. Speed is varied by varying the frequency and voltage of power to the motor. These drives allow a wide range of speeds, but they are very costly.

DC variable speed drives are widely used in industry. They employ DC motors and DC control systems.

AC Motor Nameplate Information

A motor nameplate contains information which must be noted and understood for successful motor use. Items usually appearing on a motor nameplate are listed here, and, when necessary, discussed.

1. Manufacturer's name and address.
2. Manufacturer's model number.
3. Serial number.
4. Volts. Motors are designed to operate with this voltage at the motor terminals. Induction motors can usually operate satisfactorily at voltages varying as much as ±10% from the nameplate volts. Universal motors can tolerate a ±6% variation. Some motors have two voltage ratings (see 19. Dual voltage).
5. Phase. AC power type required will be denoted by "1" (single-phase) or "3" (3-phase).
6. Horsepower. This is the power output rating of the motor. The motor is fully loaded when output is at this level.
7. RPM. The motor speed in r/min should be at this level when the motor is fully loaded, and voltage is at the rated value. Revolutions per minute is usually abbreviated as RPM.
8. Amps. The motor should be drawing this current in amperes when operating at rated output power and at rated voltage.
9. Bearings. The manufacturer may state on the nameplate the part number or the type of shaft bearings.
10. Frame. Motor frame dimensions critical for motor application are standardized among motor manufacturers by NEMA (National Electrical Manufacturers Association). Table 10.3 shows frame dimensions for various frame numbers.
11. NEMA Design. For 3-phase motors, a letter (A,B,C,D, or F) is listed here. The letter specifies the characteristics of the motor. General-purpose motors are NEMA Design B and are by far the most common and least expensive.
12. Cycles. The frequency of the power supplied to the motor in Hz is listed as "cycles." In the United States, line frequency is 60 Hz. Frequencies within ±5% (±3 Hz) will result in satisfactory motor operation. European power is 50 Hz, which is outside the specified limits. Motor damage (probably overheating) may occur if a 60-Hz motor is connected to 50-Hz power (or 50-Hz motor to 60-Hz) and fully loaded.
13. Code. The code letter predicts what amperage the motor will draw under locked rotor conditions. This is also an indication of the maximum starting current. Table 10.4 lists motor code letters and their input power ranges. Two examples will illustrate use of the table.

Example 10.2

What is the highest current that a 230-V, code L, 2-hp single-phase motor would likely draw under locked rotor conditions? From Table 10.4, code L specifies a locked rotor input of 9.0 to 9.9 kVA/hp.

$$I_{LR} = \frac{(9.9\ \cancel{k}\cancel{V}A)(2\ \cancel{hp})1000}{\cancel{hp}\qquad 230\ \cancel{V}\cancel{k}} = 86.1\ A$$

TABLE 10.3. Frame dimensions for AC motors

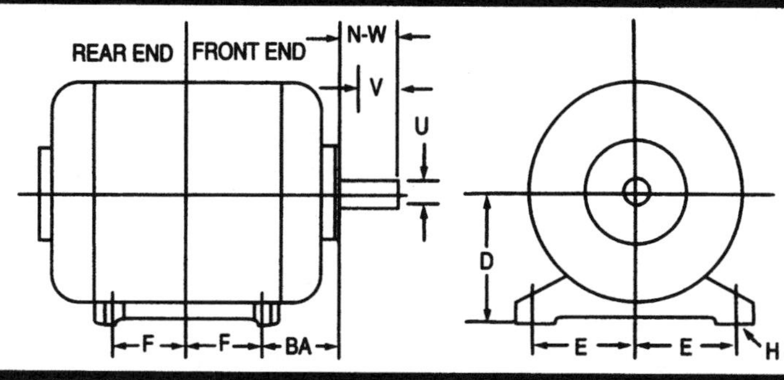

NEMA Frame	D*	2E	2F	BA	H	N-W	U	V§ Min.	Key Wide	Key Thick	Key Long
			All Dimensions in Inches								
42	2⅝	3½	1¹⁄₁₆	2¹⁄₁₆	9/32 slot	1⅛	3/8	—	—	3/64 flat	—
48	3	4¼	2¾	2½	11/32 slot	1½	1/2	—	—	3/64 flat	—
48H	3	4¼	4¾	2½	11/32 slot	1½	1/2	—	—	3/64 flat	—
56	3½	4⅞	3	2¾	11/32 slot	1⅞†	5/8†	—	3/16†	3/16†	1⅜‡
56H	3½	4⅞	3&5‡	2¾	11/32 slot	1⅞†	5/8†	—	3/16†	3/16†	1⅜†
56HZ	3½	**	**	**	**	2¼	7/8	2	3/16	3/16	1⅜
66	4⅛	5⅞	5	3⅛	13/32 slot	2¼	3/4	—	3/16	3/16	1⅞
143T	3½	5½	4	2¼	11/32 dia.	2¼	7/8	2	3/16	3/16	1⅜
145T	3½	5½	5	2¼		2¼	7/8	2	3/16	3/16	1⅜
146AT	3½	5½	5½	2¾	11/32 dia.	2¼	7/8	2	3/16	3/16	1⅜
148AT	3½	5½	7	2¾		2¼	7/8	2	3/16	3/16	1⅜
149AT	3½	5½	8	2¾		2¼	7/8	2	3/16	3/16	1⅜
1412AT	3½	5½	11	2¾		2¼	7/8	2	3/16	3/16	1⅜
182	4½	7½	4½	2¾	13/32 dia.	2¼	7/8	2	3/16	3/16	1⅜
184	4½	7½	5½	2¾		2¼	7/8	2	3/16	3/16	1⅜
182T	4½	7½	4½	2¾		2¾	1⅛	2½	1/4	1/4	1¾
184T	4½	7½	5½	2¾		2¾	1⅛	2½	1/4	1/4	1¾
182AT	4½	7½	4½	2¾	13/32 dia.	2¼	1⅛	2	1/4	1/4	1⅜
L182ACY	4½	7½	4½	2¾		2¼	7/8	2	3/16	3/16	1⅜
L182AT	4½	7½	4½	2¾		2¼	1⅛	2	1/4	1/4	1⅜
186ACY	4½	7½	7	2¾		2¼	7/8	2	3/16	3/16	1⅜
186AT	4½	7½	7	2¾		2¼	1⅛	2	1/4	1/4	1⅜
L186AT	4½	7½	7	2¾		2¼	1⅛	2	1/4	1/4	1⅜
189AT	4½	7½	10	2¾		2¼	1⅛	2	1/4	1/4	1⅜
203#	5	8	5½	3⅛	13/32 dia.	2¼	3/4	2	3/16	3/16	1⅜
204#	5	8	6½	3⅛		2¼	3/4	2	3/16	3/16	1⅜
213	5¼	8½	5½	3½	13/32 dia.	3	1⅛	2¾	1/4	1/4	2
215	5¼	8½	7	3½		3	1⅛	2¾	1/4	1/4	2
213T	5¼	8½	5½	3½		3⅜	1⅜	3⅛	5/16	5/16	2⅜
215T	5¼	8½	7	3½		3⅜	1⅜	3⅛	5/16	5/16	2⅜
219AT	5¼	8½	11	3½	13/32 dia.	2¾	1⅜	2½	5/16	5/16	1¾
2110AT	5¼	8½	12½	3½		2¾	1⅜	2½	5/16	5/16	1¾
224#	5½	9	6¾	3½	13/32 dia.	3	1	2¾	1/4	1/4	2
225#	5½	9	7½	3½		3	1	2¾	1/4	1/4	2
254#	6¼	10	8¼	4¼	21/32 dia.	3⅜	1⅛	3⅛	1/4	1/4	2⅜
254U	6¼	10	8¼	4¼	17/32 dia.	3¾	1⅜	3½	5/16	5/16	2¾
256U	6¼	10	10	4¼		3¾	1⅜	3½	5/16	5/16	2¾
254T	6¼	10	8¼	4¼		4	1⅝	3¾	3/8	3/8	2⅞
256T	6¼	10	10	4¼		4	1⅝	3¾	3/8	3/8	2⅞
284#	7	11	9½	4¾	21/32 dia.	3¾	1¼	3½	1/4	1/4	2¾
284U	7	11	9½	4¾	17/32 dia.	4⅞	1⅝	4⅝	3/8	3/8	3¾
286U	7	11	11	4¾		4⅞	1⅝	4⅝	3/8	3/8	3¾
284T	7	11	9½	4¾		4⅝	1⅞	4⅜	1/2	1/2	3¼
286T	7	11	11	4¾		4⅝	1⅞	4⅜	1/2	1/2	3¼
284TS	7	11	9½	4¾		3¼^	1⅝^	3^	3/8	3/8	1⅞^
286TS	7	11	11	4¾		3¼^	1⅝^	3^	3/8	3/8	1⅞^
324#	8	12½	10½	5¼	21/32 dia.	4⅞	1⅝	4⅝	3/8	3/8	3¾
326#	8	12½	12	5¼		4⅞	1⅝	4⅝	3/8	3/8	3¾
324U	8	12½	10½	5¼	21/32 dia.	5⅝	1⅞	5⅜	1/2	1/2	4¼
326U	8	12½	12	5¼		5¼	1⅞	5⅜	1/2	1/2	4¼
324T	8	12½	10½	5¼		5¼	2⅛	5	1/2	1/2	3⅞
326T	8	12½	12	5¼		5¼	2⅛	5	1/2	1/2	3⅞
324TS	8	12½	10½	5¼		3¾^	1⅞^	3½^	1/2	1/2	2^
326TS	8	12½	12	5¼		3¾^	1⅞^	3½^	1/2	1/2	2^
364#	9	14	11¼	5⅞	21/32 dia.	5⅝	1⅞	5⅜	1/2	1/2	4¼
364S#	9	14	11¼	5⅞		3¼	1⅝	3	3/8	3/8	1⅞
365#	9	14	12½	5⅞		5⅝	1⅞	5⅜	1/2	1/2	4¼
364U	9	14	11¼	5⅞	21/32 dia.	6⅜	2⅛	6⅛	1/2	1/2	5
365U	9	14	12½	5⅞		6⅜	2⅛	6⅛	1/2	1/2	5

(*) Dimension "D" will never be greater than the above values on rigid mount motors, but it may be less so that shims up to 1/32" thick (1/16" on 364U and 365U frames) may be required for certain machines.

(‡) Dayton motors designated 56H have two sets of 2F mounting holes—3" and 5".

(^) Standard short shaft for direct-drive applications.

(#) Discontinued NEMA frame.

(**) Base of Dayton 56HZ frame motors has holes and slots to match NEMA 56, 56H, 143T and 145T mounting dimensions.

(†) Certain NEMA 56Z frame motors have 1/2" dia. x 1½" long shaft with 3/64" flat. These exceptions are noted in this catalog.

(§) Dimension "V" is shaft length available for coupling, pinion or pulley hub—this is a minimum value.

(§§) The 2F dimension is 20.

Source: Grainger 2002.

The manufacturer states that this motor should not draw over 86.1 A during starting or under locked rotor conditions. For 3-phase motors, $\sqrt{3}$ must be added to the denominator.

Example 10.3

What is the highest locked rotor current likely to be drawn by a 5-hp, 460-V, code C, 3-phase motor? From Table 10.4, Code C: 3.55 to 3.9 kVA/hp.

For a 3-phase motor,

$$P_T = \sqrt{3}\, I_L\, E_L \text{ and } I_L = \frac{P_T}{\sqrt{3}\, E_L}$$

Thus,

$$I_{LR} = \frac{(3.9\ \cancel{kV}A)(5\ \cancel{hp})}{\cancel{hp}} \frac{1000}{\sqrt{3}(460\ \cancel{V})\ \cancel{k}} = 24.5\text{ A}$$

14. Thermal Protection or Overload Protection. If the motor is equipped with a device to open the circuit in the event of a temperature and/or current high enough to damage the motor, this is often noted on the nameplate. After the device cools down, the circuit can be closed and the motor restarted. Closing the circuit (resetting) can be done either of two ways:

    - Manual reset. If someone might be injured by automatic motor start-up, the overload device should be of the manual reset type. An example is a grain auger or a garbage disposal. A person might need to have a hand in the machine to remove the plug which caused the overload, and could be injured if the motor started automatically after a cooldown period.
    - Automatic reset. If extended shutoff would cause a problem and automatic start-up presents no safety hazards, an automatic reset device should be used. An example is a freezer compressor motor. If an overload device opens the circuit and leaves the compressor off, freezer contents may be damaged. Automatic restart allows the compressor to function periodically, but still protects the motor from overload. The cyclic operation is likely to be noticed as an indicator of a problem.

TABLE 10.4 Motor code letters (NEC Table 430.7B)

Letter Designation	kVA per Horsepower[1]	Letter Designation	kVA per Horsepower[1]
A	0–3.15	K	8.0–9.0
B	3.15–3.55	L	9.0–10.0
C	3.55–4.0	M	10.0–11.2
D	4.0–4.5	N	11.2–12.5
E	4.5–5.0	P	12.5–14.0
F	5.0–5.6	R	14.0–16.0
G	5.6–6.3	S	16.0–18.0
H	6.3–7.1	T	18.0–20.0
J	7.1–8.0	U	20.0–22.4
		V	22.4 and up

[1]Locked kVA per horsepower range includes the lower figure up to but not including the higher figure. For example, 3.14 is designated by letter A and 3.15 by letter B.

TABLE 10.5 Electric motor insulation classes

Class of Insulation	Rated Operating Temperature, °C[1]
A	105
B	130
F	155
H	180
H+	200

[1]Temperature of hottest spot in windings. The material will last, on average, for 10,000 h of continuous use or 35,000 h of normal use at this temperature.

15. Insulation or Ambient Temperature. Motor waste heat increases frame and winding temperature. The life of a motor usually is determined by how long the winding insulation lasts. It slowly oxidizes and finally breaks down, causing leakage current or a short in the windings.

    Motor insulation is classed according to its design operating temperature. Table 10.5 lists the five common classes and their design operating temperatures. Insulation in a fully loaded motor should be at its rated temperature when a motor is in a room held at the rated ambient temperature.
16. Impedance Protection. An impedance-protected motor has windings designed with impedance high enough to limit locked-rotor current to a value which will not overheat the motor, even if the rotor remains locked indefinitely. Fans on freezers and refrigerators are often impedance-protected so that if lack of cleaning and lubrication causes the rotor to stall, the motor does not overheat and no overcurrent device opens.

Higher temperature materials are more costly. Insulation life at other temperatures can be estimated using the rule which states that the rate of a chemical reaction doubles for each 10°C increase in temperature. Thus, Class F insulation will last about 5,000 h at 165°C and about 20,000 h at 145°C.

An example will illustrate the meaning of insulation and temperature ratings.

Example 10.4

A motor has this information stamped on its nameplate:

Insulation Class: B
Ambient: 40°C

a. What is the temperature of the hottest spot in the windings when this motor is fully loaded in a 40° C (104° F) room? Ans: 130°C (from Table 10.5).
b. What is the approximate motor life if the motor is fully loaded and runs continuously in a 30°C environment? Ans: 2(10,000 h) = 20,000 h.

17. Time or Duty. This rating indicates the maximum time period the motor can be operated at full load and rated ambient temperature without overheating. Most motors have a continuous duty rating. Some motors may have ratings in minutes, such as 5, 15, 30, or 60 min. Such motors must be allowed to cool after full-load operation for this time period.
18. Service Factor. The service factor is a multiplier (usually from 1.0 to 1.35) which can be applied to a motor horsepower rating to compute an allowable horsepower level which the motor can tolerate. An example will illustrate.

Example 10.5

A motor has a 5-hp rating and a service factor of 1.15. To what hp can the motor be loaded?

(5 hp)(1.15) = 5.75 hp

A motor with a service factor greater than 1.0 should be used where momentary overloads, and occasional high or low voltages, are likely.

19. Dual Voltage. Motors often have a dual voltage listing on the nameplate. This means that the motor's main (running) winding is divided into halves, which can be wired to allow operation at either of two voltages. An example will illustrate the connectors.

Example 10.6

A split-phase motor has these voltage and current listings on the nameplate:

Volts: 115/230
Amps: 9.2/4.6
How is the circuit configured?

Several points need to be made with regard to the circuit of Figure 10.25:

- For either configuration, each main winding half has 115 V connected across it.
- A split-phase motor is illustrated, but the principle applies to all motors discussed, except the synchronous and universal types.
- Note that during high-voltage operation, the starting winding is in parallel with one of the halves of the run winding.
- Full-load line current is reduced by half when operation is at high voltage rather than low voltage.
- Motor operating characteristics are the same at either voltage.
- The manufacturer will have instructions somewhere on the motor for changing voltage, in terms of wire colors or numbers.
- When there is a choice, a motor should be operated at high voltage to reduce line current.
- As connections are being changed, polarity of each half with respect to the other half must be retained. See dots at winding ends in example. If polarity is reversed at either voltage, the two halves will be bucking each other (producing torque in opposite directions) and the motor will not start.

Example 10.7

Show the circuit for each voltage rating for this dual-voltage wye-wound 3-phase motor:

Voltage: 230/460
Amps: 6.4/3.2
See Figure 10.26.

Electric Motor Selection

An electric motor selected for a particular application must be able to:

- Drive its load without being overloaded.
- Reliably start its load.
- Withstand the environment in which it is placed.

The best motor choice is the cheapest, simplest motor that will meet all these criteria.

Drive Load without Overloading

The motor hp rating should be selected so that once the load is started, the motor operates near, or somewhat below, its nameplate hp (Figure 10.11). When this is the case, the motor will be operating at, or slightly above, its nameplate r/min, and at or slightly below its nameplate current. Any of several methods can be used to estimate the motor hp necessary for a load:

a. For low-voltage operation

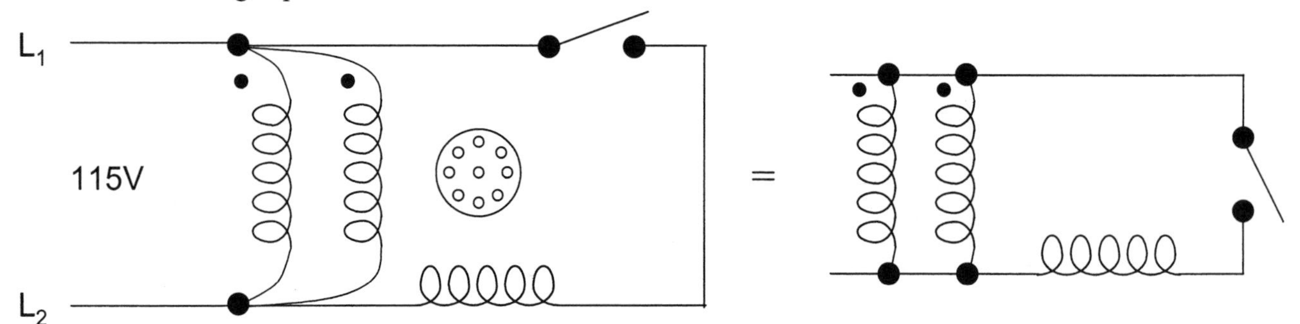

b. For high-voltage operation

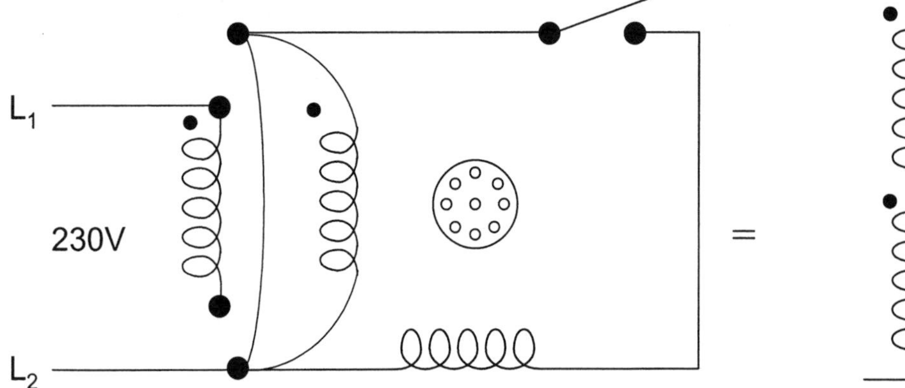

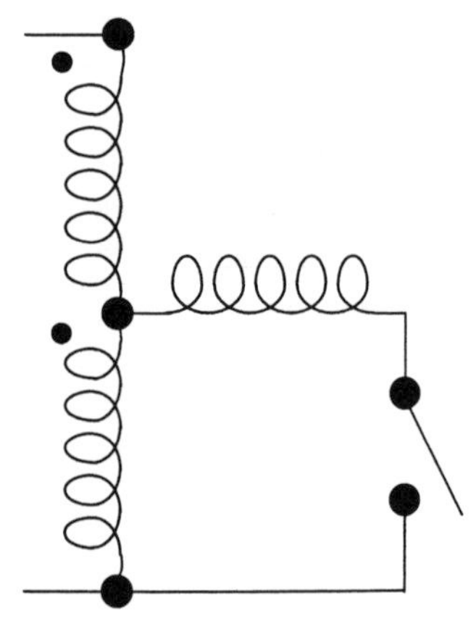

Fig. 10.25. Single-phase dual-voltage motor circuit.

a. Low voltage

L_1
L_2
L_3

b. High voltage

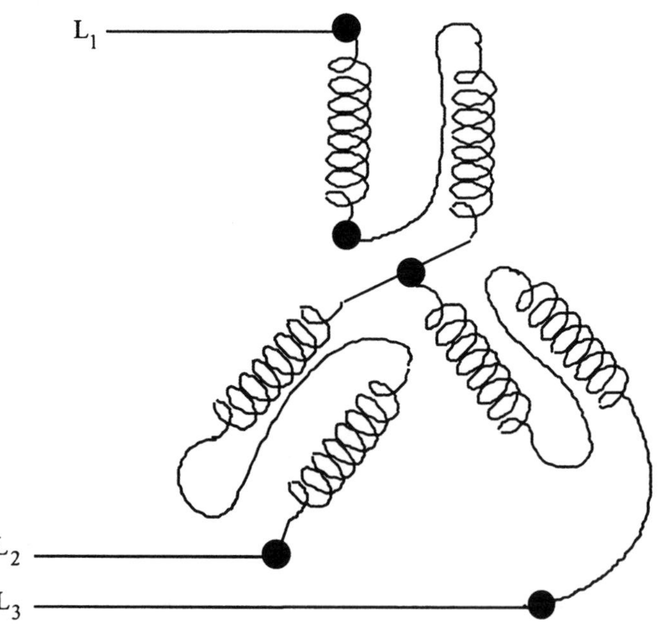

Fig. 10.26. Dual-voltage 3-phase wye-wound motor circuit.

- Use the hp size recommended by the load's manufacturer. Literature for the load may recommend a motor size to be used, or the load may have been fitted with a motor that is being replaced. The manufacturer of the load might be contacted for a recommended motor size.
- Use a reference source. Tables, such as Appendices A and B of this text, can be used to estimate motor hp for some loads. Handbooks and extension publications often include power requirement tables.
- Calculate hp required. Using principles of physics, power necessary to drive a load such as a fan, a pump, or a conveyor can be calculated as the sum of the power needed to do the work, plus the power necessary to overcome friction and other losses in the load. It is usually not desirable to select a motor so large that it always operates at half of rated load or less because efficiency will be low (Figure 10.13d) and the original cost of the motor will be needlessly high.

If a belt or chain drive connects the motor to the load, it will result in a power loss of about 7% and motor power must be increased to account for this. Once a hp figure is calculated, a standard motor hp rating at, or slightly above, this value can be selected. Table 10.6 lists standard hp ratings.

TABLE 10.6 Full-load currents for single-phase AC motors

Standard Motor Power Ratings, hp	Full-Load Current, A Single-phase 115 V	230 V
1/6	4.4	2.2
1/4	5.8	2.9
1/3	7.2	3.6
1/2	9.8	4.9
3/4	13.8	6.9
1	16	8
1 1/2	20	10
2	24	12
3	34	17
5	56	28
7 1/2	80	40
10	100	50

Source: Based on NEC Table 430.148.

Reliably Start Load

Select a motor type which has torque adequate to start the load. Once the power rating of the motor has been selected, the motor type needs to be selected so that starting torque is adequate to start the load turning from rest and to accelerate it up to its proper operating speed. This is illustrated in Figure 10.27. In the figure, torque-speed curves are drawn for a hypothetical load and four hypothetical motors having the same hp ratings. We assume here that the motor is

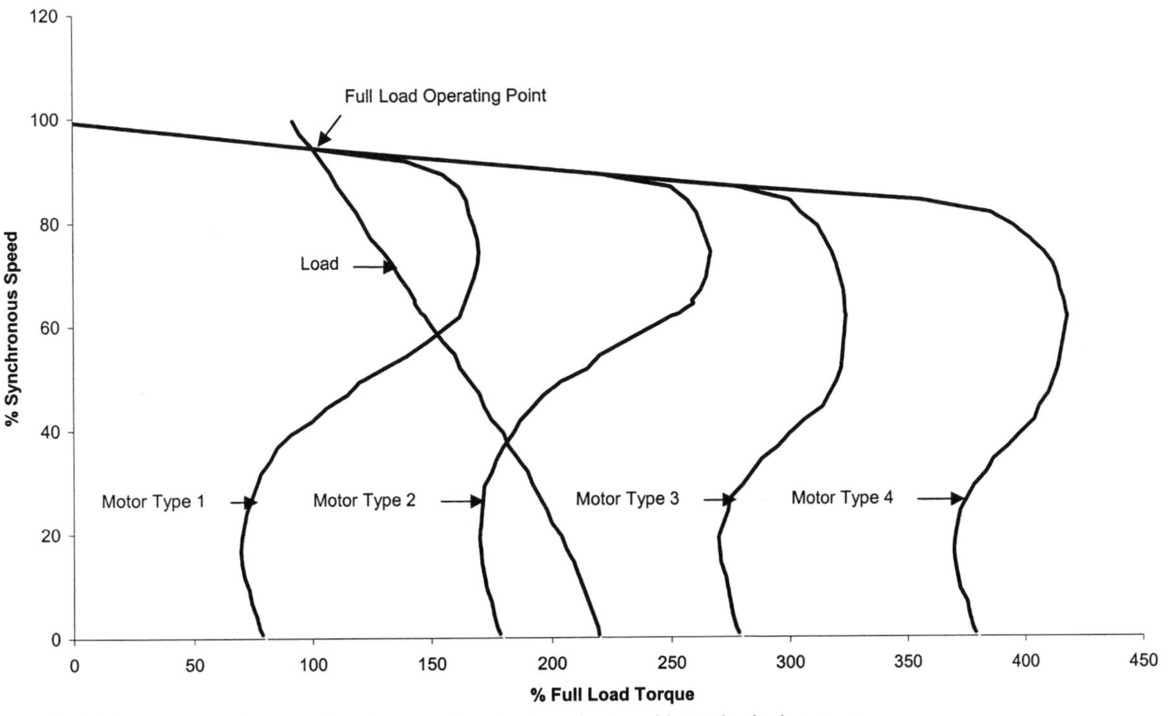

Fig. 10.27. Torque-speed curves for a hypothetical load and several hypothetical motors.

direct-connected to the load, so the load and the motor turn at the same speed.

The desired operating point of the motor-load system is the point where the load curve crosses the motor curve (100% of full-load torque and 96% of synchronous speed). Since the curves cross here, the motor will operate at its proper nameplate full-load r/min. Note that the curves for all the motors are the same between the synchronous speed point and the full-load speed point.

In order for the selected motor to be able to start the load from rest and accelerate it up to its operating point, the motor must have more torque than the load requires at all speed below the operating speed. But purchase price tends to increase with starting torque. Looking at the figure, we see that motor types 1 and 2 would not be capable of starting the load, although they have the correct hp rating. Motor types 3 and 4 are both capable of starting the load from rest and accelerating it up to its proper operating speed.

Because motor price will be higher for higher starting torques, motor type 3 can be chosen as the least expensive design which will reliably start the load. In reality, precise torque-speed curves for motors are available from manufacturers but are almost never available for loads. Loads are broadly classified as to the torque required to start them (high, medium, low). Numerical ranges are assigned to these classification as part of motor descriptions (see footnote 1 of Table 10.2).

Withstand Motor Environment

Selectable motor features related to environment include bearings, mounting method, and enclosure.

Bearings

Ball bearings and sleeve bearings are used in electric motors. Motors larger than 1 hp are likely to have ball bearings. Each type of bearing has advantages compared to the other (Table 10.7).

Mounting Method

The simplest and cheapest motor mounting method is the *solid mounting method* in which the motor mounts are integral with the motor frame. When the motor stator is cast iron, the mounts are part of this casting (Figure 10.14c). When the stator is steel, a steel mounting bracket is welded or bolted to the stator (Figures 10.16c, 10.19c).

Sometimes, motor vibration is a problem. Inaudible vibrations generated in the motor can be transmitted through the mounting and can excite audible resonant frequencies in other parts of the installation. A common example is an air conditioning installation where a motor drives a fan and motor vibrations can resonate with panels of the air ducting. In such cases, a *resilient mounting* can be used to reduce transmission of motor vibration. A common type of resilient mounting is the resilient cradle base mount in which the motor mounting bracket is attached to the motor through rubber donut-like bushings which fit around the bearing mounts on both ends of the motor (Figure 10.14c). A cradle base mount increases motor cost.

Enclosure

The motor enclosure must protect the motor from its environment, and must also be designed to transfer internally generated motor heat to air around the motor. There are two categories of motor enclosures:

- *Open motors* have holes which allow circulation of air through the motor. This is the cheapest way to cool the motor. However, circulating air can also carry in dirt, moisture, and other contaminants to damage the motor.
- *Enclosed motors* have no openings, but they must be designed so that internal heat is transferred to the outer surface and then carried away in an airstream.

TABLE 10.7 Bearing comparison (adapted from Bodine 1978)

Characteristic	Rank (1 = best)		Comments
	Sleeve Bearing	Ball Bearing	
Simplicity	1	2	Sleeve has no moving parts.
Friction loss	2	1	Rolling friction lower than sliding friction.
Noise	1	2	Rolling action makes noise.
Cost	1	2	Simplicity makes sleeve type cheap.
Need for relubrication	2	1	Ball bearings require less lubricant.
Corrosion resistance	1	2	Steel balls rust.
Ability to withstand loads	2	1	Axial thrust is a problem for sleeve bearings.

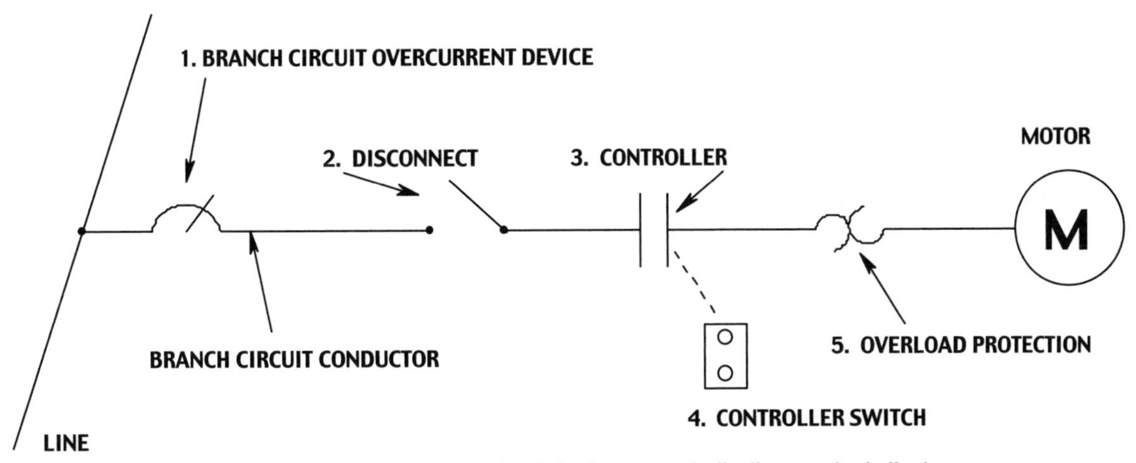

Fig. 10.28. Motor circuit components (the motor circuit is shown symbolically as a single line).

Following are descriptions of common motor enclosure types, listed in order of increasing cost:

- Open. The motor has numerous openings in the stator and end bells to allow circulation of cooling air. It is usually equipped with an internal fan to circulate air. This type of motor enclosure is suitable for use in clean, dry, non-hazardous environments (Figure 10.20c).
- Open Drip-Proof (ODP). The motor has openings in the stator and end bells, but they are located so that dirt and water falling straight down or at an angle of up to 15° from vertical does not fall into the motor. Use in clean, non-hazardous environments (Figure 10.16c).
- Totally Enclosed Air-Over (TEAO). This motor is enclosed to stand use in dusty, moist, and outdoor locations. The "air-over" designation means that the motor has no external fan for cooling and depends on airflow from another source for cooling. TEAO motors are common on fans.
- Totally Enclosed Fan-Cooled (TEFC). This enclosed motor can withstand dirty, moist, and outdoor locations. An external fan on one end of the motor moves air over the stator for cooling. The stator is often equipped with fins to aid in cooling (Figures 10.14c, 10.19c).
- Explosion-Proof. An explosion-proof motor is totally enclosed and designed for use in hazardous locations where dust or gas present may be explosive. The motor enclosure must withstand an explosion within itself without igniting the dust or gas outside. It must be selected to withstand a specific National Electrical Code explosive atmosphere classification.

Checking a Motor Application

After a motor has been selected and installed, some tests can be carried out to determine if the application is appropriate.

Starting

The motor should start its load quickly. Unless the load has a very high inertia, the load should be up to speed within less than 3 seconds.

Voltage

With the load started and running, voltage at the motor terminals should be near the motor nameplate value. It must be within 10% of the nameplate value for satisfactory operation (see the Motor Nameplate Information section).

Rotational Speed

A good indicator of proper motor operation is motor rotational speed, which can be measured with a rotational or stroboscopic tachometer. While making this measurement, it is very important to take appropriate precautions to avoid any entanglement in rotating shafts. The rotational speed should be at or slightly above the nameplate value for full load speed. Recall from the Motor Operation section that this corresponds to about 4.2% slip. If slip exceeds 4.2% (Speed drops below 1,725 r/min for a 4-pole motor), the motor is overloaded. This assumes the motor is operating in it rated environmental temperature and has a service factor of 1.0. Review these sections under Motor Nameplate Information. Overloading can cause abnormally high motor winding temperatures and premature motor failure.

If the measured speed shows slip to be 2% or less (rotational speed $\geq$ 1,764 r/min for a 4-pole motor), the motor is underloaded. Operating with the motor underloaded means that its efficiency and power factor are well below full-load values (see Figure 10.13, d and e) and that a smaller, cheaper motor might have been a better choice. Some amount of extra hp rating may be desirable if periodic load increases are possible, or if the extra starting torque afforded by the oversized motor is needed for reliable starting.

Current

Another good indicator of proper motor operation is motor current. This can be easily measured by placing a current clamp around one motor conductor somewhere between the panel and the motor. These conductors are energized to motor voltage, so appropriate safeguards for personal safety must be employed.

If measured current exceeds the nameplate value, the motor is overloaded or motor windings are shorted. If a 3-phase motor is being tested, current in all 3 conductors should be nearly the same.

Motor Circuits

Design of a motor circuit follows, for the most part, practices discussed in chapter 5. However, extra requirements are necessary because of the characteristics of motors as electrical loads.

Generic Form

The discussion of motor circuits here is necessarily generic in nature because it is not practical to consider all cases and exceptions described and allowed in the National Electrical Code (NEC). *Consult a current edition of the National Electrical Code before finalizing a design!*

Motor Circuit Components

The NEC requires every motor circuit to have the 5 elements shown in Figure 10.28 identifiable. Each will be discussed separately in following sections.

Table 10.8 lists motor current definitions as they are used in this section.

TABLE 10.8 Motor current definitions

I = motor current, amps
I_{LR} = motor locked-rotor current, amps
I_{FL} = motor full-load current, amps
Overload: $I_{FL} < I < I_{LR}$
Short or ground fault: $I > I_{LR}$

TABLE 10.9 Available size ratings of inverse-time circuit breakers and time-delay fuses

Inverse-time circuit breaker ratings, A: 10, 15, 20, 30, 35, 40, 45, 50, 60, 70, 80, 90, 100, 110, 125, 150, 175, 200, 225, 250, 300, and larger
Time-delay fuse ratings, A: 1, 3, 6, 10, plus all breaker sizes

Circuit Conductors

Motor circuit conductors must be designed to have these minimum ampacity values:

- One motor on circuit: Design for ampacity of at least 125% I_{FL} (use I_{FL} from Table 10.6, not from motor nameplate).
- >1 motor on circuit: Design for ampacity of at least 125% I_{FL} of largest motor +100% of sum of the I_{FL}s of the other motors.

A 2% design voltage drop is common.

Branch Circuit Overcurrent Device (1)

The primary purpose of the branch circuit overcurrent device is to protect the circuit (conductors, control apparatus, motor) from ground faults or short circuits ($I > I_{LR}$). The device must be sized to provide protection (open very quickly due to a short or ground fault) but also to allow the motor to start reliably without opening. If the device opens when the motor starts, this is called a "nuisance trip" because the short duration of the motor starting current presented no hazard to the circuit.

Appropriate time-trip characteristics are available with use of either a time-delay fuse or an inverse-time circuit breaker. Available sizes are listed in Table 10.9.

Maximum size for the branch circuit overcurrent device as a percentage of I_{FL} is listed in Table 10.10. The branch circuit overcurrent device may not be larger than this percentage of I_{FL}. I_{FL} must be as listed in Table 10.6 even if the nameplate value on the motor is different. A minimum size rating of 115% of I_{FL} is recommended (McPartland 1985). If the computed value falls between rated overcurrent device sizes, the next larger size may be used.

TABLE 10.10 Maximum rating of branch circuit overcurrent protection device for single-phase motors, % I_{FL}

Motor Code Letter	Time-Delay Fuse	Inverse-Time Breaker
Any code letter	175%	250%

Source: Based on NEC Table 430.52.

Good design consists of using the smallest overcurrent device, within these bounds, which will reliably start the motor. An example will illustrate the procedure.

Example 10.8

Specify the minimum and maximum overcurrent device ratings for use on a circuit supplying a single 5-hp, 230-V, code H single-phase motor.

$I_{FL} = 28$ A (Table 10.6)

Assume that an inverse-time circuit breaker is to be used.

Minimum size rating: $(1.15)(28) = 32.2$ A
select: 35 A rating (Table 10.9)

Maximum size rating: $(2.5)(28) = 70$ A
select: 70 A rating (Table 10.9)

The designer uses experience and judgement in selecting a size between 35 A and 70 A. For motors requiring protection in a size smaller than available inverse larger breakers, a breaker can be used for switching and a time-delay fuse of proper size can be installed in series.

Disconnect (2)

The purpose of the disconnect is to allow disconnection of the motor and controller from all ungrounded conductors for protection of persons working on the motor and controller.

Requirements of the device are:

- Device must indicate on and off.
- Device must be located within 50 feet and within sight of controller.
- No individual poles of the device may operate independently.
- The switch hp rating must equal or exceed total motor hp.
- The switch current rating must equal or exceed 115% of I_{FL}.

A receptacle and plug can serve as the disconnecting means for motors not permanently fixed in place (NEC 430.109).

Controller and Control Switch (3 and 4)

The purpose of the controller is to start and stop the motor.

Requirements of the device are:

TABLE 10.11 Motor horsepower and voltage ratings for standard NEMA size motor starters

NEMA Size	Single Phase		Three Phase
	115 V	230 V	230 V
00	1/3	1	1 1/2
0	1	2	3
1	2	3	7 1/2
1P	3	5	--
2	3	7 1/2	15
3	7 1/2	15	30
4	--	25	50
5	--	50	100
6	--	--	200
7	--	--	300
8	--	--	450

Source: Adapted from NEMA 1978.

- Each motor must have its own controller, except if motors are in one machine or system.
- The controller cannot double as the disconnect. (Exception: a hand-controlled knife switch sized as a disconnect can be used as a controller.)
- The controller is only required to open enough conductors to stop the motor.
- The controller must be located within sight and within 50 ft of the motor. (Exception: does not apply if controller can be locked off.)
- The current rating of the controller must equal or exceed I_{FL} of the motor.

Overload Protection (5)

The purpose of the motor overload protection device is to protect the motor from overload ($I_{FL} < I < I_{LR}$).

- This device might not operate fast enough to provide protection from shorts and ground faults, since the branch circuit overcurrent device will do this.
- General rule: size at 1.15 I_{FL}.
- If hp >1, motor starts automatically, motor ambient temperature = 40°C, and SF ≥ 1.15, overload protection can be sized at 1.25 I_{FL}.
- If hp ≤ 1, motor is started manually, is within sight and within 50 ft of controller, and installation is permanent, the branch circuit overcurrent device can serve as the overload protector.

Controllers are rated by NEMA according to hp and voltage (Table 10.11).

References

Bodine, C. 1978. Small motor, gearmotor, and control handbook. 4th ed. Bodine Electric Company, Chicago, IL.

Bureau of Naval Personnel. 1969. Basic electricity. Rate Training Manual NAVPERS 10086-B. U.S. Gov. Printing Office, Washington, DC.

Grainger, W. W., Inc. 2002. Grainger Catalogue No. 393, 2002-2003. W. W. Grainger, Inc., Lake Forest, IL.

McPartland, J. F. 1985. Design and layout of modern motor circuits. Electrical Construction and Maintenance, New York, NY.

NEMA 1978. NEMA publication ICS 2-1978. National Electrical Manufacturers Association, Washington, DC.

NFPA 2002. National electrical code 2002. National Fire Protection Association, Inc., Quincy, MA.

Problems

10.1. Prove that the direction of field rotation in a 3-phase stator is reversed by exchanging two of the leads. Do this using Figure 10.6. Map the field orientation at successive times with phase C connected at the top, phase A on the right, and phase B on the left. (This corresponds to reversing phases B and A.)

10.2. An induction motor stator has 10 poles.

(a) How many power cycles are there per revolution of the rotating field?

(b) What is its synchronous speed on a 60-Hz power system?

(c) What is its synchronous speed on a 400-Hz aircraft system?

(d) What is the highest synchronous speed attainable by any induction motor on this 400-Hz system?

10.3. A certain motor is rated at 1,150 r/min when operated on 60-Hz power.

(a) How many poles does it have?

(b) How many r/min of slip does it have when at full load?

(c) What is its percent slip when at full load?

10.4. An electronic device for varying induction motor speed uses line power, and supplies power to an induction motor. Induction motor speeds up to 4,000 r/min are attainable with 2-pole motors. What power frequency does the device supply?

10.5. The full-load speed on the nameplate of a 4-pole, 60-Hz, induction motor is usually 1,725 r/min.

(a) At full load, how many r/min of slip is there?

(b) At full load, what is the percent slip?

10.6. Is it possible for an induction motor with a squirrel cage rotor to run at synchronous speed? Explain.

10.7. The nameplate of a 3-phase general purpose motor gives this information: 100 hp, 3,450 r/min. Compute values for these torques in lb·ft units:

(a) T_{FL}

(b) T_{LR}

(c) T_{BD}

10.8. A certain load is to be started by direct connection to a 6-pole 3-phase, general-purpose induction motor. The load requires 51 lb·ft of torque to start it turning. What is the smallest motor size which will start it?

10.9. Derive an equation for T_{FL} in lb·ft units, in the form $T_{FL} = (\text{constant})(f(HP_{FL}, P))$ of an induction motor operating on 60-Hz power.

10.10. The locked rotor torque of a 4-pole motor is measured to be 45.7 lb·ft. This type motor is supposed to develop a locked rotor torque of 150% of full-load torque. What is the hp rating of this motor?

10.11. Explain four different means of achieving a rotating field effect in an induction motor stator.

10.12. Assume one of each of these types of motors was displayed for inspection with the nameplate covered. Assuming that you can take apart the motor as far as you need to, how would you identify each type? (State only what is needed to identify a particular type among the others.)

(a) Shaded-pole
(b) Split-phase
(c) Capacitor start-induction run
(d) Permanent split capacitor
(e) 3-phase
(f) Repulsion start-induction
(g) Universal

10.13. (a) Calculate the full-load torque for a 2-hp two-value capacitor motor with a full-load speed of 3,450 r/min.

(b) About how much starting torque (lb·ft) could be expected from this motor?

10.14. (a) Draw a schematic of a dual-voltage two-value capacitor motor wired for high voltage.

(b) Draw a schematic of the same motor wired to run in the opposite direction, at low voltage.

10.15. Say we have a 120/240-V single-phase, split-phase-type motor. Removing the cover plate reveals six leads, with DC resistances between as shown:

Red to blue = 2.3 Ω
Black to yellow = 2.3 Ω
White to green = 2.5 Ω
Indicating colors of leads,

(a) Draw a schematic with leads connected for 120-V operation.

(b) Assume when connected as in (a) motor hums but does not start when connected to power. Draw a schematic showing your remedy.

(c) Draw a schematic of the motor as in (b), except wired to run in the reverse direction.

(d) Draw a schematic of the motor wired for 240-V operation.

10.16. A motor has this information on the nameplate:

$^1/_2$ hp, 115 V, 8 A, Code K
Efficiency = 75%
This motor is placed 100 ft from a 120-V source. All wiring is inside. Specify:

(a) Conductor type and material

(b) Size of conductor

(c) Maximum starting amps

(d) Voltage at motor under full load

(e) Voltage at motor with rotor locked

(f) Power factor at full load (use nameplate voltage)

10.17. A split-phase motor has this information on its nameplate:

$^3/_4$ hp	Code U	13.8/6.9 A
1,725 rpm	115/230 V	Full-load power factor: 0.70
		Insulation class B

The motor is wired for low-voltage operation.

(a) What is the synchronous speed of this motor?

(b) What is the maximum locked-rotor current it should draw if wired for 115-V operation?

(c) What is the full-load torque (lb·ft)?

(d) What is the highest starting torque this motor could be expected to deliver (lb·ft)?

(e) At what speed would this motor run if output was 0.5 hp?

(f) Compute the full-load efficiency of this motor.

(g) What is the temperature of the hottest spot in the windings when this motor is fully loaded in a 40°C room?

10.18. Assume you are replacing a 4-hp gasoline engine on a burr mill with an electric motor. The motor will be mounted under the mill, and will not be readily accessible. Power available is 120/240-V single-phase. The motor will run 8 h every day. Specify voltage, motor type, hp rating, enclosure, bearings, mounting method.

10.19. A motor is required to drive a 25-ft long × 10-in. diameter horizontal dry corn auger. The auger turns at 450 r/min, driven by a belt from a 1,725-r/min motor. The auger is stationary and is to have a start-stop station as near as possible. Single-phase 115/230-V service is available at a panel about 120 ft from the motor. Conductors are to be run in a conduit. The auger must start under all load conditions. Starting reliability is important. Conductors are to be copper. Need not specify grounding conductor.

(a) Select motor. Specify motor (type, hp, voltage, poles, bearings, mounting method, enclosure).

(b) Design motor circuit (Sketch and label components on one sheet, put computations on another page):

1. Conductor type, size, length.

2. Branch circuit over-current device (number of poles, minimum and maximum amp rating).

3. Controller (hp rating, number of poles).

4. Overload (heater ratings).

5. Disconnect (number of poles, hp rating).

10.20. Design a branch circuit to supply a 10-hp, code D, 230-V single-phase motor 100 ft (including connection) from a panel board. Conductors are 75°C copper. The motor is to have a controller and one integral start-stop station.

Specify: Conductor size
Branch circuit breaker rating
NEMA size of controller
Disconnect

10.21. A branch circuit is to be run to a single-phase, 7.5-hp (code C, 230-V, 40°C, 1.15 service factor) motor which is 120 ft from the service entrance panel. The circuit is to be run underground, with direct-burial conductors. The operator will control the motor from one 115-V push-button station at the motor. Overload protection is to be in the controller.

Draw a complete pictorial (not ladder diagram) circuit. On the drawing, specify:

(a) Branch circuit over-current device (number of poles, amps rating [min and max]).

(b) Disconnect (number of poles, min hp and current rating, placement requirement).

(c) Controller (number of poles, NEMA size).

(d) Overload protection (current rating).

(e) Branch circuit conductors (conductor or cable type, conductor material, size for ampacity, size for voltage drop, size needed, length needed).

10.22. An electric motor is needed to drive a 25-ft long 6-in. diameter auger conveyor operating at 500 r/min and carrying dry corn up a 60° slope. Service is 115/230-V single phase, and the conveyor is outside. The motor must have a manual reset overload protector. Specify motor type, hp, full-load speed, mountings, bearing, and enclosure. Choose a motor from the Grainger catalog or another source and specify it by stock number, page number, and price.

10.23. An electric motor is needed to drive a pump in an inside, clean location. The pump's manufacturer states that the load will be about 4.8 hp, and is to be driven at 3,450 to 3,550 r/min. The pump requires a 184 T motor frame. Power available is 3-phase, 230-V. Motor overload protection is to be done by the magnetic motor starter.

(a) Choose a 3-phase motor from the Grainger catalog or from another source. Specify hp, motor type, enclosure, mounting, full-load speed, and bearings. State catalog number, page, stock number, and price.

(b) Choose a single-phase motor with the same specifications and state catalog information.

Chapter 11

Programmable Logic Controllers

Introduction

Programmable Logic Controllers (PLCs) are widely used in machine and process control. This chapter will help you to understand what PLCs are, how they work, how they are programmed, a few of the many things they can do, and what some common applications of PLCs are.

PLC Fundamentals

What is a PLC?

A PLC is an industrial computer that accepts inputs from switches and sensors, evaluates these in accordance with a stored program, and generates outputs to control machines and processes (GE Fanuc Automation 1995). Figure 11.1 is a block diagram of a PLC's components. Input devices provide input to the input/output system, which in turn provides information to an input table, which feeds information to a user program. The user program communicates with the programming device (a microcomputer), data storage, and the output table. The output table feeds information to the input/output system, which finally sends information to the output devices. Table 11.1 contains the definitions of these terms.

Hardware

An example of a low-end PLC is the GE Fanuc Series 90 Micro PLC. The Series 90 Micro PLC is physically a box 3 in. × 3.25 in. × 4.5 in. which can be attached to a standard DIN rail (Figure 11.2). It uses 120-V line power. The power supply produces DC voltages needed to run the microprocessor (an H8), and for I/O devices. Screw terminals for input and output wires are located on the top and bottom, respectively. Programs are loaded through the RS-422 port on the front.

Principles of Operation

The operating principles of a PLC are illustrated in Figure 11.3. Think of the box as the PLC. There are two inputs shown on the left. These inputs consist of wires extending from a +24 VDC bus, through switches to input terminals %I0002 and %I0005. %I0002 and %I0005 are also addresses within the computer memory. Note that input %I0002 is open and input %I0005 is closed, and that in the status column only the closed input has a 1.

This box also shows two outputs (%Q0001 and %Q0006). (You've probably noted that input addresses begin with %I and output addresses begin with %Q). Both outputs are closed, since there are 1s in the status column for each. This means

TABLE 11.1 PLC Definitions (GE Fanuc Automation 1995)

Memory	The part of a programmable controller where data and instructions are stored either temporarily or semipermanently. The control program is stored in memory.
Inputs	Any input device connected equipment that will supply information to the central processing unit, such as switches, buttons, limit switches, sensors, etc.
Outputs	Any output device connected to equipment that will receive information or instructions from the central processing unit, such as control devices like motors, solenoids, lights, alarms, etc.
I/O	Abbreviation for Inputs and Outputs
Address	An alphanumeric value that uniquely identifies where data is stored. (Ex. %10001 is input 1, %Q0003 is output 3, and %R0100 is Register 100)
High-Speed Counter	Allows rapid pulses that are faster than the scan time of the PLC to be counted independently.
Encoder	A rotary device which transmits position information. The pulses are fed into a high speed counter on the PLC.
Status Table	The status table is part of the CPU that stores the status of all the inputs (%1000x), Outputs (%Q000x), Registers (%R000x), etc.
Normally-open Contact	A ladder logic symbol that will allow logic continuity (flow) if the reference input is a logic "1" when evaluated.
Normally-closed Contact	A ladder logic symbol that will allow logic continuity (flow) if the reference input is a logic "0" when evaluated.
PID	Proportional-Integral Derivative. A mathematical formula that provides a closed loop control of a process. Inputs and outputs are continuously variable and typically will be analog signals.
Scan Time	The time required by the processor to read all inputs, execute the control program, update I/O, and evaluate and execute the control logic. The program scan repeats continuously while the processor is in the run mode.
Register	A temporary storage device for various types of information and data (e.g., Timer/Counter values). In the PLC a register is normally 16 bits wide or holds a number up to 32,768.
Flash Memory	Provides a non-volatile user-program storage and system firmware. Does not require battery backup.
PWM	Pulse Width Modulation, used in motion control where position is critical.
SNP	Series Ninety Protocol is the protocol required to talk to the GE Fanuc Series 90 Family of PLCs.
P/S	Power supply.
Bit	One binary digit. The smallest unit of binary information. A bit can have a value of "1" or "0".
Byte	A group of adjacent bits usually operated upon as a unit, such as when moving data to and from memory. There are 8 bits in a byte.
Word	The unit of binary digits (bits) operated on at a time by the central processing unit when it is performing an instruction or operating on data. A word is usually composed of a fixed number of bits. One word is equal to 2 bytes or 16 bits.
AND	A Boolean operation that yields a logic "1" output if all inputs are "1" and a logic "0" if any input is a "0".
OR	A Boolean operation that yields a logic "1" output if any inputs are "1" and a "0" if all inputs are "0".

that the run light and the motor are both on. (The PLC can't supply enough power to run the motor. The output is really powering the coil of the relay which switches the motor on and off). They are both on because the user program (Figure 11.1) existing in the PLC has scanned the inputs and determined that under these conditions the program in memory dictates that outputs %Q0001 and %Q0002 should be on.

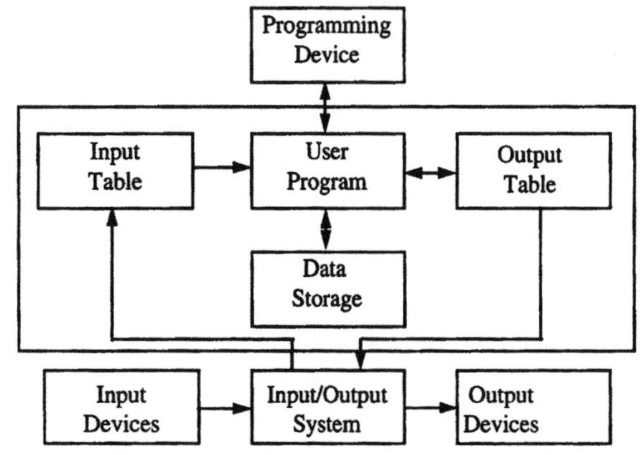

Fig. 11.1. Block diagram of a PLC (GE Fanuc Automation 1995; reprinted with permission.)

The program therefore turns on these two outputs. They remain on until the user program determines they should be off. Read the definition of scan time from Table 11.1.

Ladder Diagrams

The original users of PLCs in the 1960s were familiar with ladder diagrams because they drew electromechanical control systems by this approach. It was therefore logical for the first PLC manufacturers to develop high-level computer languages which allowed PLC programming by means of ladder diagrams. Chapter 8 describes ladder diagram methods. Table 11.2 shows ladder logic symbols which will be used.

Program Example

Figure 11.4 is a PLC with one input and one output and programmed with a simple one-rung PLC program. Contacts are always placed on the left and coils are placed on the right in the program. The assumption is made that the left rail (vertical segment) of the ladder is energized, and that the right rail is

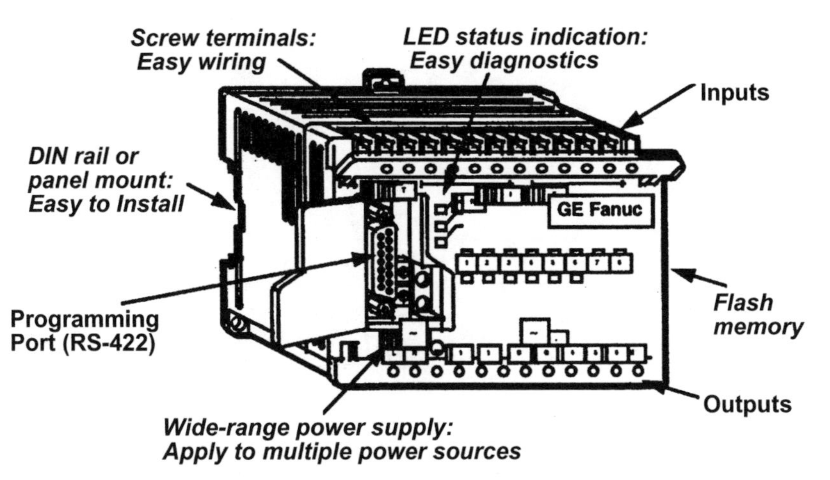

Fig. 11.2. Series 90 PLC (GE Fanuc Automation 1995; reprinted with permission.)

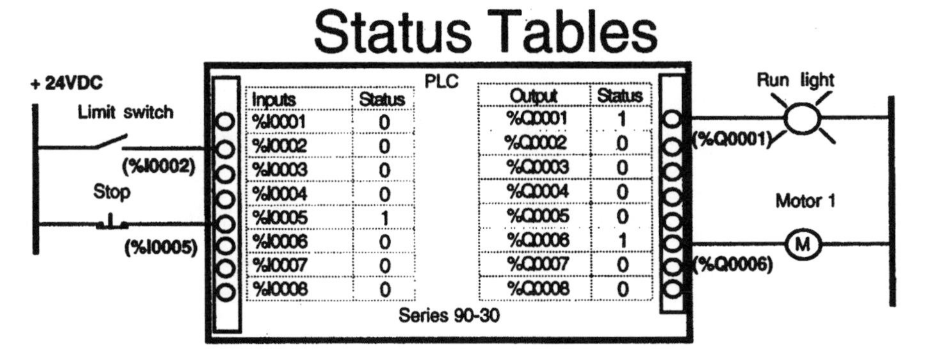

Status tables: An area in the PLC memory where the status of inputs outputs and data registers are stored. The status of the inputs and outputs are represented by "1"s and "0"s. One represents a true condition or a valid signal present. Zero represents a false condition or a invalid signal present. This information is later used to solve ladder logic programs.

In the above example, the Limit switch is not made (no power flow to input 2 on the Series 90-30) therefore a "0" is placed in the input status table for input 2 (%I0002). The Stop switch is passing power to input 5 therefore a "1" is placed in input status table 5 (%I0005).

Logic inside the PLC is writing a "1" to both output 1 (%Q0001) and output 6 (%Q0006). Therefore the Run light and the Motor will be turned on by switching the appropriate voltage to the devices.

Fig. 11.3. PLC operating principles (GE Fanuc Automation 1995; reprinted with permission.)

TABLE 11.2 Ladder logic symbols

Type of Symbol	Symbol	Operation
Normally-open contact	─┤ ├─	Follows state of its input
Normally-closed contact	─┤/├─	Opposite to state of its input
Coil	─( )─	Its output follows the coil's state
Negated coil	─(/)─	Its output is in the state opposite to that of the coil

grounded or neutral. Then coils are drawn using good wiring practice with one terminal connected to neutral and the other terminal energized through the contacts.

In Figure 11.4, normally-open contact %I0001 is connected in series with coil %Q0001. The program is written so that whenever input %I0001 is open, or de-energized, output %Q0001 is off, and whenever input %I0001 is closed, output %Q0002 is on. Note that the input and output (which are wired to terminals on the PLC) have the same addresses as the contact and the coil (which are program instructions).

Figure 11.5 has the same elements, but now contact %I0001 is programmed as normally-closed and control logic is reversed. When input %I0001 is energized or closed, %Q0001 is off, and when %I0001 is de-energized or off, %Q0001 is on. This is a difficult

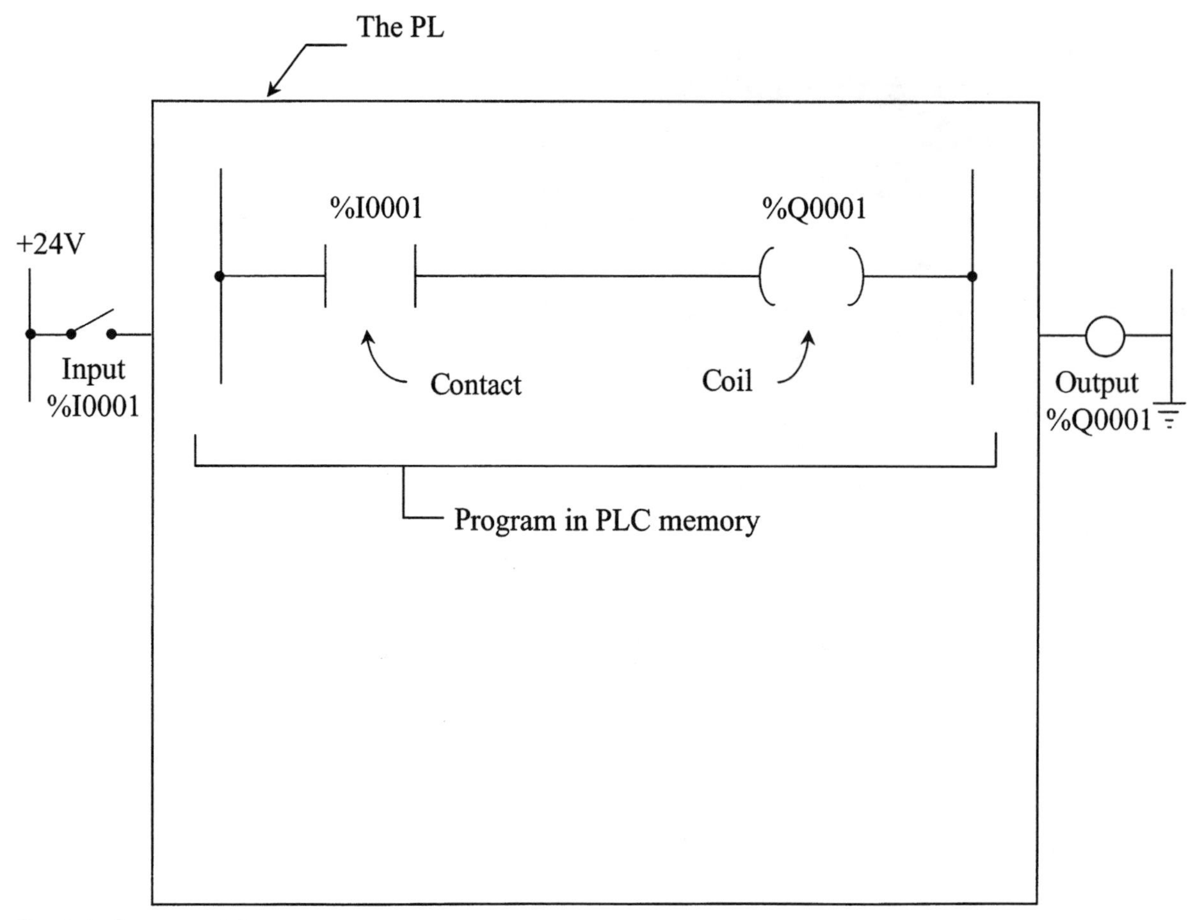

Fig. 11.4. One-rung PLC program.

Fig. 11.5. Normally closed contact program example.

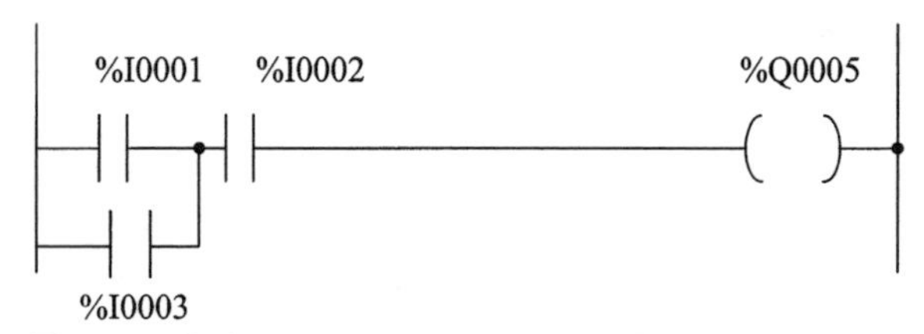

Fig. 11.6. Series–parallel arrangement of contacts.

and tricky concept. Be sure you understand this before you go further! The central issue is that Figure 11.5 represents *the program instruction* and not the state of the actual input wire. Table 11.3 may help you understand the concept.

Contacts can be arranged in series, parallel, or in combination series–parallel circuits in order to accomplish whatever logic is required for the problem. In Figure 11.6, for example, the rung is programmed so that coil %Q0005 will be on if input %I0002 is on, and input %I0001 or input %I0003 is on.

Using Contacts More than Once

A contact can be used more than once in a program. (Coils are usually used only once.) This can be a valuable capability in many instances. For example, look at Figure 11.7. When input %I0005 is on, coil %Q0003 is on and coil %Q0004 is off. When input %I0005 is off, coil %Q0003 is off and coil %Q0004 is on.

TABLE 11.3 Contact/Input/Output truth table

Contact	Power flow to output? Input wire is: on	off
—| |—	yes	no
—|/|—	no	yes

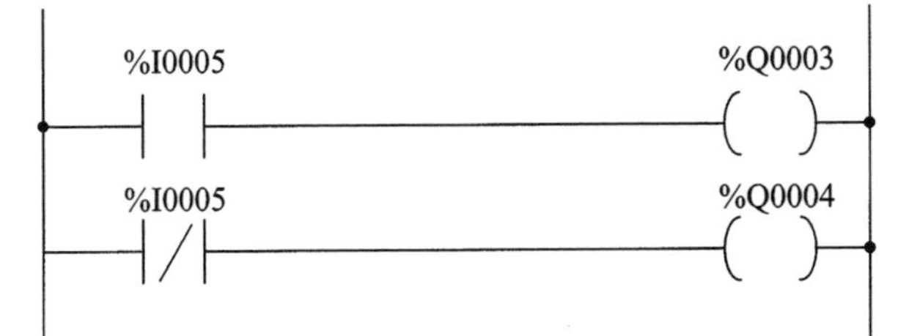

Fig. 11.7. Using an input more than once.

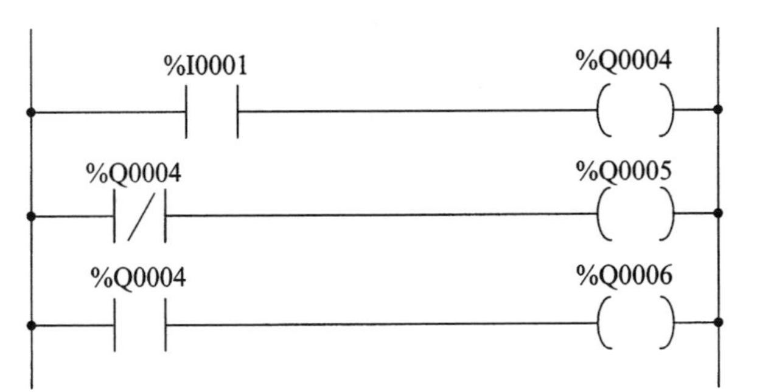

Fig. 11.8. Using a coil address for a contact.

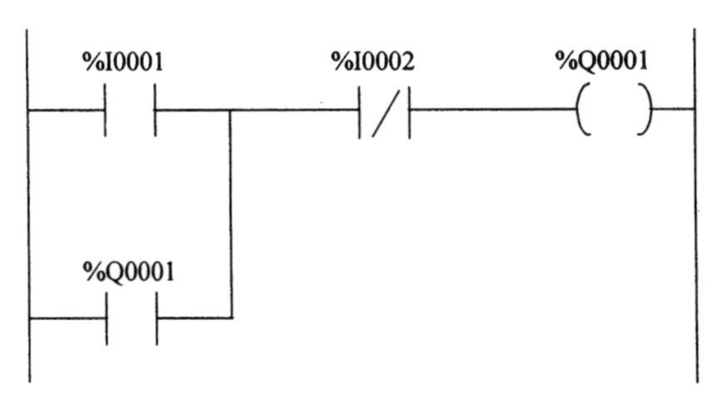

Fig. 11.9. Latching circuit.

Using a Coil as a Contact

The address of a coil can be used for a contact, and the contact then follows the state of the coil. Figure 11.8 illustrates such a use. When address %Q0004 is given to a contact, that contact follows the state of coil %Q0004. In the figure, when %Q0004 is on, coil %0005 is off and coil %Q0006 is on.

Latching

The use of coils as contacts allows outputs to turn on and stay on regardless of the state of the controlling input. This "latching" function is illustrated in Figure 11.9. Assume that inputs %I0001 and %I0002 are both open. If input %I0001 is momentarily closed, and then opened, output %Q0001 is turned on, and will stay on because contact %Q0001 closes when coil %Q0001 closes. When input %I0001 opens, closed contact %Q0001 provides a parallel path and output %Q0001 stays on and is said to be *latched.* If input %I0002 is closed and then opened, coil %Q0001 is de-energized and contact %Q0001 opens. The rung is now *unlatched* because contact %Q0001 opens when coil %Q0001 goes off.

Negated Coil or Out-Not

A coil drawn with a slash inside is a negated coil (Table 11.2). A negated coil's output will be energized when the rung to the left is off and de-energized when the rung to the left is on. In other words, it is in the opposite logic state from a conventional coil.

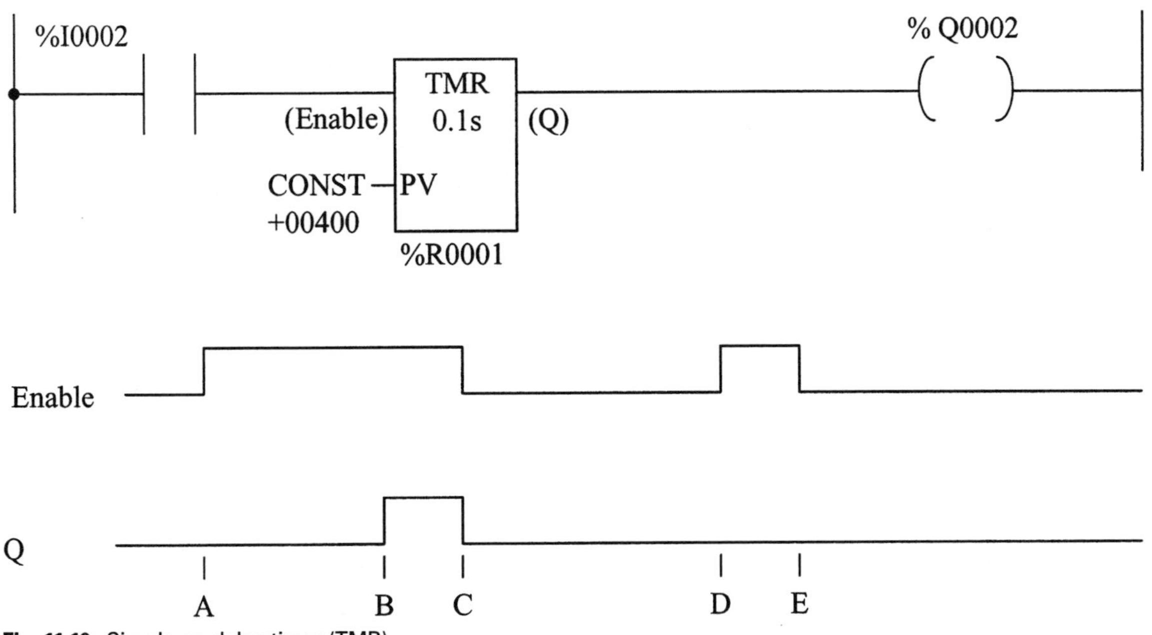

Fig. 11.10. Simple on-delay timer (TMR).

Timers

Timers can be programmed into PLC ladder diagrams to accomplish many different time-dependent control operations

TMR

TMR is an on-delay timer (Figure 11.10). TMR counts time in selectable increments. Increments available are 0.1 s (default, used in Figure 11.10), 0.01 s, and 0.001 s. TMR turns on after CONST x increment seconds have elapsed. In the example, output Q turns on after (400)(0.1) = 40 s have elapsed. CONST can be any value up to 32,767. %R0001 identifies the timer.

Time lines at the bottom of Figure 11.10 illustrate operation of TMR. Enable goes high at time A. At time B, the programmed time has passed, so Q goes high and stays high until time C when enable goes low. At time C, Q also goes low. Enable goes high at time D and stays high until time E. Since the time from D to E is less than the programmed time of 40 s, Q never goes high.

Programming and Running PLCs

PLC programs can be written using a personal computer (PC) which has software from the PLC manufacturer installed. Programming consists of drawing a ladder diagram containing rungs which associate PLC inputs with PLC outputs. Problems at the end of this chapter deal with such ladder diagrams.

When the program is finished, it is downloaded from the PC to the PLC through a cable. The GE Fanuc Series 90 PLC has an RS-422 port on the front for this purpose. The PLC can be put into its run mode from the PC keyboard. The ladder diagram program can be viewed on the PC screen while it runs, and changes in the brightness of rung segments indicate operation of contacts and coils. Energized segments are noticeably brighter than non-energized segments. This feature is an aid in troubleshooting since open and closed circuit segments are indicated by brightness of rung segments on the screen.

The PC can be disconnected from the PLC during normal operation, since its functions are limited to programming and troubleshooting.

References

GE Fanuc Automation. 1995. Series 90-30 Workshop student guide. GE Fanuc Automation, Charlotesville, VA.

GE Fanuc Automation. 1996. Series 90-30/20/Micro reference manual. GE Fanuc Automation, Charlottesville, VA.

Problems

11.1.

(a) Input wire %I0002 is off. Therefore output %Q0007 is ________.

(b) When input wire %I0002 is turned on (becomes hot), output %Q0007 is turned ________.

11.2.

%I0002 %Q0007

(a) Input wire %I0002 is off. Therefore output %Q0007 is ________.

(b) When input wire %I0002 is turned on, output %Q0007 is turned ________.

11.3. Output %Q0006 is to be on only if input %I0003 is on and input %I0004 is off. Draw the circuit.

11.4. Output %Q0006 is to be off if input %I0003 is off or if input %I0004 is on or if both conditions are true. Draw the circuit.

11.5. Output %Q0006 is to be off only if input %I0003 is off and input %I0004 is on. Draw the circuit.

11.6. Output %Q0005 is to be on if inputs %I0001 and %I0002 are on or if inputs %I0003 and %I0004 are off. Draw the circuit.

11.7. Describe the response of these circuits.

a.

%I0001 %Q0001

%I0001

b.

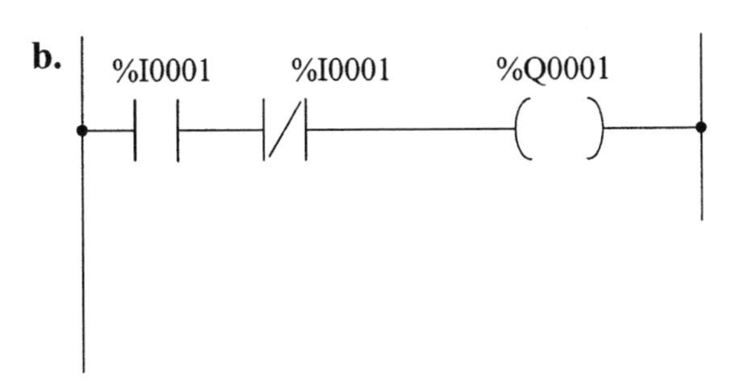

11.8. Describe the response of this circuit.

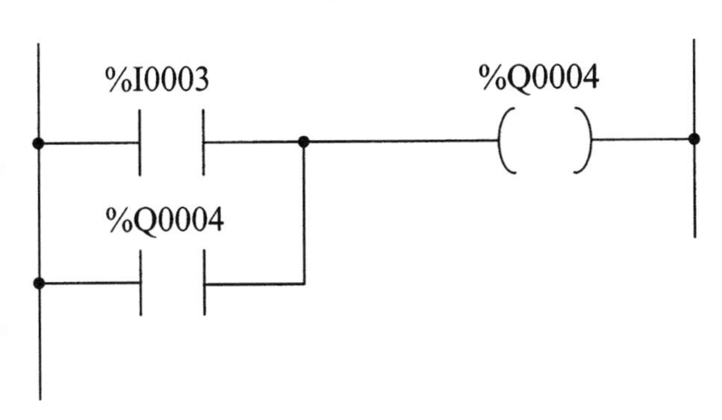

11.9. Inputs %I0001 and %I0002 are push-button momentary contact normally open switches. Draw a circuit where output %Q0006 turns on and stays on (that is, latches) when %I0001 is momentarily closed, and then later, output %Q0006 turns off and stays off when %I0002 is momentarily closed. Draw the circuit.

11.10. Input %I0001 is a push-button momentary contact normally open switch and input %I0002 is a push-button normally closed momentary open switch. Draw a circuit where output %Q0006 turns on and stays on when %I0001 is momentarily closed, and then later, output %Q0006 turns off and stays off when %I0002 is momentarily opened. This circuit operates like a 3-wire start-stop station on a magnetic motor starter.

11.11. A push-button station consists of a START button which is a normally open, momentary contact switch, and a STOP button, which is a normally closed momentary open switch. The two switches in this station are inputs %I0004 (the START button) and %I0007 (the STOP button) to a PLC. Output %Q0004 powers the coil of a motor driving a grinder. Through a horrible mistake in programming, the motor starts when the STOP button is pushed and stops when the START button is pushed. Draw the circuit of the program.

11.12. Output %Q0007 is to be on when input %I0003 is off, and off when input %I0003 is on. Draw the circuit and use a negated coil in your solution.

11.13.

%I0001 %Q0001

%Q0001 %Q0002

In order for %Q0002 to be on, what must be the state of %I0001?

11.14. Output %Q0001 is to turn on 1 minute after input %I0001 turns on. It is to stay on until input I0001 turns off. Draw the circuit of the program, using TMR %R0002.

11.15. Output %Q0001 is to turn off 2 minutes after input %I0001 turns on. It is to stay off until input %I0001 turns off. Draw the circuit of the program, using TMR %R0002.

Chapter 12

LIGHTING

Introduction

Light is visually evaluated radiant energy:

- It is "energy" because it heats any object it falls upon.
- It is "radiant" because it moves from its source in a straight line.
- It is "visually evaluated" because it allows us to see (Boylan 1987).

Light makes vision possible. Light from Thomas Edison's masterpiece, the electric filament lamp, changed life by making high-quality artificial light readily available for the first time. On farms, electric light lengthened the day so that chores and household activities could be carried on safely and conveniently after dark or before dawn.

The earliest applications of electrical energy were for lighting. In fact, many companies selling electrical energy were called "light companies." Some of these names still exist in Iowa. In Preston, Iowa, electrical energy is furnished by Preston Light and Power. Fredericksburg, Iowa has a Municipal Light Department. In 1896, the Ames Iowa Municipal Electric System based its customer bills for electrical energy solely on the number of light bulbs in the customer's home.

In this chapter, some physical principles of light will be explained. Sections on lighting devices and lighting systems design follow.

Physical Principles

Light makes up only a narrow band in the electromagnetic spectrum (Figure 12.1). Different types of radiation are specified by frequency (Hz), and wavelength (m). These units are used because one of the ways to describe electromagnetic radiation is as a wave phenomenon. A wave has velocity, wavelength, and frequency. These are related as in Equation 12.1.

$$v = \lambda \nu \qquad \textbf{(12.1)}$$

where v = velocity, m/s
λ = wavelength, m
ν = frequency, Hz or cycles/s

In air or free space, the velocity of electromagnetic radiation (the speed of light) is about 2.30×10^8 m/s (186,000 miles/s). The velocity is slower in water or glass. In water, it is about 2.25×10^8 m/s. Frequency is constant for a given kind of radiation. Therefore, for Equation 12.1 to hold, wavelength must be different for each medium.

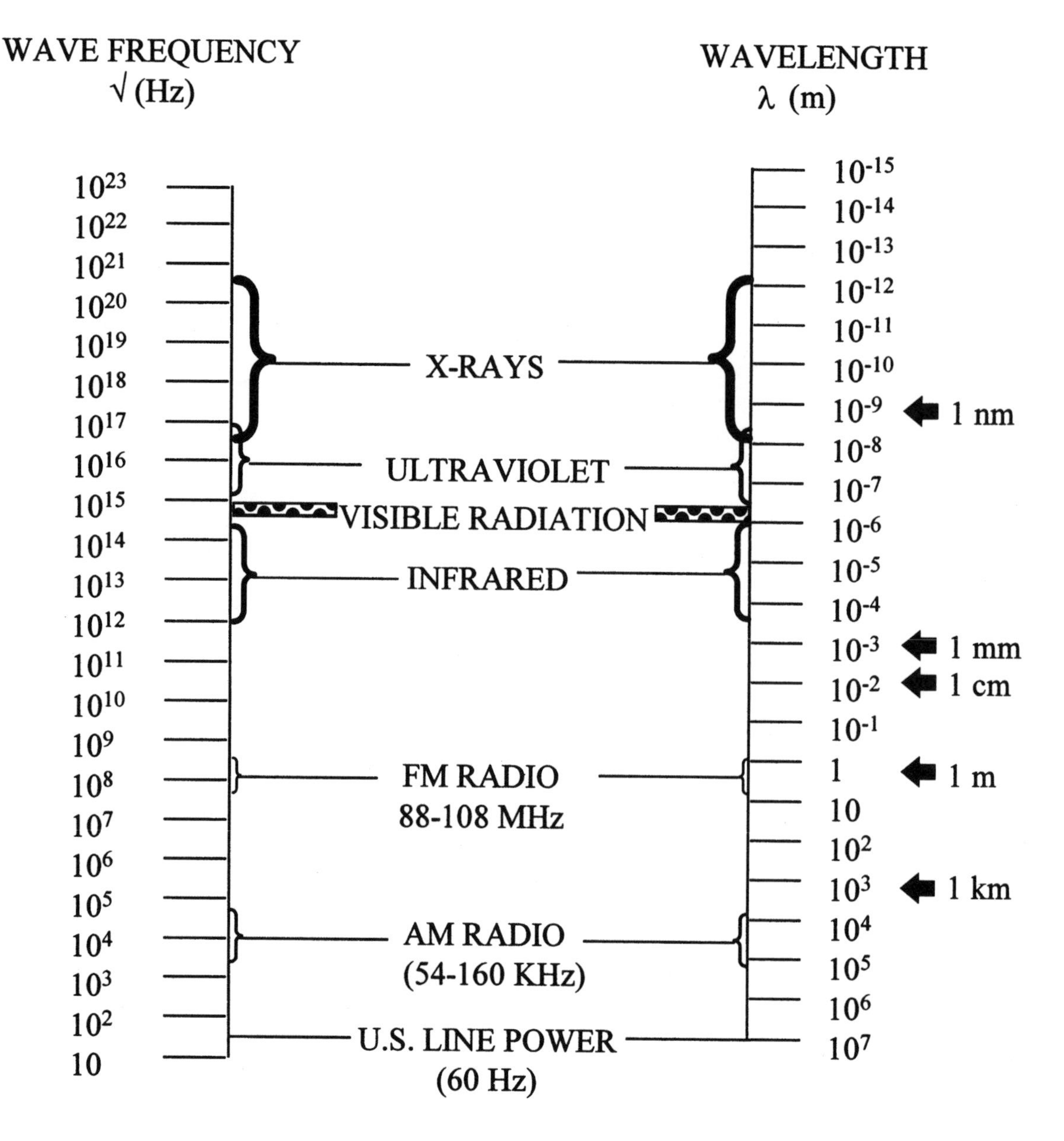

Fig. 12.1. Electromagnetic spectrum.

Within the electromagnetic spectrum, the visible region extends only over the wavelength range from 380 to 780 nm (1 nm $= 10^{-9}$ m). Figure 12.2 specifies colors within this band.

Fundamental Photometric Quantities

Luminous Flux (Lumens)

Luminous flux is the time rate of flow of light and the unit is the lumen (lm). Light output of lighting devices is specified in lumens. Lumen outputs of some common bulbs and tubes are listed in Table 12.1. For example, a 100-W inside-frosted incandescent bulb has an approximate initial output of 1,710 lm. Initial lumens are specified because output decreases with bulb or tube age.

Eye Sensitivity

Luminous flux is what stimulates the sense of sight, but radiant flux is what flows from the radiation source. The ratio of these two quantities is a measure of the eye's efficiency at sensing radiation. This efficiency varies drastically over the visible spectrum. See the eye sensitivity curve on Figure 12.3. Eye sensitivity peaks at a wavelength of about 555 nm in the

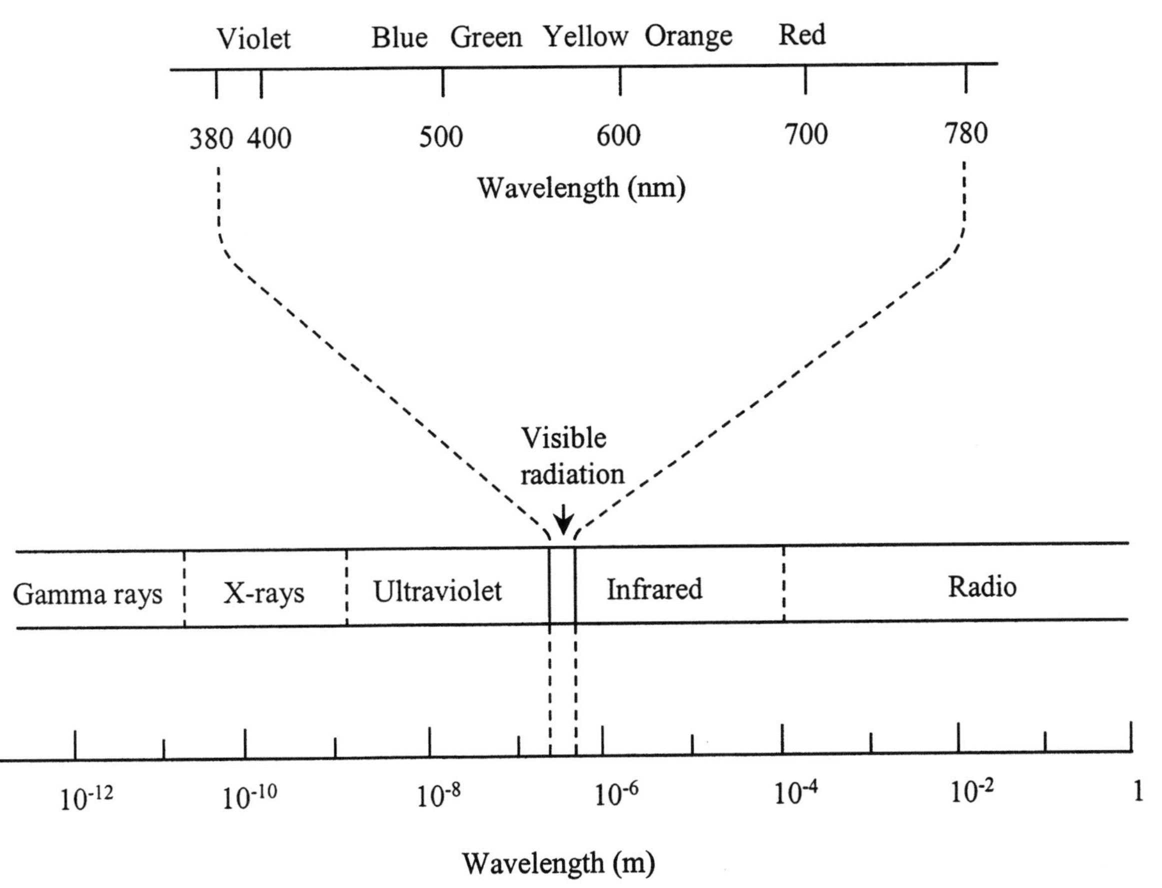

Fig. 12.2. Visible radiation.

yellow-green color band, and drops off drastically above and below this value. At very low levels of radiation, the eye sensitivity shifts about 48 nm to the left, toward the blue band. This shift occurs when the eye shifts from using cones (photopic vision) to rods (scotopic vision) for sensing light. Lighting devices emitting distinctly yellow light use this sensitivity to advantage.

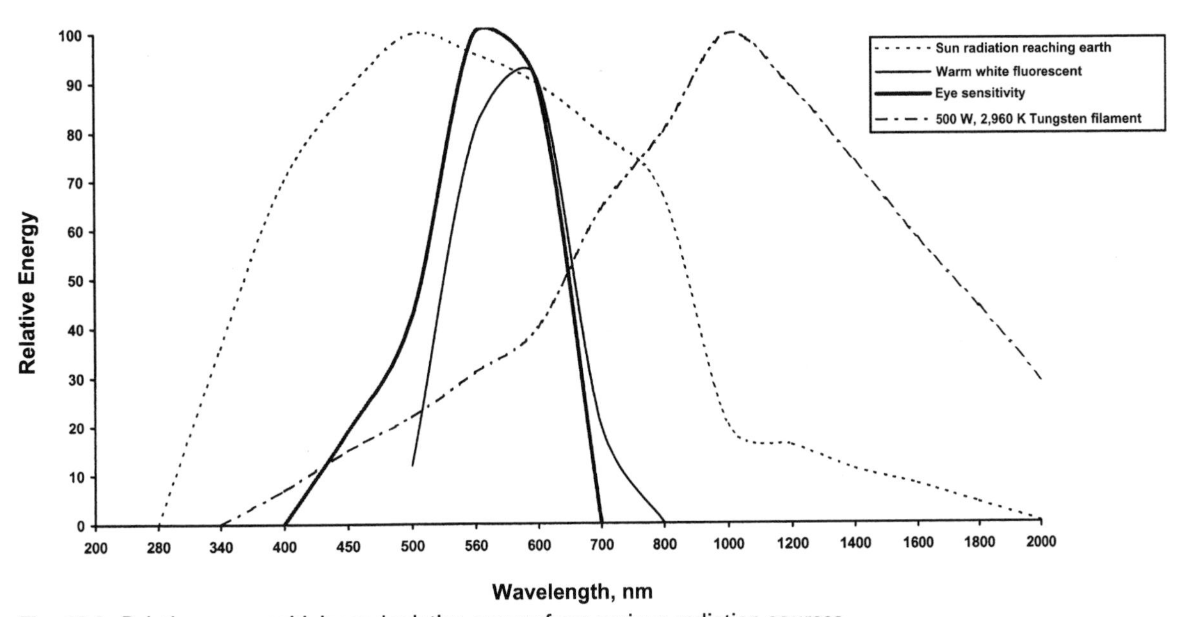

Fig. 12.3. Relative eye sensitivity and relative energy from various radiation sources.

TABLE 12.1 Operating characteristics of some lighting devices

	Voltage, V	Power use, W[1]	Initial	Mean Luminous flux, lm	Efficacy, lm/W[1]	Rated av. life, h	CRI[2]
Candle				12.6	0.1		
Oil lamp					0.3		
Edison's 1879 carbonized cotton thread incandescent					1.4		
60-W carbon filament (1905)					4.0		
100% energy conversion to visible (white light) energy					200.0		
100% energy conversion to 555 nm (yellow green) light					673.0		
Theoretical tungsten filament limit (tungsten melting point)					53.0		
Incandescent							
100-W rough-service incandescent	120	100	960	893	9.6	1,000	100
60-W incandescent, inside frosted	120	60	850	791	14.2	1,000	100
100-W incandescent, inside frosted	120	100	1,710	1,590	17.1	750	100
150-W incandescent, inside frosted	120	150	2,850	2,651	19.0	750	100
300-W incandescent, inside frosted	120	300	6,200	5,766	20.7	750	100
10,000-W incandescent	120	10,000	330,000	306,900	33.0	75	100
Fluorescent							
48-in., 40-W fluorescent, daylight deluxe	120	50	2,250	1,910	45.0	20,000	84
48-in., 40-W fluorescent, cool white	120	50	3,200	2,464	64	20,000	70
48-in., 40-W fluorescent, high efficiency	120	50	3,350	2,580	67	20,000	82
Mercury vapor							
100-W mercury vapor, deluxe white	120	125	4,000	2,600	32.0	24,000	50
400-W mercury vapor, deluxe white	120	500	22,600	14,400	45.2	24,000	50
Metal halide							
100-W metal halide, clear	120	125	8,100	5,800	64.8	15,000	75
400-W metal halide, clear	120	500	44,000	33,900	88	20,000	65
High-pressure sodium							
400-W high-pressure sodium, clear	120	500	51,000	45,000	102	15,000	22
100-W high-pressure sodium, clear	120	125	9,500	8,550	76	24,000	22
Low-pressure sodium							
180-W low-pressure sodium, clear	120	225	32,000	27,800	142	16,000	0
90-W low-pressure sodium, clear	120	115	12,750	11,095	111	16,000	0

[1]Ballast power included if device requires a ballast.
[2]Color-rendering index.

Luminous Intensity (Candelas)

Luminous intensity of light sources is specified in candelas (cd). One candela is very nearly the output of one wax candle. More concisely, a candela is 1 lumen per steradian or 1 cd = 1 lm/sr. A steradian encloses an area of r^2 on the surface of a sphere of radius r. Since the surface area of a sphere is $4\pi r^2$, the area totals 4π steradians. Since a 1-candela light source emits 1 lumen through each steradian, its total emission is 4π lm or about 12.6 lm.

Luminous Efficacy (Lumens/Watt)

Luminous efficacy, in units of lm/W, describes the energy efficiency of a light source. Column 5 of Table 12.1 lists the efficacy of some light sources. A candle operates at about 0.1 lm/W. A hypothetical light source capable of converting all input energy to 555-nm yellow light would operate at 673 lm/W. Practical electric light sources operate in the range from about 10 to under 150 lm/W.

Illumination (Footcandles or Lux)

Illumination in footcandles (fc) is the density of luminous flux on a surface. Luminance is an alternate term for this quantity. One fc is the illumination produced from one lumen of flux uniformly distributed on a 1-ft^2 surface, or 1 fc = 1 lum/ft^2. In the SI system, the unit of illumination is the lux (lx), and 1 lux = 1 lm/m^2. Thus, 1 lx is the illumination resulting from 1 lumen of flux uniformly distributed on a 1-m^2 surface. Since 1 m^2 = 9.29 ft^2, 10 lx is approximated as 1 fc. In practice, at least 2 lm/ft^2 is required to produce 1 fc. This will be discussed later in the section on lighting design.

Illumination levels for design are specified in footcandles (fc). Some approximate footcandle levels are stated in Table 12.2. These will help you get a feel for the unit. As can be seen from the table, the amazing human eye allows visual function over a range of illumination levels from 0.01 to 10,000 fc. The illuminance of bright sunshine is a million times that of moonlight.

TABLE 12.2 Approximate illumination levels

Location	Approximate fc	Illumination, lux
Moonlight	0.01	0.1
Dimly lit bar	1	10
Bright room	100	1,000
Bright sunlight	10,000	100,000

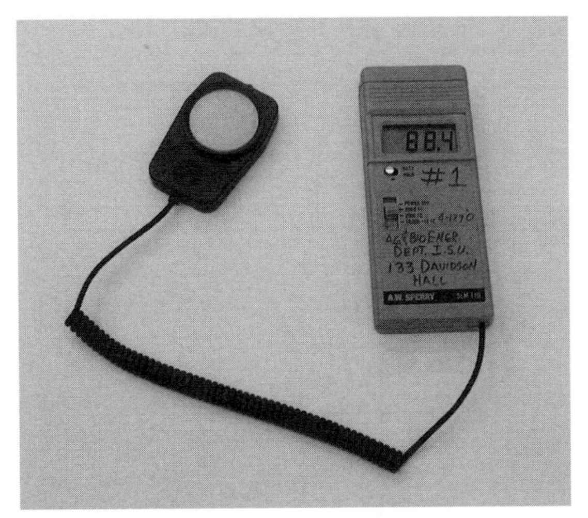

Fig. 12.4. Light meter.

Light Meters

A hand-held light meter can be used to measure illumination (Figure 12.4). The sensor is placed horizontally, at workplace height. A photovoltaic cell in the sensor converts incident visible radiation to an electrical voltage, which is related to illumination. The meter reads out footcandles or lux.

Lighting Sources

Most devices for artificial lighting use electricity as a source of energy. The lighting device converts electrical energy to visible radiation or light. Common electric lighting devices can be classified into incandescent or gaseous discharge types (Table 12.3).

In this section, each of the common lighting devices will be discussed. Discussion includes an explanation of how the device works, how much energy it requires, and other characteristics useful for design.

Incandescent Tungsten Filament Lamp

Incandescent filament lamps utilize the I^2R heating effect of an electric current to heat a wire filament until it reaches incandescence—the state of emitting visible radiation. Thomas Edison demonstrated the first practical incandescent lamp in 1879. The filament in Edison's lamp was carbonized cotton thread,

TABLE 12.3 A classification of electric lighting devices

Incandescent	Gaseous Discharge
Tungsten filament	Fluorescent
Conventional	High-intensity discharge
Tungsten Halogen	Mercury
	Metal-halide
	High-pressure sodium
	Low-pressure sodium

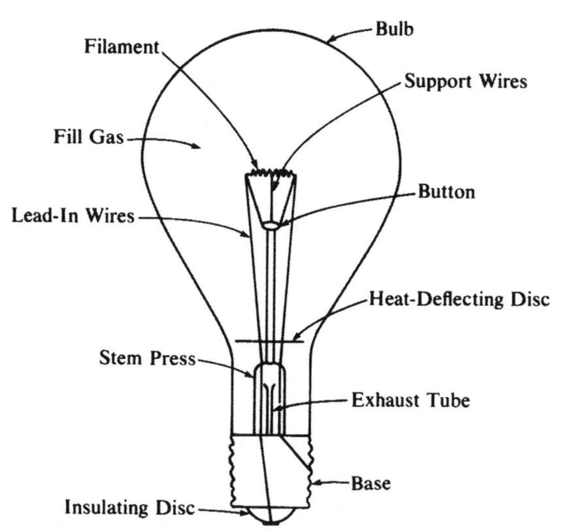

Fig. 12.5. Components of an incandescent lamp (Murdoch 1985).

which he later found to be inferior to tungsten. Components of a modern incandescent lamp are shown in Figure 12.5.

Filament

The filament is a coil of tungsten wire. Tungsten's melting point (3,655 K) is higher than all other elements except carbon. Coiling increases luminous efficacy. A coiled-coil filament has an efficacy about 10% higher than a coil because of improved concentration of heat. A 60-W, 120-V GE bulb has a $^5/_8$-in. long filament which is a coiled-coil design using 21 in. of 0.0018-in. diameter tungsten wire.

Filling Gas

The bulb contains a mixture of nitrogen and argon. This mixture prevents oxidation and evaporation of the filament and subsequent bulb blackening. Bulb pressure approximately equals atmospheric pressure during operation. Bulbs smaller than 40 W are usually evacuated because of heat loss through any filling gas.

Lead-In Wires and Stem Press

The lead-in wires complete the circuit between base and filament. Each of the three sections is made of a different material. The stem press holds the lead wires and forms the seal around them. Within the stem press, Dumet wires are used. This wire has a nickel–iron alloy core in a copper sleeve and has a coefficient of thermal expansion very near that of the glass stem press to minimize stresses during heating and cooling. Wires below the stem press are copper, while those above are nickel.

Exhaust Tube

Air is exhausted and a filling gas is inserted through the exhaust tube.

Support Wires and Button

Support wires for the filament are usually molybdenum. They are fused into a glass button atop a glass rod for support. Their number is kept at a minimum since they reduce lighting efficiency.

Heat-Deflecting Disc

The heat-deflecting disc protects the lower parts from excessive temperature by limiting convective flow of gases within the lamp.

Fuse

The fuse wire opens the circuit quickly if the filament shorts or opens and begins to arc. This fuse reduces the incidence of bulb breakage and the opening of circuit overcurrent devices.

Base

The base contains the connection terminals. It is usually made of aluminum or brass with glass insulation between the terminals. The center contact is energized (hot). The rim (threaded portion) is connected to neutral.

Bulb

The bulb is glass. Clear bulbs have the highest efficacy, but may cause glare (annoying brightness) in the region of the filament. Finishes to the bulb interior, such as etching or a coat of silica powder, spread the high brightness of the filament. A bulb with such a finish is said to be *frosted.*

Operating Characteristics

When voltage is applied to an incandescent lamp, current surges as the cool, low-resistance filament heats. In a 200-W bulb, for example, the cold resistance is 4.6 Ω. When 120 V is applied, current reaches 40 A, but within 0.13 s, drops to 1.7 A as resistance climbs to 72 Ω. Within 0.22 s, light output is 90% of normal. This current surge, although greater than the starting current of a motor, is not troublesome because it subsides so fast.

The tungsten filament operates near 4750°F. Light output varies from 96 to 104% of normal, due to the AC current variation at 60 Hz. This variation is greater at low frequencies. The color rendering index (CRI) is high, indicating the color looks like it would in sunlight (Table 12.1). For example, a 6-W bulb on 25-Hz

TABLE 12.4 Lamp efficacy

Lamp Type	Efficacy, lm/W
Incandescent	10–40
Halogen	20–45
Mercury vapor	50–60
Fluorescent	35–100
Metal halide	80–125
High-pressure sodium	90–140
Low-pressure sodium	100–180

power varies in light output from 31% to 169% of normal due to current variation during each cycle (General Electric Co., circa 1970).

Typical energy output of a tungsten filament is shown in Figure 12.3. Note that most of the energy output from the tungsten filament lamp is at wavelengths above the visible range. The incandescent lamp is, thus, not very energy efficient as a producer of visible radiation. Also, the large amount of nonvisible radiant energy received in order to achieve a certain light level can limit use of incandescent lamps for some applications. A 100-W incandescent bulb radiates 10% of the input energy in the visible range, while 72% is radiated as infrared. Heat lost from the bulb due to convection and conduction account for the remainder of the energy.

Energy Use

Operating characteristics of several light-producing devices are shown in Table 12.1. A 100-W inside-frosted bulb produces 17.1 lumens/W. Compared to other lighting devices, incandescents are lowest in efficacy (Table 12.4). Notice how much incandescent efficacy has increased since Edison's first bulb in 1879. Notice also that efficacy goes up with voltage rating and down with rated life. Longer life is achieved by using a stronger, cooler, heavier filament. The 10,000-W bulb has an efficacy of about 63% of the theoretical limit for tungsten filaments. Concerns about energy conservation, along with advances in other types of lighting devices, are resulting in continual replacement of incandescent lamps by other types. However, incandescent lamps have so many advantages for some applications that they will not go completely out of use.

Voltage Effects

Incandescent lamps are sensitive to voltage. Figure 12.6 shows effects of voltage on life, light

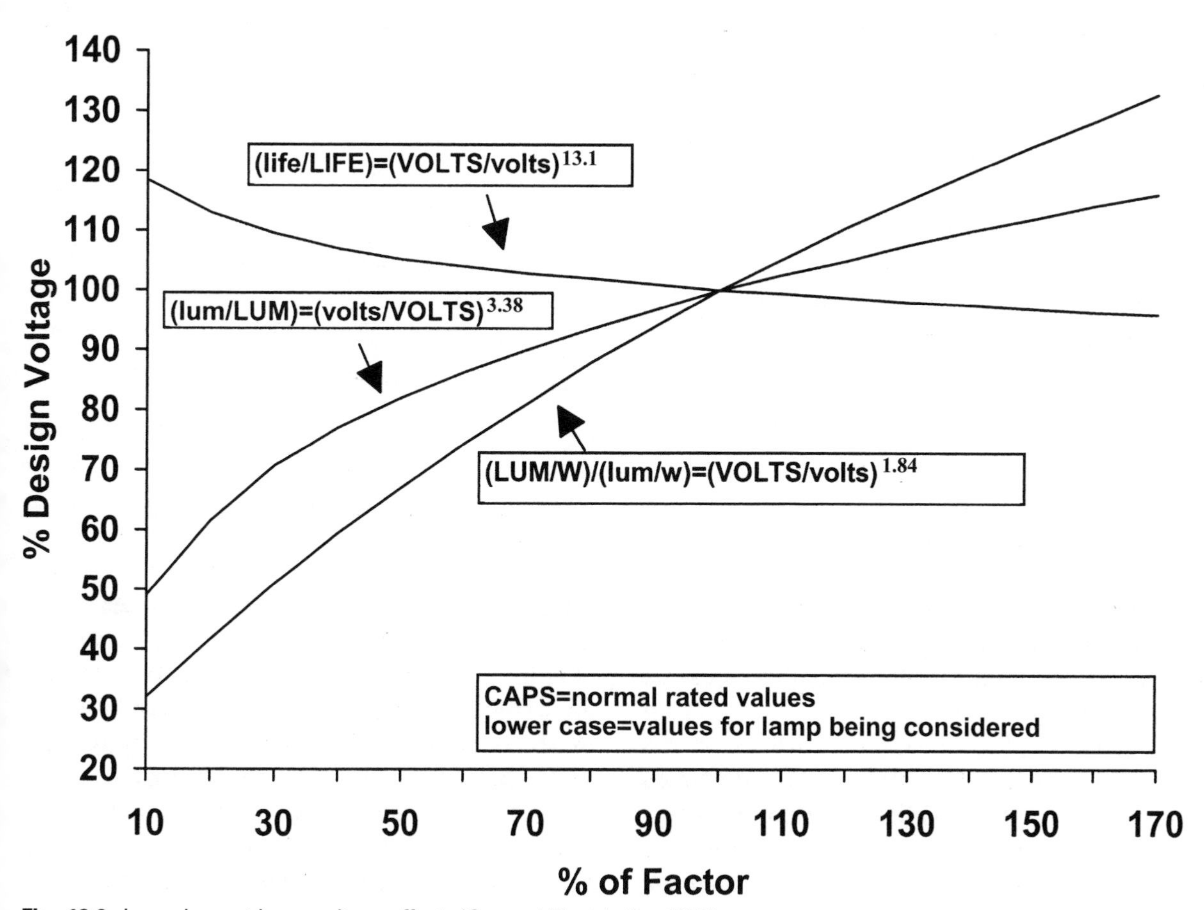

Fig. 12.6. Incandescent lamp voltage effects (General Electric Co. 1970).

output, and efficacy. Operating above design voltage increases light output and efficiency, but drastically reduces lamp life. Well-designed lamps have filaments designed to maximize lumen hours of light per dollar of cost, taking into account lamp replacement cost and electrical energy cost.

Example 12.1

A 300-W inside-frosted incandescent bulb is rated at 120 V, 750 h, and 6300 lm. Predict its life, output, and power use if it is operated at 132 V. The bulb is operated at 115% of rated voltage.

$$\frac{132\text{ V}}{120\text{ V}}(100) = 110\%$$

life: From Figure 12.6, at 110% voltage, life is about 30% of rating:

$$0.3(750\text{ h}) = 225\text{ h}$$

output: From graph, lumens will be about 140% of rating:

$$1.4(6{,}300\text{ lm}) = 8{,}820\text{ lm}$$

efficacy: From the graph, lm/W will be about 120%.

$$\text{Rated efficacy} = \frac{6{,}300}{300} = 21\,\frac{\text{lm}}{\text{W}}$$

$$1.2(21) = 25.2\,\frac{\text{lm}}{\text{W}}\text{ then}$$

$$\text{Power} = \frac{8820\,\cancel{\text{lm}}\,\text{W}}{25.2\,\cancel{\text{lm}}} = 350\text{ W}$$

This problem can also be solved using equations listed in Figure 12.6.

Example 12.2

A 60-W inside-frosted bulb costs $0.24, and a 150-W bulb costs $1.79. If these bulbs have the operating characteristics shown in Table 12.1, and are operated at rated voltage, which one produces the most light per dollar of cost? (Neglect cost of the luminaire which holds the bulb, and the ownership cost of the bulb. Assume electrical energy costs $0.08/kWh.)

Equation 12.2 can be used to compute lumen hours per dollar cost:

$$\text{LH} = \frac{(\text{LUM})(\text{LIFE})}{C_B + C_E} \quad \textbf{(12.2)}$$

where LH = hours of light per unit cost, lm h/$
LUM = bulb output, lm
LIFE = bulb life, h
C_B = cost of bulb, $
C_E = cost of energy, $

For the 60-W bulb:

$$\text{LH} = \frac{(850\text{ lm})(1000\text{ h})}{\$0.24 + (0.06\,\cancel{\text{kW}})(1000\,\cancel{\text{h}})\,\frac{(\$0.08)}{\cancel{\text{kWh}}}}$$

$$= 168{,}651\,\frac{\text{lmh}}{\$}$$

For the 150-W bulb:

$$\text{LH} = \frac{(2850\text{ lm})(750\text{ h})}{\$1.79 + (0.15\,\cancel{\text{kW}})(750\,\cancel{\text{h}})\,\frac{(\$0.08)}{\cancel{\text{kWh}}}}$$

$$= 198{,}100\,\frac{\text{lmh}}{\$}$$

The 150-W bulb provides over 17% more light per dollar than the 60-W bulb.

Failure

If the bulb breaks, the filament oxidizes and burns off very quickly. During use, the incandescent filament slowly sublimes and deposits, causing a gradual blackening of the inside surface of the bulb. Eventually, the bulb fails when the filament breaks or burns through at the thinnest point. Filaments, which are very weak during use, can be broken by shock or vibration.

Tungsten-Halogen Lamp

Tungsten-halogen (TH) lamps are incandescent lamps which use different materials to achieve longer life, higher efficacy, and smaller size than conventional tungsten filament lamps. A halogen gas (for example iodine) is used as part of the fill gas. As tungsten particles sublime, they combine with halogen molecules to form a tungsten halide. When the halide comes in contact with the incandescent filament, it disassociates into tungsten (which redeposits on the filament) and halogen gas. This allows a higher operating temperature and efficacy, and a longer life. The bulb is quartz glass to withstand the higher temperature.

Advantages and Disadvantages of Incandescent Lamps

Incandescent lamps have these advantages:

1. First cost is low.
2. Circuitry is simple and complete within the lamp (no ballast).
3. Operation is normal in low ambient temperatures.
4. Bulb life is not reduced by a large number of on–off cycles.
5. Lamps operate near unity power factor.
6. Warm-up takes only a fraction of a second.
7. Lamps are available in a wide variety of wattages.
8. Control is easy.

Incandescent lamps have these disadvantages:

1. Their life is short.
2. Efficacy is low.
3. Their high infrared output is detrimental in some applications.
4. The high-temperature filament is subject to breakage due to shock or vibration.
5. Life, light output, and efficacy are drastically affected by voltage variations.

Fluorescent (Gaseous Discharge) Lamp

Fluorescent lamps consist of long tubes which emit light. They overcome, to varying degrees, all the disadvantages of incandescent lamps, but have a few of their own. The invention of the fluorescent lamp actually came before Edison's 1879 incandescent lamp. The French physicist Becquerel demonstrated the fluorescent lamp principle in 1867 (Alphin 1959). However, practical fluorescent lighting did not appear until 1938.

Operating Principles

A fluorescent lamp consists of a glass tube containing an inert gas such as argon, and a few mg of mercury. The tube has electrodes at each end. The lamp and circuit of Figure 12.7 illustrate a simple preheat type circuit. When the start switch is closed, the electrodes heat up to incandescence and ionize nearby argon atoms, thereby lowering their electrical resistance. When the start switch is opened, the ballast transformer opposes the change in current and causes an increase in voltage between the electrodes. This high voltage establishes an arc between the electrodes. Some of the electrons flowing through the arc strike electrons of vaporized mercury atoms in the tube (see Figure 12.8). The collision causes these valence electrons to be diverted out of their normal orbits. When these diverted electrons snap back into place, they emit energy in the form of ultraviolet radiation, mostly at a wavelength of 253.7 nm in the ultraviolet band, which strikes phosphor crystals on the inside of the tube, causing them to "fluoresce," and emit visible radiation. The start switch remains open during operation. To shut off the lamp, the stop switch is opened long enough for the arc to be extinguished.

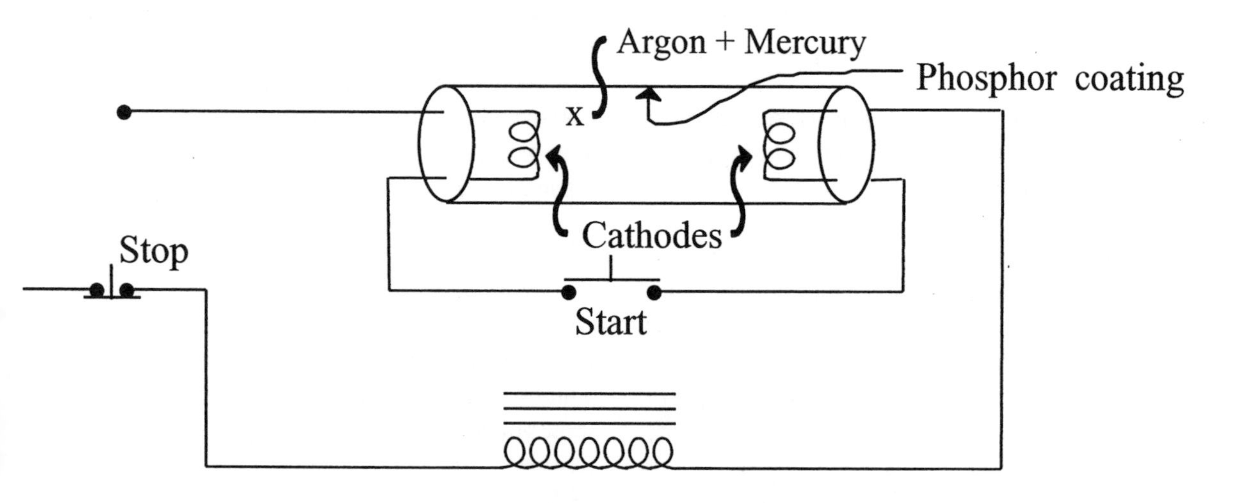

Fig. 12.7. Preheat fluorescent lamp circuit.

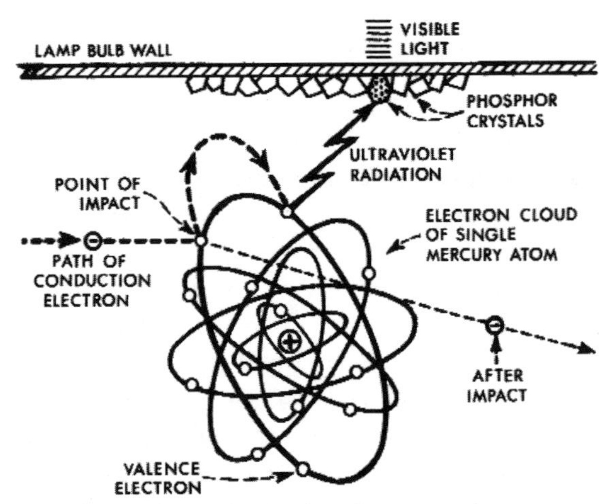

Fig. 12.8. Light production by a fluorescent lamp (Kaufman and Christensen 1984; reprinted with permission.)

The ballast transformer (or ballast) consists of an iron core with numerous turns of conductor around it. Unlike the transformer of chapter 1 (Figure 1.24), a ballast transformer has no secondary winding. A fluorescent ballast transformer produces a voltage increase for starting, and also serves as a current-limiting device during lamp operation. Since arc resistance decreases with time after starting, a current-limiting circuit component is required. Ballast transformers are usually encapsulated in tar. Ballasts consume power, so the power of a fluorescent luminaire (lamp plus fixture) must account for this. Ballast power varies considerably among lamp types. We will estimate ballast power at 20% of lamp power. Thus the power use of a 40-W fluorescent lamp will be figured as 48 W. If a ballast has power factor correction components, the luminaire (ballast plus lamp) will operate at a power factor of about 0.95. Without power-factor correcting, the luminaire will operate at about 0.60. Electronic ballasts are available which operate lamps at power frequencies above 20 kHz. These ballasts increase lamp output and reduce ballast power use.

Light Emission

The light emitted by a fluorescent lamp, and also the efficacy of the lamp, depends on the chemical makeup of the phosphor. The output of a warm white tube is included in Figure 12.3. Notice that most of this lamp's emission is in the range of eye sensitivity and infrared emission is minimal. Lamp emissions (and therefore color and efficacy) depend on the lamp's phosphor composition, and there are many types available.

Lamp Circuits

The preheat circuit of Figure 12.7 was one of the earliest. It is still in use, but other circuits (rapid-start, instant start, trigger start) are used more now. Lamps are generally not interchangeable among circuit types.

Fluorescent Lamp Characteristics

Temperature Effects

Fluorescent lamps are temperature-sensitive. Starting, light output, and power use all vary with temperature. Lamp starting becomes increasingly troublesome below 50°F and reliable starting at temperatures below 32°F requires special starting circuits and ballasts. Ideal lamp temperature during operation is between 100°F and 120°F and luminaries are designed so this temperature is reached during operation in still air at 77°F. Lumen output decreases when lamp temperature is above or below this range (North American Philips 1984). Conventional fluorescent luminaires are not suitable for use in unheated spaces in climates where there is a significant winter. Special cold-weather luminaires are available at higher costs.

Humidity Effects

The electrostatic charge on the outside of fluorescent lamps affects the voltage necessary for starting. Humid surrounding air can form a film which affects this charge and makes higher starting voltages necessary. Lamps for use in high humidity areas can be equipped with high-voltage starters to ensure reliable starting.

Voltage Effects

Fluorescent lamp operation is influenced by voltage, but not to the extent of incandescent lamps. Low voltage and high voltage reduce efficacy and lamp life. In addition, low voltage may cause starting problems and high voltage may overheat ballasts and also cause premature lamp failure.

Number of Starts

Each time a fluorescent lamp starts, some electron emission material "boils off" the electrodes, shortening lamp life. Published life ratings generally assume 3 hours of operation per start. Multipliers in Table 12.5 show effects of hours per start on lamp life.

Efficacy

Fluorescent lamp efficacy tends to be higher than that of incandescent lamps, but generally lower than

TABLE 12.5 Typical life extension multipliers for fluorescent lamps

Hours per Start	Standard Life Multiplier
3	1.00
6	1.25
12	1.60
continuous	2.50

that of other gaseous discharge lamps (see Table 12.4). Figure 12.9 illustrates the energy distribution in the emission of a typical cool white fluorescent lamp. About 21% of the input energy is emitted as visible light.

Table 12.1 contains a column showing efficacy in lumens per watt. Note that fluorescent lighting, with 67 lumens/W available from a 48-in. high efficiency tube, has more than twice the efficacy of any incandescent lamp. This is, however, far below the 673 lm/W theoretical limit of a lighting device.

Color

Fluorescent lamp color depends on the material mix of the inside phosphor coating. Six standard "white" lamps are used in industry:

Cool white	(CW)
Warm white	(WW)
White	(W)
Cool-white deluxe	(CWX)
Warm-white deluxe	(WWX)
Daylight	(D)

Choosing lamp color necessitates a compromise between efficacy and color rendering characteristics. See Table 12.1. CW has high efficacy but is weak in the red portion of the spectrum and has a low CRI. D provides better color rendition with a lower efficacy.

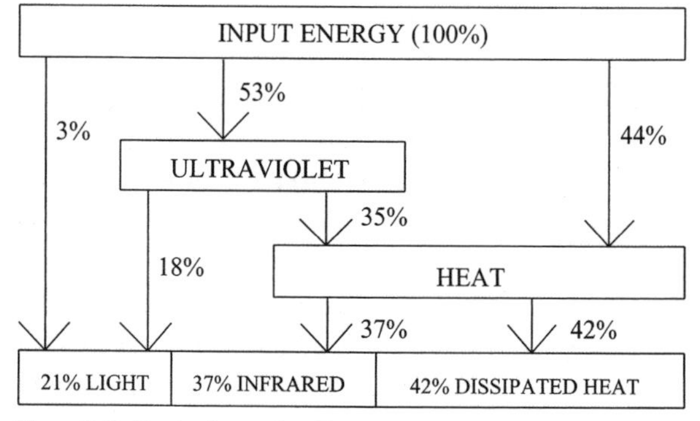

Fig. 12.9. Typical energy distribution from a cool-white fluorescent lamp.

CW or WW are common choices for offices and institutions since they combine satisfactory appearance, low cost, and good availability.

Advantages and Disadvantages of Fluorescent Lighting

Fluorescent lighting has these advantages:

1. Moderately high efficacy.
2. Relatively long tube life.
3. Relatively low infrared component.
4. Low surface brightness.
5. Moderate purchase price.

Fluorescent lighting has these disadvantages:

1. Temperature sensitivity.
2. Poor power factor.
3. Humidity sensitivity.
4. Life sensitivity to number of starts.
5. Possible AM radio interference.

High-Intensity Discharge

Besides fluorescent, another type of gaseous discharge lamp in use is categorized as high-intensity discharge (HID). There are four common types (mercury, metal-halide, high-pressure sodium, and low-pressure sodium). These lamps, like the fluorescent, utilize emission from an electrical arc, but with several differences. Operation of the mercury-vapor lamp (Figure 12.10) will be explained. The lamp contains a pressurized inner bulb of quartz, and an outer glass envelope. The quartz arc tube contains argon pressurized to 1 to 10 atm, along with mercury. When power is turned on, an arc is established between the starting electrode and a working electrode. Soon, an arc is established between the working electrodes. As with the fluorescent lamp, electrons in the arc collide with valence electrons of the mercury atoms, forcing them to a higher energy state. As they snap back to their normal orbits, they emit energy as radiation. The wavelength of the emissions is mostly shifted into the visible region by selecting the pressure in the arc-tube. The outer glass envelope is designed to filter out ultraviolet radiation emitted. This is necessary for safety, and operating an HID lamp with the outer envelope broken may expose people to radiation dangerous to eyes and skin.

Mercury vapor lamp efficacy is usually in the range of 50 to 60 lm/W, which means they are between

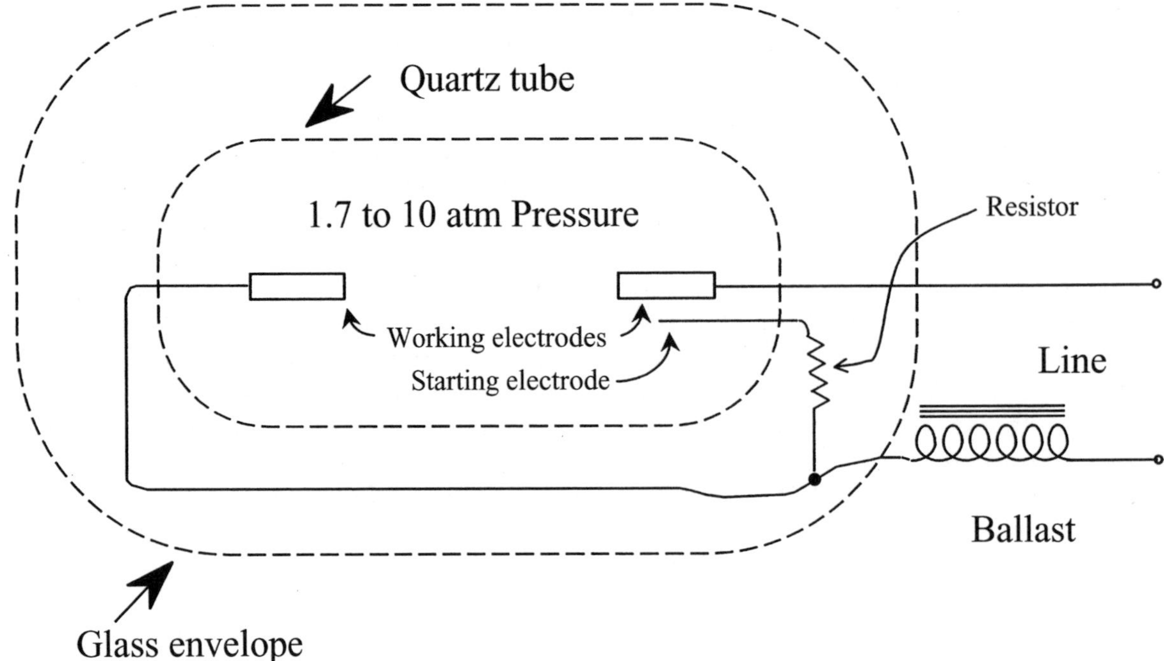

Fig. 12.10. Mercury-vapor lamp.

incandescent and fluorescent in efficiency (Table 12.4). All HID lamps require ballasts, and all require several minutes to reach full output after being turned on. This characteristic must be considered in applications where persons need to turn on lights as they enter a dark room.

Metal Halide, High-Pressure Sodium, Low-Pressure Sodium

The other HID lamp types are similar in construction to the mercury vapor lamp, but each type has features which influence the emission. In metal halide lamps, metal iodides such as scandium iodide and sodium iodide are placed in the arc tube. These compounds cause emissions in additional wavelength bands causing their light to be blue-white. Efficacy and color rendering index are significantly improved (Table 12.1).

Sodium lamps employ sodium vapor as the main light-producing gas in an arc tube. Sodium vapor produces a monochromatic (one color) emission at 590 nm (yellow) at low pressure. At high pressure, the spectrum broadens to include most of the visible range. Sodium lamps have the highest efficacy of all common light devices (Table 12.4). However, their color rendering index tends to be low, which means they are usually not suitable for lighting work or living areas (Table 12.1).

Advantages and Disadvantages of Gaseous Discharge Lamps

Advantages:

1. High efficacy.
2. Long life (up to 24,000 h).
3. Not temperature or humidity sensitive.

Disadvantages:

1. A 2- to 10-min warm-up time is required at starting.
2. A large and possible noisy ballast is required.
3. Color rendering may be poor.
4. Relatively high purchase price.
5. Burn position may affect operation.

Lighting Design

If a lighting system has been well designed, attention has been paid to both the quality and the quantity of illumination.

Quality of Illumination

Quality of illumination pertains to the distribution of brightness within the whole visual field. Quality is affected by walls, ceiling, floor, and furnishings as well as by the luminaire.

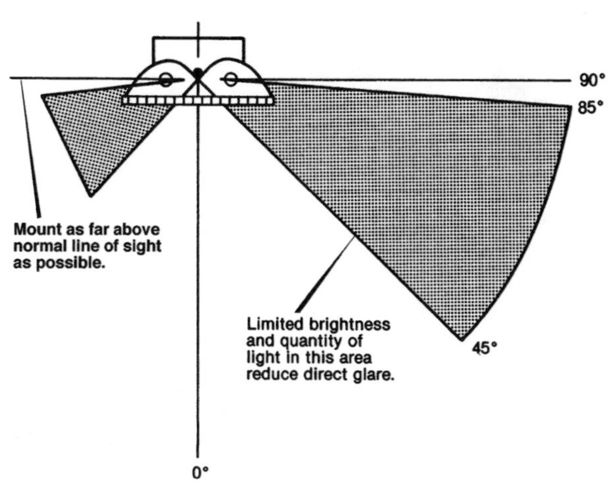

Fig. 12.11. Glare from a light source (Campbell and Thimijan 1981).

Glare

Glare is any brightness in the field of view which causes discomfort, annoyance, interference with vision, or eye fatigue. Glare is generally undesirable because it can be annoying, fatiguing, painful, and even dangerous. The causes of glare include:

1. Bright light source.
2. Low field brightness.
3. Location of exposed light sources within the line of vision.

Frost inside incandescent bulbs and fins on fluorescent luminaires tend to reduce glare. Some designs use glare to advantage. Used car lots are often lighted with glare-producing clear bulbs so that car finishes reflect and sparkle at night. Dark walls and surroundings can make glare more noticeable.

Figure 12.11 shows that light emissions are most likely to cause glare in the 40° angle from 45° to 85°. Although extreme contrasts in brightness are undesirable, some contrast is needed for easy eye focusing and orientation to a room.

The quality aspect of lighting design is a highly developed art, and discussion in this chapter is necessarily brief. Many good references on lighting quality are available.

Quantity of Illumination

The quantity of illumination (*illuminance*) can be specified as the footcandle or lux level on a plane. Response of the eye to changes in illumance is approximately geometric rather than arithmetic. For example, suppose illuminance on a plane is 10 fc or 100 lux. An increase of 10 fc or 100 lux causes a "significant increase in visual effect." If the initial level is 100 fc or 1000 lux, an increase of about 100 fc or 1000 lux is required for a "significant increase in visual effect." Human responses to sound, taste, and drugs are similar.

Table 12.6 allows estimation of proper illuminance according to the nature of the task, the room, and the persons present. From Table 12.6, a category is selected according to activity. Three fc levels are specified. From Table 12.6, weighting factors are assigned. The sum of the weighting factors determines which fc level should be used. If the sum is −2 or −3, use the lowest value. If the sum is −1, 0, or +1, use the middle value. If the sum is +2 or +3, use the highest value. An example will illustrate use of the tables.

Example 12.3

What illumination should be selected for the general lighting in a farm shop? Some users will be over 55 years of age.

From Table 12.6, this area is estimated to be Category D (20-30-50 fc). From Table 12.6, occupant age is over 55 (+1), speed and/or accuracy is important (0), reflectance of task background is 30 to 70% (0). The sum is +1, so the recommended level is 30 fc. Note that using the table requires judgement, and results are not always clearcut. Table 12.7, which lists suggested illuminance values, also specifies 30 fc.

Zonal Cavity Method

The zonal cavity method is a simple and fast approach to designing a lighting system which will provide a desired illumination level on a workplane. Using this method, the total lumen output required of all lamps in the room (lamp lumens) is calculated using Equation 12.3:

$$LL = \frac{(I)\,(A)}{(CU)\,(LLF)} \tag{12.3}$$

where LL = lamp lumens required, lm
I = desired illumination on workplane, fc
A = area of workplane, ft^2
CU = coefficient of utilization, dimensionless
LLF = light loss factor, dimensionless

Earlier, 1 footcandle was defined as the illumination resulting from 1 lumen of flux falling uniformly

TABLE 12.6 Illuminance values and weighting factors for interior activities

Illuminance categories and values			
Category	Activity	Footcandles	Lighting Type
A	Public spaces, dark surrounds	2-3-5	
B	Short, temporary visits to space	5-7.5-10	General lighting throughout
C	Visual tasks performed only occasionally	10-15-20	
D	Visual tasks of high contrast	20-30-50	
E	Visual tasks of medium contrast or small size	50-75-100	
F	Visual tasks of low contrast or very small size	100-150-200	Illumination on task obtained by a combination of general and local (supplementary) lighting
G	Category F over a prolonged period	200-300-500	
H	Very prolonged and exacting visual tasks	500-750-1000	
I	Very special visual tasks of extremely low contrast and small size	1000-1500-2000	

Categories A through C				Categories D through I			
	Weighting Factor				Weighting Factor		
Characteristics	−1	0	+1	Characteristics	−1	0	+1
Occupant age	Under 40	40–55	Over 55	Occupant age	Under 40	40–55	Over 55
				Speed and/or accuracy	Not important	Important	Critical
Room surface reflectances	Over 70%	30–70%	Under 30%	Reflectance of task background	Over 70%	30–70%	Under 30%

Source: Kaufman and Christensen 1981.

on a 1-ft^2 area. That is, 1 footcandle equals 1 lumen per square foot. Equation 12.3 reduces to this relationship if CU = LLF = 1. CU and LLF are factors which account for characteristics of the luminaire and room which prevent total lamp lumens from reaching the workplane. Procedures for estimating these factors are described in lighting design handbooks (Kaufman and Christensen 1984).

Abbreviated Zonal Cavity Method

A rule of thumb in the lighting industry says that half the light gets to the workplace (Boylan 1987). For this rule of thumb to hold, (CU) (LLF) = 0.5 in Equation 12.3. This assumption greatly simplifies the design procedure and experience shows that, for many applications, the procedure gives acceptable designs. The equation for the abbreviated zonal cavity method is:

$$LL = \frac{(I)(A)}{0.5} \qquad \textbf{(12.4)}$$

Example 12.4 illustrates use of this procedure.

Example 12.4

Design a lighting system for a 20-ft × 60-ft feeding–inspection area of a poultry laying house.

1. Select illumination level.
   From Table 12.6, the recommended illuminance is 20 fc.
2. Calculate required lamp lumens.
   $$LL = \frac{(20\text{ fc})(21\text{ ft})(63\text{ ft})}{0.5} = 52{,}920\text{ lm}$$
3. Select lamps (Table 12.1).
   Incandescent: 150-W inside frosted 2,651 lm (mean)
   or
   Fluorescent: 48-in., 40-W, cool white 2,464 lm (mean)
   $$\text{Incandescent: } \frac{52{,}920\text{ lm lamp}}{2{,}651\text{ lm}} = 20\text{ lamps}$$

Possible luminaire layouts using approximately 20 lamps include 7 × 3, 10 × 2, 5 × 4, 4 × 5. A 7 × 3 layout gives the most nearly equal lamp-to-lamp spacing:

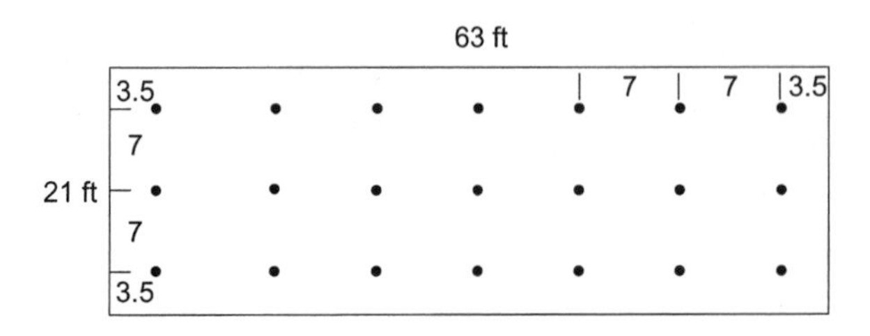

Luminaire-to-wall spacing is one-half of luminaire-to-luminaire spacing.

TABLE 12.7 Minimum recommended illuminance levels for various agriculture-related tasks areas

Areas and Visual Tasks	Illuminance (fc)
Milking operation	
general, milk parlor	20
cow's udder	50
Milk handling	
general, loading platform	20
washing area, bulk tank interior	100
Feed storage & handling	
haymow, silo, grain bin	5
feed preparation, processing	10
hay inspection area, stairs, silo room, feed alley, pens, loose housing, feed areas	20
charts and records	20
Livestock housing	10
Poultry brooding & laying houses	
feeding, inspection and cleaning	20
read charts and records	50
read controls	50
Hatcheries	
general and loading platform	20
inspect and clean inside incubators	50
dubbing station	200
sexing	1000
Egg handling, packing, & shipping	
loading platform, egg storage	20
general cleaning, inspecting, processing for food	100
Fowl processing plant	
unloading and killing	20
general lighting (clean-up)	100
government inspecting and grading	100
Machine storage	10
Farm shop	
active storage area	10
general shop lighting	50
rough bench-machine work	50
Miscellaneous	
pumphouse	20
restrooms	20
farm office (bookkeeping, etc.)	100
Exterior	
general inactive areas	0.5
barn lots, paths, rough storage	5
service areas, building entrances	3

Source: ASAE 1997.

$$\text{Fluorescent:} \quad \frac{52{,}920 \not{\text{lm}} \text{ lamp}}{2{,}464 \not{\text{lm}}} = 21.5 \text{ lamps}$$

Once again a 7 × 3 layout (as shown in the figure above) appears to be a good choice. The fluorescent luminaires are 4 ft long. If the luminaires are mounted in rows across the width of the room, luminaire-to-luminaire spacing will be 3 ft, and luminaire-to-wall spacing will be 1.5 ft.

References

Alphin, W. 1959. Primer of lamps and lighting. Chilton Company, New York, NY.

American Society of Agricultural Engineers. 1997. Lighting for dairy farms and the poultry industry–EP344.2 ASEA Standards. ASEA, St. Joseph, MI.

Boylan, B. R. 1987. The lighting primer. Iowa State University Press, Ames.

Campbell, L. E., and R. W. Thimijan. 1981. Farm lighting, Farm Bulletin 2243. USDA, Washington, D.C.

General Electric Company, 1970. The General Electric lumen calculator. General Electric Company, Cleveland OH. (Not copyrighted)

Kaufman, J. E., and J. F. Christensen. 1981. IES Lighting handbook, Applications volume. Illuminating Engineering Society of North America, New York, NY.

Kaufman, J. E., and J. F. Christensen. 1984. IES Lighting handbook, Reference volume. Illuminating Engineering Society of North America, New York, NY.

Murdoch, J. B. 1985. Illumination engineering–From Edison's lamp to the laser. Macmillan Publishing Co., New York, NY.

North American Philips Lighting Corp. 1984. Philips lighting handbook. North American Philips Lighting Corp., Bloomfield, NJ.

Problems

12.1. What is the illumination level on the inside surface of a 3-ft-radius sphere containing at its center a 9-cd source which emits equally in all directions? Area of a sphere = $4\pi r^2$.

12.2. At what distance from a single wax candle is the illumination level about the same as moonlight?

12.3. An illuminance photometer reads 10 fc when aimed at a light source 25 feet away which emits equally in all directions. What is the lumen output of the light?

12.4. Compute the luminous intensity for a new 100-W inside-frosted incandescent lamp, assuming its intensity is equal in all directions.

12.5. A 100-W inside-frosted bulb costs $1 to replace and electrical energy costs 8 ¢/kWh. The bulb is rated at 120 V and has the characteristics shown in Table 12.1 (use mean lumen output). Assuming lighting cost can be estimated by lm·h/$, use a spreadsheet to determine what voltage in the range from 90 V to 140 V provides the least expensive light. Ignore the cost of the fixture. Use equations from Figure 12.6.

12.6. Using Figure 12.9 compute the energy conversion efficiency of the phosphor in a cool-white fluorescent tube.

12.7. Use the abbreviated zonal cavity method to design the lighting system for a 30 × 40-ft livestock feed preparation area. Luminaires are simple lamp holders and 100-W frosted incandescent bulbs. Show luminaire placement.

12.8. A 100-W incandescent bulb is rated at 130 V, 750 h, and 1,600 lm. Compute life, efficacy, light output, and power use.

(a) If bulb is operated at 117 V.

(b) If bulb is operated at 143 V.

12.9. Estimate light cost for several types of lighting devices. From a store or catalog, obtain data for 10 types of lighting devices and record the data on the following table. Consider several types of devices (incandescent, fluorescent, halogen, etc.) and different sizes within a type (several wattages of incandescent, for example). Assume electrical energy costs $0.08/kWh. Neglect cost of luminaire.

Source of information:

PROBLEM 12.9 Estimating lighting costs

Type of Lighting Device	Output, lm	Life, h	Lamp Cost, $	Power, W	Energy Cost, $	Total $	lm·h/$

Note: Increase power of fluorescent and gaseous discharge types by 10% to account for ballast power.

Chapter 13

DIGITAL LOGIC FOR CONTROL

Introduction

Electrical control discussed up to now has involved specifying and arranging mechanical or electromechanical switches in circuit conductors carrying current to loads or to relay coils. Control of many electrical loads by this approach is simple, low-cost, and reliable, and will continue in wide use for many applications. The hardware is developed, standardized among brands, and readily available in many forms.

In applications where the logic is complicated, some form of digital system may be appropriate. Integrated circuit (IC) technology allows logic of the switching to be done by small solid-state devices operating at low DC voltage levels. This logic for control is illustrated in Figure 13.1. Input signals (one or several) enter the logic system where they are processed, and where output signals (one or several) originate. This chapter provides an introduction to digital control and deals with inputs, the logic system, and outputs for agricultural application.

Digital and Analog

The control logic of this chapter is digital rather than analog. Digital devices have discrete or incremental output levels, whereas the output of analog devices is continuous and the number of possible output levels is not limited. Table 13.1 gives some examples. A room light switch has two discrete positions: on and off. However, a dimmer switch for the same light allows an unlimited number of output levels between off and fully on. Brightness of the light is analogous to the position of the switch knob. The other devices listed have similar characteristics. The availability of low-cost reliable digital ICs has made digital logic systems practical for control applications in many areas, including agriculture.

Binary Number System

The binary number system uses only two digits, 1 and 0. The word *binary* means "consisting of two things or parts." Since most digital logic devices operate at one or the other of two states, binary numbers find extensive use.

Binary numbers operate with a base 2 instead of the base 10 as used in the decimal system. See Table 13.2 for a listing of common number systems. The octal and hexadecimal systems are widely used in computer technology.

The position of a digit represents its weight in a positional number system. Equation 13.1 illustrates this. The equation is applicable to any number system. In the equation, d represents a digit and b represents the base of the system. A binary digit is called a *bit*.

$$d_n \ldots d_3d_2d_1d_0 = d_n b^n + \ldots d_3b^3 + d_2b^2 + d_1b^1 + d_0b^0 \qquad \textbf{(13.1)}$$

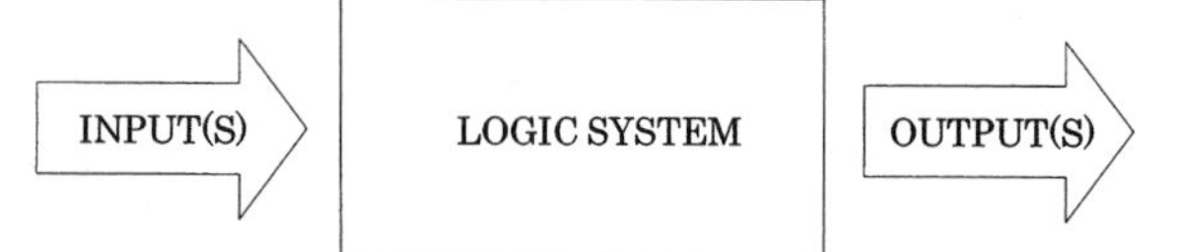

Fig. 13.1. Block diagram of logic system.

Use of the equation is illustrated by Example 13.1. A subscript after a number defines its base.

Example 13.1

Show the meaning of the decimal number 3458 by use of Equation 13.1.

$$3458_{10} = 3 \times 10^3 + 4 \times 10^2 + 5 \times 10^1 + 8 \times 10^0$$
$$= 3{,}000 + 400 + 50 + 8$$
$$= 3458_{10}$$

Similarly, the decimal equivalent of a binary number can be computed:

Example 13.2

What is the decimal equivalent of 10100110?

$$10100110_2 = 1 \times 2^7 + 0 \times 2^6 + 1 \times 2^5 + 0 \times 2^4 + 0 \times 2^3 + 1 \times 2^2 + 1 \times 2^1 + 0 \times 2^0$$
$$= 128 + 0 + 32 + 0 + 0 + 4 + 2 + 0$$
$$= 166_{10}$$

To find the binary number which represents a decimal number, the procedure illustrated in Example 13.3 can be used.

Example 13.3

What binary number represents 1453_{10}?

2^{11}	2^{10}	2^9	2^8	2^7	2^6	2^5	2^4	2^3	2^2	2^1	2^0
2048	1024	512	256	128	64	32	16	8	4	2	1

Select largest number smaller than 1453_{10} and subtract from 1453:

$$\begin{array}{r} 1453 \\ -1024 \\ \hline 429 \end{array}$$ • eleventh number from right is 1

$512 > 429$ • tenth number is 0

$$\begin{array}{r} 429 \\ -256 \\ \hline 173 \end{array}$$ • ninth number is 1

$$\begin{array}{r} 173 \\ -128 \\ \hline 45 \end{array}$$ • eighth number is 1

$64 > 45$ • seventh number is 0

$$\begin{array}{r} 45 \\ -32 \\ \hline 13 \end{array}$$ • sixth number is 1

$16 > 13$ • fifth number is 0

$$\begin{array}{r} 13 \\ -8 \\ \hline 5 \end{array}$$ • fourth number is 1

$$\begin{array}{r} 5 \\ -4 \\ \hline 1 \end{array}$$ • third number is 1

$2 > 1$ • second number is 0

$$\begin{array}{r} 1 \\ -1 \\ \hline 0 \end{array}$$ • first number is 1

Thus we see that $1453_{10} = 10110101101_2$.

TABLE 13.1 Digital and analog devices

Digital	Analog
Room light switch (ON, OFF)	Dimmer switch for room light
Digital clock	Conventional rotating-hand clock
Stairway	Escalator

TABLE 13.2 Number systems

Number System	Base	Digits
Binary	2	0, 1
Octal	8	0, 1, 2, 3, 4, 5, 6, 7,
Decimal	10	0, 1, 2, 3, 4, 5, 6, 7, 8, 9
Hexadecimal	16	0, 1, 2, 3, 4, 5, 6, 7, 8, 9, A, B, C, D, E, F

Digital logic circuits are composed of logic elements interconnected in a way which will yield proper binary output(s) for any combination of binary input(s).

Transistor–Transistor Logic

The logic elements to be discussed in this section are available in the transistor–transistor logic (TTL) integrated circuit family. TTL is one of several families available. CMOS (complimentary metal–oxide–silicon) is another popular family. TTL is the most widely used because it combines low cost, ease of use, adequate performance characteristics, and good interfacing capability.

A typical TTL logic element, or *gate*, consists of 5 transistors, 4 resistors, and a diode. It operates from a +5-V DC power supply. Inputs and outputs correspond to about +2.4-V DC for binary 1 and about 0-V DC for binary 0. The TTL logic family will operate in the range of +2.0 to 5.0-V DC for binary 1 and in the range from 0 to +0.8-V DC for binary 0.

Logic Elements

Solid Conductor

A solid conductor (Figure 13.2a) can be thought of as a simple logic element. It, of course, is complete in itself and contains no additional electronic components. It will be used here to introduce the concept of a logic element.

The logical function of a solid conductor is illustrated in the truth table, Figure 13.2b. The truth table lists outputs for all possible input combinations. It is a very useful tool in working with digital circuits because it can verify the behavior of the circuit. The number of lines in the truth table is a function of the number of inputs (Equation 13.2).

$$L = 2^N \qquad \textbf{(13.2)}$$

where: L = number of lines in truth table
N = number of inputs to the circuit

In the case of the switch, there is one input and since $2^1 = 2$, there are two possible input–output combinations. For a switch the output Q corresponds exactly with the input, A. The relay equivalent of a switch is a normally open (NO) relay (Figure 13.2c). In these examples we will assume that Q = 0 if it is not connected to the battery. We cannot make this assumption with actual TTL devices.

Figure 13.2d is the Boolean algebra equation describing the function of a solid conductor in terms of the input and output signals. Boolean algebra is a discrete algebra developed by 19th-century English clergyman and mathematician George Boole. In this algebra, which Boole described in 1847, variables can have only values of 0 and 1. It is the foundation of digital design work since its rules and theorems (which will be explained as needed) allow manipulation of binary variables. In the case of a solid conductor, the equation states simply that the output, Q, always corresponds to the input, A.

Inverter

The simplest electronic logic element is the inverter. Its symbol is shown in Figure 13.3a. The small circle represents the inverting function. The inverter has the property of always having an output opposite to the input. This is illustrated in the truth table, Figure 13.3b. Its relay equivalent is a normally closed (NC) relay (Figure 13.3c). The equation (Figure 13.3d) says that Q, the output, equals $\overline{A}$ (read A NOT).

The inverter also functions as a buffer and the triangle represents the buffer function. A buffer is an amplifier which provides an increase in current level in the gate's output. Each device in a digital circuit supplies power in its output to provide input signals to downstream devices. If too many inputs are fed from one output, the device can stop functioning or become unreliable. Insertion of a buffer expands the drive capability of a device. The device we call an inverter

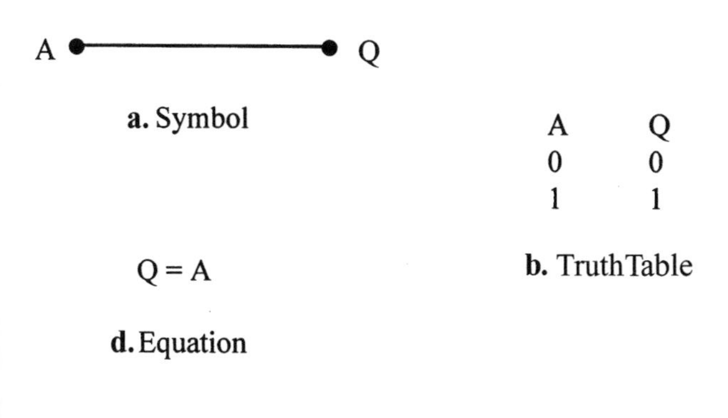

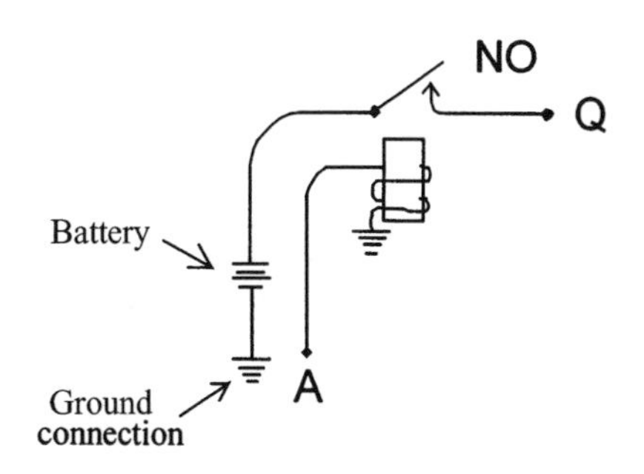

Fig. 13.2. Solid conductor as logic element.

Q

a. Symbol

A	Q
0	1
1	0

b. Truth Table

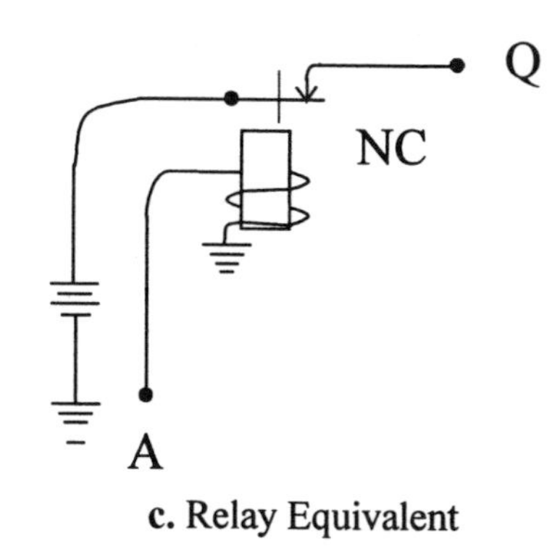

c. Relay Equivalent

$Q = \overline{A}$

d. Equation

Fig. 13.3. Inverter.

may be called an inverting buffer. Non-inverting buffers (triangles with no circles) are also available.

AND Gate

The AND gate (Figure 13.4a) is a logic element having two or more inputs and one output. The truth table (Figure 13.4b) shows that the output, Q, is high or true only when both A and B are true or high. Notice that since there are 2 inputs the truth table has $2^2 = 4$ lines. Its relay equivalent (Figure 13.4c) consists of two NO relays with their load poles in series. The descriptive equation (Figure 13.4d) is read "Q equals A and B". The A and B might also be shown with a dot between to indicate ANDed inputs: A·B. In symbolic notation ANDing in Boolean algebra corresponds to multiplication in conventional algebra.

AND gates, as well as the other gates to follow, are also available with more than two inputs.

A
B
Q

a. Symbol

A	B	Q
0	0	0
0	1	0
1	0	0
1	1	1

b. Truth Tab

NO NO Q
A B

c. Relay Equivalent

Q = AB

d. Equatio

Fig. 13.4. Two-input AND gate.

OR Gate

The OR gate (Figure 13.5a) also has two or more inputs and one output. Its truth table (Figure 13.5b) shows that the output is high whenever one or more inputs are high. The relay equivalent (Figure 13.5c) consists of two NO relays with the load poles in parallel. The OR function in Boolean algebra corresponds to the + sign in conventional algebra. Figure 13.5d, the equation for a 2-input OR gate, is read as "Q equals A or B."

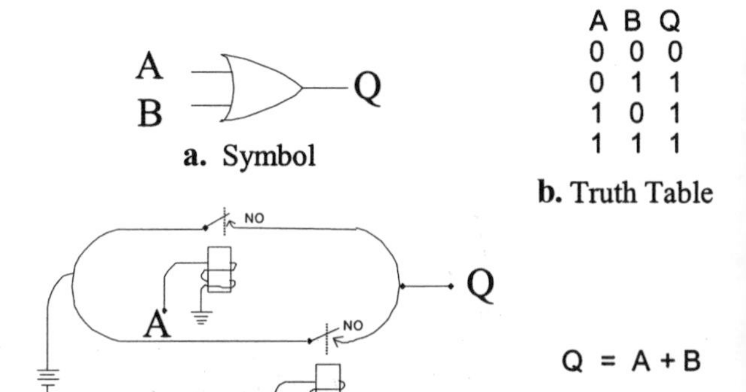

A	B	Q
0	0	0
0	1	1
1	0	1
1	1	1

b. Truth Table

Q = A + B

d. Equation

c. Relay Equivalent

Fig. 13.5. Two-input OR gate.

NAND and NOR Gates

The NAND and the NOR gates are variations of the AND or OR gates. Functionally and symbolically they consist of an inverter placed behind an AND or an OR gate, as shown in Figures 13.6 and 13.7.

In equation form, they consist of NOTed AND and OR functions, respectively.

XOR Gate

The XOR (exclusive OR) gate (Figure 13.8) differs from the OR gate in that its output is high only

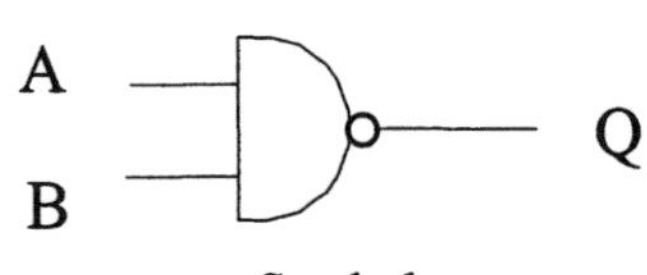

a. Symbol

A	B	Q
0	0	1
0	1	1
1	0	1
1	1	0

b. Truth Table

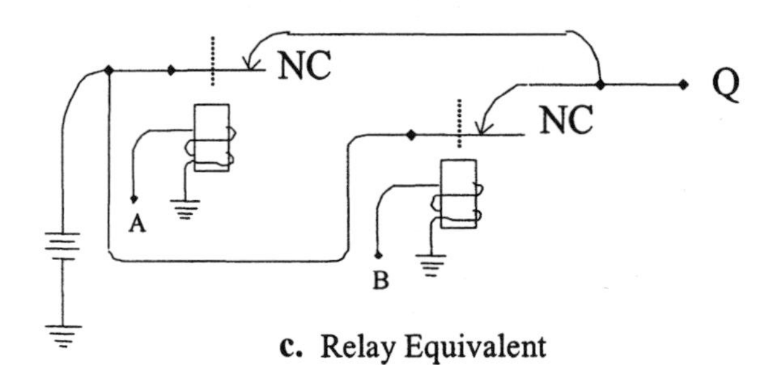

c. Relay Equivalent

$Q = \overline{AB}$

d. Equation

Fig. 13.6. Two-input NAND gate.

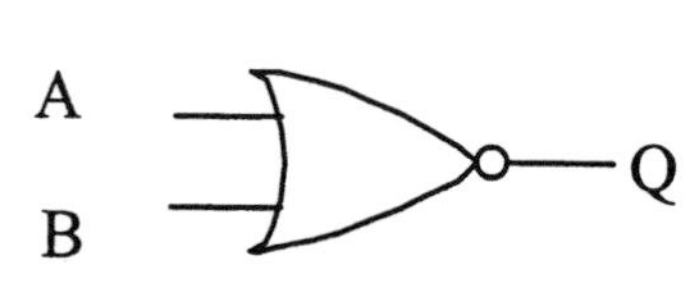

a. Symbol

A	B	Q
0	0	1
0	1	0
1	0	0
1	1	0

b. Truth Table

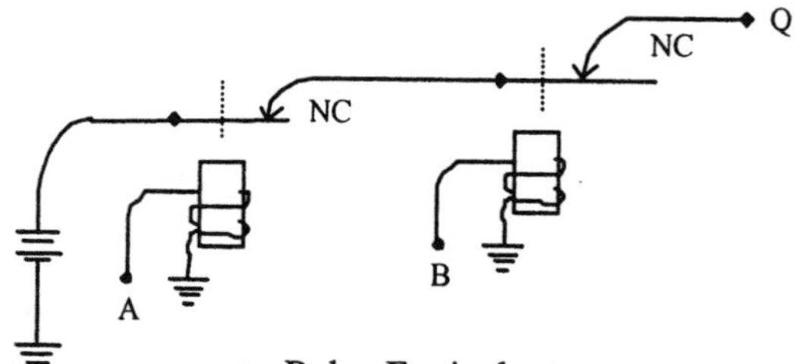

c. Relay Equivalent

$Q = \overline{A + B}$

d. Equation

Fig. 13.7. Two-input NOR gate.

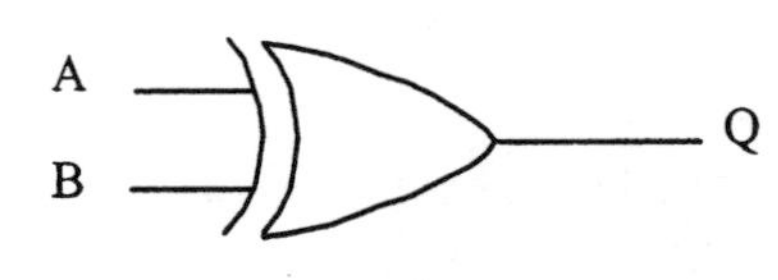

a. Symbol

A	B	Q
0	0	0
0	1	1
1	0	1
1	1	0

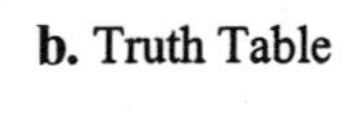

b. Truth Table

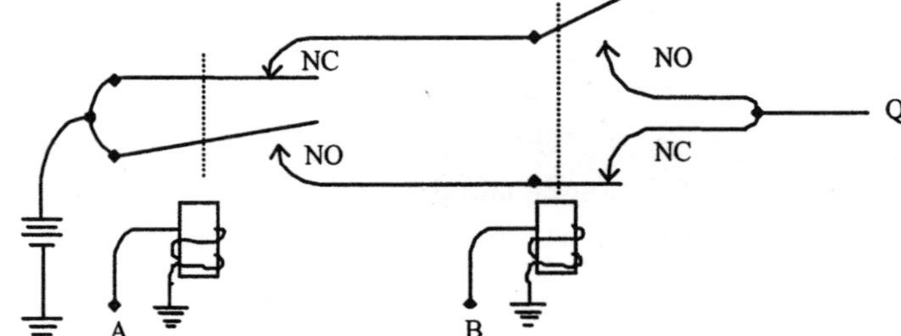

$Q = A \oplus B$ or $Q = A\overline{B} + \overline{A}B$

d. Equation

c. Relay Equivalent

Fig. 13.8. Two-input XOR gate.

when one of the inputs is high. The exclusive feature is specified in the function equation by the circle around the + sign (Figure 13.8d).

Combinations of Gates

Combinations of gates are often required in order to produce a desired output signal from a set of inputs. Example 13.4 illustrates writing a truth table and an equation for a 4-input combination circuit.

Example 13.4

Write the truth table and the equation for this 4-input, 1-output circuit:

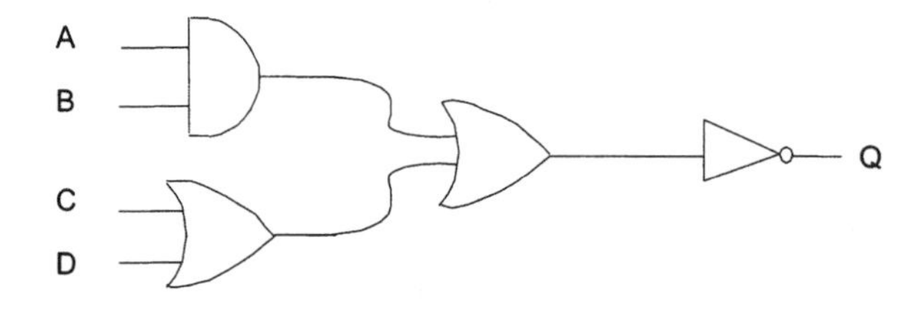

First, it's convenient to write in the terms describing each output:

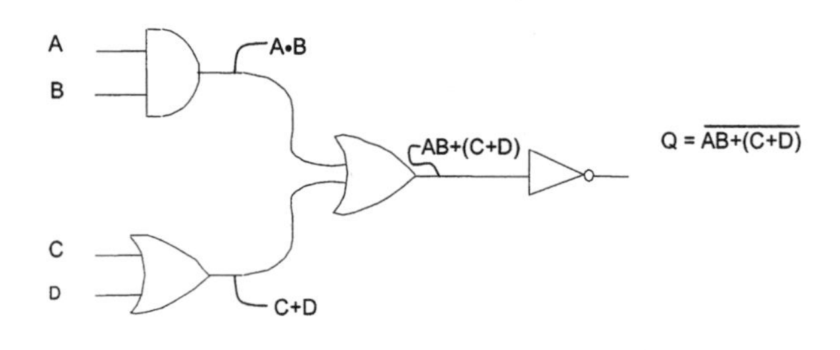

The Q term is then the output of the last gate.

With four inputs, the truth table contains 16 lines. First we write in the A, B, C, and D columns. Notice how the A, B, C, and D columns are written to include all possible combinations in an orderly way. As done in the example, it is often desirable to build the truth table in several intermediate steps with columns written for intermediate points in the circuit. In succession, columns for AB, C + D, AB + (C + D), and $\overline{AB + (C + D)}$ are written in. This step-by-step procedure decreases the possibility of an error in the final Q column.

A	B	C	D	AB	C + D	AB + (C + D)	$\overline{AB + (C + D)} = Q$
0	0	0	0	0	0	0	1
0	0	0	1	0	1	1	0
0	0	1	0	0	1	1	0
0	0	1	1	0	1	1	0
0	1	0	0	0	0	0	1
0	1	0	1	0	1	1	0
0	1	1	0	0	1	1	0
0	1	1	1	0	1	1	0
1	0	0	0	0	0	0	1
1	0	0	1	0	1	1	0
1	0	1	0	0	1	1	0
1	0	1	1	0	1	1	0
1	1	0	0	1	0	1	0
1	1	0	1	1	1	1	0
1	1	1	0	1	1	1	0
1	1	1	1	1	1	1	0

Example 13.5 illustrates going from a problem statement to a circuit.

Example 13.5

A drying fan is to turn on if 4 vent doors are all closed, or if an override switch S is closed. The fan comes on when Q is high. Inputs D_1, D_2, D_3, and D_4 are high when the corresponding vent door is closed. Draw a logic circuit to satisfy this logic using 2-input gates.

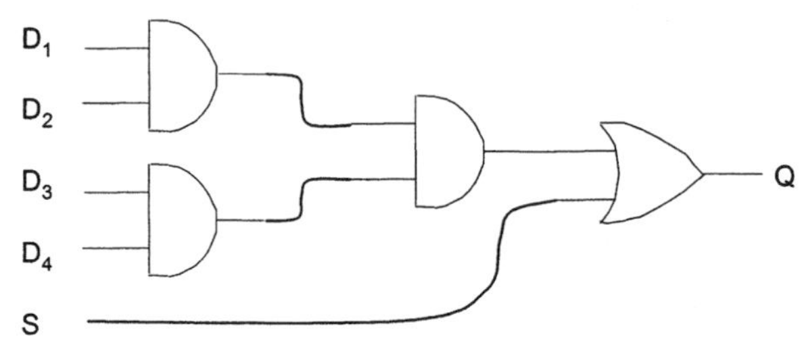

As drawn, Q is high if D_1 and D_2 and D_3 and D_4 are high, or if S is high. The reader can verify by means of a 32-line truth table that Q is 1 or high whenever S is high, or D_1 and D_2 and D_3 and D_4 are high.

Boolean Algebra

In dealing with combinations of logic gates, and their function equations, Boolean algebra theorems and laws are very useful for manipulation and simplification.

Theorem of idempotency: $A + A = A$ **(13.3)**

$AA = A$ **(13.4)**

These (and all) Boolean algebra theorems and laws can be verified by use of a truth table. This is illustrated in Example 13.6.

Example 13.6

Prove, by use of a truth table, that A + A = A. Construct a truth table for A + A = Q. Since there is one input, the truth table has two lines:

A	A	Q = A + A
0	0	0
1	1	1

Since the A + A column is identical to the A column, we have proven that A + A = A.

Theorems of union and intersection:

$$A + 0 = A \quad \textbf{(13.5)}$$

$$A + 1 = 1 \quad \textbf{(13.6)}$$

$$A1 = A \quad \textbf{(13.7)}$$

$$A0 = 0 \quad \textbf{(13.8)}$$

Theorem of complementation:

$$A + \overline{A} = 1 \quad \textbf{(13.9)}$$

$$A\overline{A} = 0 \quad \textbf{(13.10)}$$

Theorem of commutation:

$$AB = BA \quad \textbf{(13.11)}$$

$$A + B = B + A \quad \textbf{(13.12)}$$

Theorem of association:

$$A(BC) = AB(C) \quad \textbf{(13.13)}$$

$$A + (B + C) = (A + B) + C \quad \textbf{(13.14)}$$

Theorem of distribution:

$$AB + AC = A(B + C) \quad \textbf{(13.15)}$$

$$(A + B)(A + C) = A + BC \quad \textbf{(13.16)}$$

Theorem of redundancy or absorption (Holdsworth 1982):

$$A + AB = A \quad \textbf{(13.17)}$$

$$A(A + B) = A \quad \textbf{(13.18)}$$

DeMorgan's theorem:

$$\overline{AB} = \overline{A} + \overline{B} \quad \textbf{(13.19)}$$

$$\overline{A + B} = \overline{A}\,\overline{B} \quad \textbf{(13.20)}$$

DeMorgan's theorem is very useful in changing ANDed terms to ORed terms and vice versa. It can be conveniently applied as follows.

Change AB to its ORed form:

1. Make the change (retaining any NOT lines): $A + B$
2. Compliment (NOT) each term: $\overline{A} + \overline{B}$
3. Compliment (NOT) the whole expression: $\overline{\overline{A} + \overline{B}}$

Use of truth tables can prove that:

$$AB = \overline{\overline{A} + \overline{B}}$$

Since 2 NOTs cancel, we can get the form shown in Equation 13.19:

$$\overline{AB} = \overline{\overline{\overline{A} + \overline{B}}} = \overline{A} + \overline{B}$$

The same procedure works for an ORed expression.

Change A + B to its ANDed form:

1. Make the change: AB
2. Compliment (NOT) each term: $\overline{A}\,\overline{B}$
3. Compliment (NOT) the whole expression $\overline{\overline{A}\,\overline{B}}$

As can be verified by a truth table: $A + B = \overline{\overline{A}\,\overline{B}}$

Likewise: $\overline{A + B} = \overline{\overline{\overline{A}\,\overline{B}}} = \overline{A}\,\overline{B}$, as shown in Equation 13.20.

Integrated Circuit Packaging

The most common way to package ICs is in dual in-line packages (DIPs). DIPs consist of a package containing the circuitry, with 2 rows of pins along the sides (Figure 13.9). They come with 8 to 40 pins (half on each side).

The DIP can be connected to a circuit board by simply pushing it down over a set of matching pin receiver holes. The DIP can be removed and re-connected many

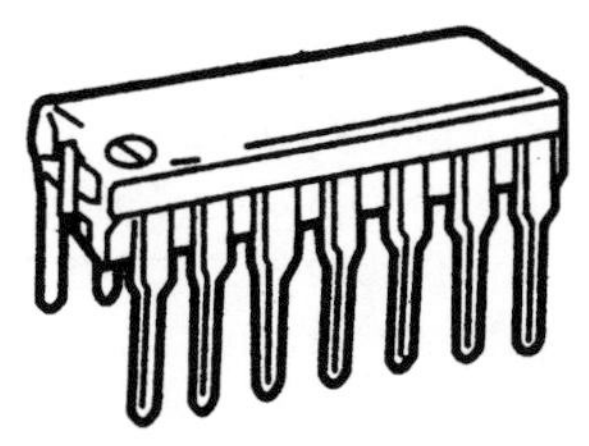

Fig. 13.9. A 14-pin dual in-line package (DIP).

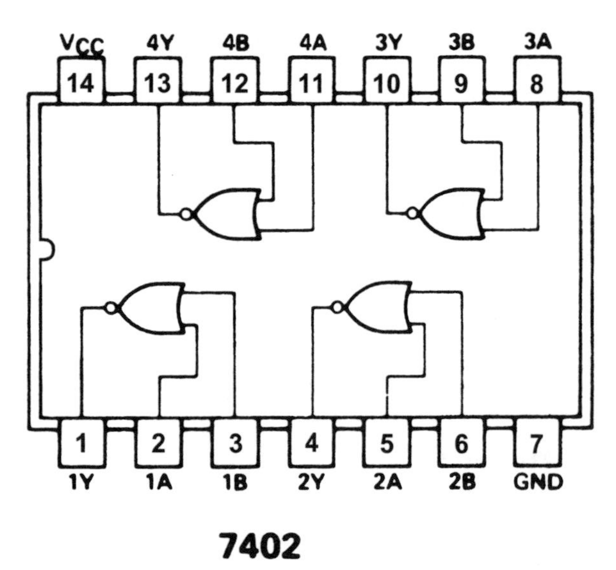

Fig. 13.10. Pin connections for a TTL 7402 (quad 2-input NOR) IC.

times. When installing DIPs, care must be taken to have all pins aligned with the pin receiver holes. During removal, care must be taken to avoid bending pins.

The connection pins of the DIP are internally wired to inputs and outputs of gates contained in the IC, and to power terminals. Figure 13.10 is the connection diagram (top view) for a TTL 7402 quad 2-input NOR. *Quad* means there are 4 gates in the IC. The number 7402 specifies the device (regardless of manufacturer) to be a quad 2-input NOR.

At all times that the IC is in operation, a DC power supply must furnish about +5-V DC to pin 14 (designated V_{CC}) and a ground connection to pin 7. A notch or a dot on top at one end of the IC is a reference mark which ensures correspondence with a pin connection diagram. This reference mark is always placed on the left, as shown in Figure 13.10.

Control Applications

When a control circuit is to be designed for a specific control application, these steps need to be completed:

1. A description of the control function to be carried out is written (in words).
2. This description is translated to a truth table.
3. An equation relating output to inputs is derived from the truth table and simplified.
4. A circuit is built to satisfy the equation.

An example will be used to illustrate this procedure.

Example 13.7

The pump for a sprinkler irrigation system is to be controlled by 4 inputs.

1. Description

   These are the 4 inputs:

   A. Electrical power demand sensor
      A = 1 during periods of low electrical demand
      A = 0 during periods of high electrical demand

   B. Soil moisture sensor I
      B = 1 when soil is *dry* or *very dry*
      B = 0 when soil is *wet*

   C. Soil moisture sensor II
      C = 1 when soil is *very dry*
      C = 0 when soil is *dry* or *wet*

   D. Wind sensor
      D = 1 when wind velocity is high
      D = 0 when wind velocity is low

   The pump is to operate (QP = 1) when the soil is wet if electrical demand is low and wind velocity is low. The pump is to operate during conditions of dry soil if demand is low and wind is low, or if one of these (but not both) is low and the other high. If the soil is very dry, the pump is to operate regardless of wind or electrical demand.

2. Truth table

   Below is a truth table version of the problem description.

A	B	C	D	QP	QA
0	0	0	0	0	0
0	0	0	1	0	0
0	0	1	0	Impossible	1
0	0	1	1	Impossible	1
0	1	0	0	1	0
0	1	0	1	0	0
0	1	1	0	1	0
0	1	1	1	1	0
1	0	0	0	1	0
1	0	0	1	0	0
1	0	1	0	Impossible	1
1	0	1	1	Impossible	1
1	1	0	0	1	0
1	1	0	1	1	0
1	1	1	0	1	0
1	1	1	1	1	0

The truth table is a concise statement of the description. Four conditions cannot occur if the soil sensors are properly working. If the sensors are working properly, it is not possible for sensor I to be indicating "wet", while sensor II is indicating "very dry". Therefore all four combinations having B = 0 and C = 1 indicate a sensor problem. We can form a QA column and turn on an alarm if this occurs.

3. Equation

Each line of the truth table for which QP = 1 is, in fact, an AND statement. For example, the bottom line can be read: "QP = 1 if A and B and C and D are 1 . . ." continuing this statement to the next to bottom line ". . . or if A and B and C are 1 and D is not 1." Saying D is not 1 is the equivalent of saying $\overline{D}$ is 1. Therefore, when there is a zero, we NOT the input and include it in the OR statement. Then, the second line up reads ". . . or if A and B and C and $\overline{D}$ are 1."

Following this procedure, we can write a complete equation of QP. For each line where Q = 1, we NOT the 0 inputs, then AND all inputs on that line, then OR all of the ANDed inputs:

$$QP = \overline{A}B\overline{C}\overline{D} + \overline{A}BC\overline{D} + \overline{A}BCD + A\overline{B}\overline{C}\overline{D} + AB\overline{C}\overline{D} + AB\overline{C}D + ABC\overline{D} + ABCD$$

The corresponding equation for QA is:

$$QA = \overline{A}\overline{B}C\overline{D} + \overline{A}\overline{B}CD + A\overline{B}C\overline{D} + A\overline{B}CD$$

4. The Circuit

A circuit of logic gates could be constructed which would satisfy the QP equation in step 3; however, the resulting circuit would be rather large and complex. A mapping procedure can be used to simplify the QP equation.

Karnaugh Mapping

Karnaugh (pronounced car-NO) maps provide a fast, systematic approach to Boolean algebra simplification for systems of up to six variables. The procedure will be illustrated using the problem of Example 13.7.

A general map structure for four variables is laid out as shown below.

	$\overline{A}\overline{B}$	$\overline{A}B$	AB	$A\overline{B}$
$\overline{C}\overline{D}$	$\overline{A}\overline{B}\overline{C}\overline{D}$	$\overline{A}B\overline{C}\overline{D}$	$AB\overline{C}\overline{D}$	$A\overline{B}\overline{C}\overline{D}$
$\overline{C}D$	$\overline{A}\overline{B}\overline{C}D$	$\overline{A}B\overline{C}D$	$AB\overline{C}\overline{D}$	$A\overline{B}\overline{C}D$
CD	$\overline{A}\overline{B}CD$	$\overline{A}BCD$	$ABCD$	$A\overline{B}CD$
$C\overline{D}$	$\overline{A}\overline{B}C\overline{D}$	$\overline{A}BC\overline{D}$	$ABC\overline{D}$	$A\overline{B}C\overline{D}$

Each cell in the box represents possible product term. For example, the box second from the left and third from the top is $\overline{A}BCD$. This represents the product of $\overline{A}B$ (the column heading) and CD (the row heading). The peculiar arrangement is purposely chosen so that the sum of product terms from any two adjacent boxes is independent of one of the variables. For example, consider adjacent cells ABCD and $A\overline{B}CD$:

$$ABCD + A\overline{B}CD = ACD(B + \overline{B}) = ACD$$

The cells are seen to be independent of B. In the up direction,

$$ABCD + AB\overline{C}D = ABD(C + \overline{C}) = ABD$$

This adjacency also extends beyond all map edges to the opposite edge:

$$A\overline{B}\overline{C}D + \overline{A}\overline{B}\overline{C}D = \overline{B}\overline{C}D(A + \overline{A}) = \overline{B}\overline{C}D$$

Going further, a subcube of four adjacent cells represents a sum of products which is independent of two variables.

$$\begin{aligned} & AB\overline{C}\overline{D} + A\overline{B}\overline{C}\overline{D} + AB\overline{C}D + A\overline{B}\overline{C}D \\ &= A\overline{C}\overline{D}(B + \overline{B}) + A\overline{C}D(B + \overline{B}) \\ &= A\overline{C}\overline{D} + A\overline{C}D \\ &= A\overline{C}(\overline{D} + D) = A\overline{C} \end{aligned}$$

You will also find that a 4 × 2 block or a 2 × 4 block is independent of three variables.

Using Karnaugh Mapping for Simplification

To use Karnaugh mapping to simplify a sum of products, proceed as follows.

1. Map the function by placing 1s in all the squares that correspond to terms in the expanded sums of squares:

$$QP = \overline{A}B\overline{C}\overline{D} + \overline{A}BC\overline{D} + \overline{A}BCD + A\overline{B}\overline{C}\overline{D} + AB\overline{C}\overline{D} + AB\overline{C}D + ABC\overline{D} + ABCD$$

	$\overline{A}\overline{B}$	$\overline{A}B$	AB	$A\overline{B}$
$\overline{C}\overline{D}$		1	1	1
$\overline{C}D$			1	
CD		1	1	
$C\overline{D}$		1	1	

2. Locate any isolated 1s and draw square boxes around them. (There are none.)
3. Draw the largest possible boxes around all the remaining adjacent 1s until every 1 is in at least 1 subcube. Overlapping is accepted. Boxes need to contain 2, 4, or 8 1s.

	$\bar{A}\bar{B}$	$\bar{A}B$	AB	$A\bar{B}$
$\bar{C}\bar{D}$		1	1	1
$\bar{C}D$			1	
CD		1	1	
$C\bar{D}$		1	1	

4. Identify the reduced terms represented by the subcubes and find their logical OR sum. Keep going until every 1 is represented in a reduced sum.

4-square subcube, lower half, center = BC

4-square subcube, upper center, lower center = $B\bar{D}$

4 in AB column = AB

2 in upper right corner = $A\bar{C}\bar{D}$

Every 1 has now been accounted for.

$$QP = BC + B\bar{D} + AB + A\overline{CD}$$
$$= B(C + \bar{D} + A) + A\bar{C}\bar{D}$$

It is left to the reader to verify that this QP equation also yields a truth table identical to the original problem statement of Example 13.7.

The circuit can be implemented as follows (Figure 13.7).

This simplified circuit requires:

2 inverters
3 OR gates
3 AND gates

The complete equation for QA from the truth table is:

$$QA = \bar{A}\bar{B}C\bar{D} + \bar{A}\bar{B}CD + A\bar{B}C\bar{D} + A\bar{B}CD$$

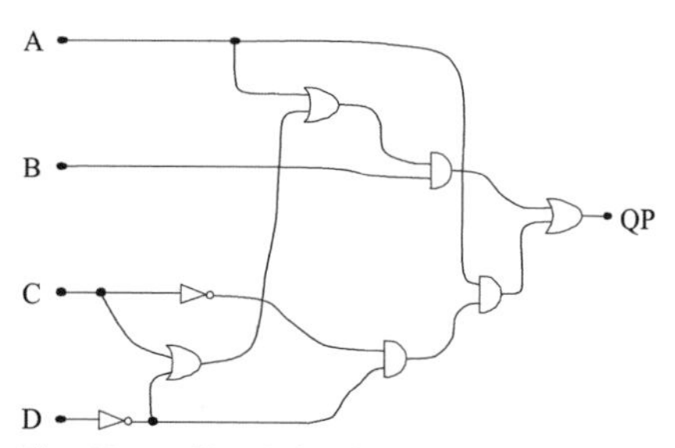

Fig. 13.11. Circuit for QP in Example 13.7.

This equation can be simplified to:

$$QA = \bar{B}C$$

It is again left to the reader to verify that the truth table for this equation corresponds exactly to the QA column in the original truth table. If QA is high, an alarm can be energized indicating a soil moisture sensor malfunction.

Applying TTL Circuits

A discussion is appropriate here about building the circuit of Example 13.7 for actual use.

TTL Input Requirements

TTL and its close relative, TTL/LS (low-power Schottky) chips have these general requirements (Mimms III 1986):

- V_{CC} (positive supply to pin 14) must never exceed + 5.25-V DC.
- Input voltages must never exceed V_{CC} nor fall below ground.
- Connect unused inputs to V_{CC}. They may otherwise pick up stray signals.
- Force outputs of unused gates to the high-state voltage to save power.

TTL inputs should not be left unconnected. When an input is left unconnected, the gate will probably behave as if this input is at logic 1; however, the designer should not rely on this (Mimms III 1982).

Switch Inputs

A problem with switch inputs is contact bounce. The phenomenon of contact bounce is illustrated in Figure 13.12. When the position of a simple input switch is changed, the input state may change several times before definitely assuming the other state. Some switches are worse than others. It has been observed that a switch bounced 563 times before settling down to a logic 0 state. The problem with bounce is that the TTL circuitry will react to each bounce as a change of state.

Input switches on designer boards are usually "debounced," that is, equipped with circuitry which assures only one change of state when the switch position is changed. Since there is no timing associated with the circuit of Example 13.7, bounce will be ignored.

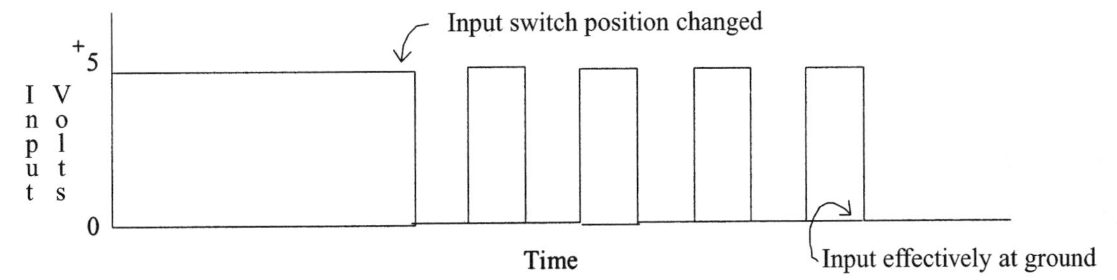

Fig. 13.12. Contact bounce.

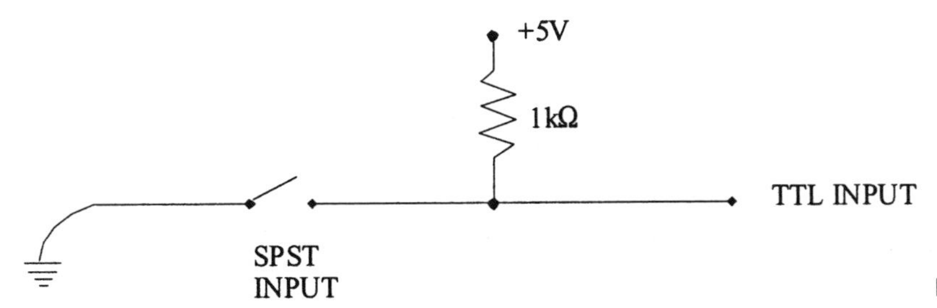

Fig. 13.13. Input circuit.

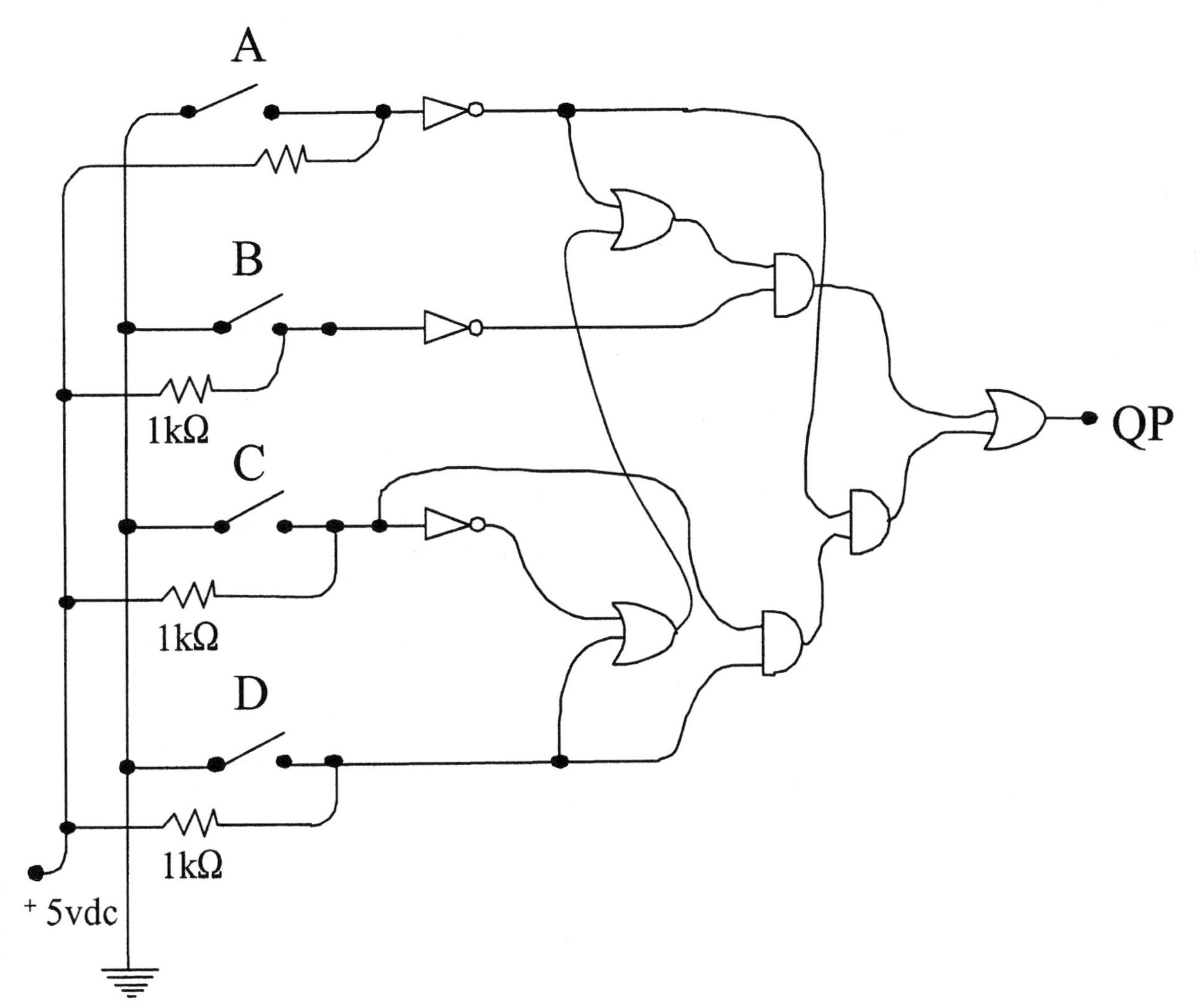

Fig. 13.14. Circuit for Example 13.7 with inverted inputs and additional inverters.

The inputs for Example 13.7 are likely to originate with SPST switches. Therefore, the input circuits need to be constructed in some way to avoid having unconnected inputs. One approach is shown in Figure 13.13. With the switch closed, voltage at the TTL input is very nearly 0 because of the voltage drop through the resistor. With the switch open, voltage at the TTL input is near 5 V because current through the resistor is very low. This switch circuit, though not debounced, allows an SPST switch to provide actual logic 1 and logic 0 inputs.

As the circuit of Figure 13.13 is constructed, an open switch gives a logic 1 and a closed switch gives a logic 0 input. Since this is the opposite of the assumptions at the start of Example 13.7, we must invert each input signal from what is shown in Figure 13.11. The modified circuit is shown in Figure 13.14. Making this change does not require any additional inverters.

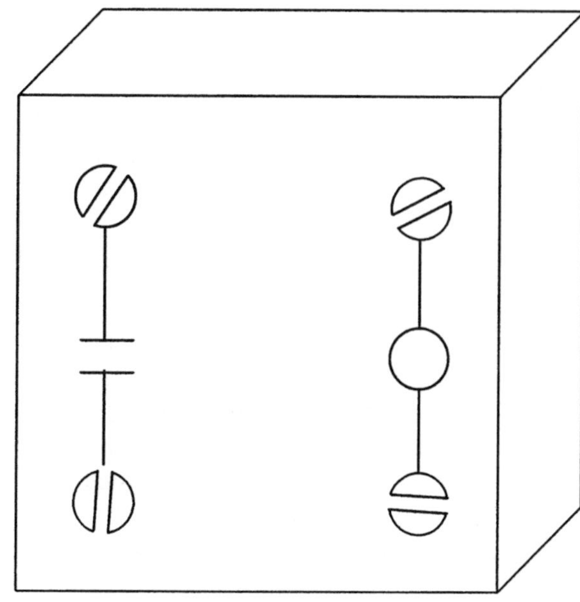

Fig.13.16. Solid-state relay.

Output Interface

If the logic system is to control an AC load like a motor or heater, a suitable system must be incorporated which will allow a change in the TTL output signal to turn on or to turn off an AC load. This is shown schematically in Figure 13.15.

Solid-State Relay

The electromechanical relay was discussed in chapter 8. The solid-state relay can perform the same function without any moving parts (Figure 13.16). This feature makes solid-state relays usable in dirty environments, and in locations where there cannot be any arcing due to contacts opening or closing.

For clarity, solid-state relays use coil and contact notation and terminology of electromechanical relays (Figure 13.16) although contacts and coils are not present. Switching is done by transistors, and other solid state devices. The link between coil and contacts can be optical, thus avoiding any electrical connection between the two. Resistance through open solid-state relay contacts is not infinite like that of electromechanical relays, and a small leakage current continues to flow while they are open.

Solid-state relays are available in a large range of contact currents and DC or AC coil voltages. They are applicable in interfacing TTL logic circuits with electrical loads because coils rated at TTL voltages operate at currents low enough to allow being driven by TTL circuit outputs.

Interface Circuit

Assuming that the load is a 1-hp AC motor, the interface could be designed as in Figure 13.17. The TTL output is fed to a solid-state relay.

The relay chosen must operate on the output of the TTL system, and must be able to switch a load consisting of the 120-V magnetic motor starter relay. One such device is an Elec-Trol part number SA 10014106, which closes for a coil voltage over 3 V DC and opens for a coil voltage under 1.0 V DC.

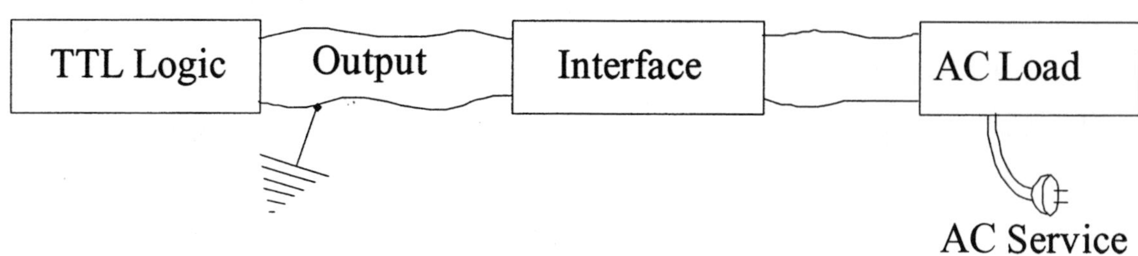

Fig. 13.15. Output interface.

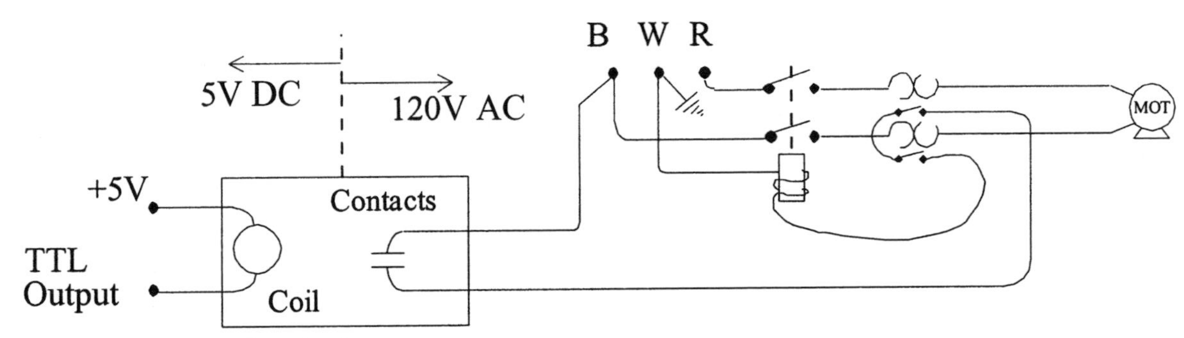

Fig. 13.17. Output interface circuit.

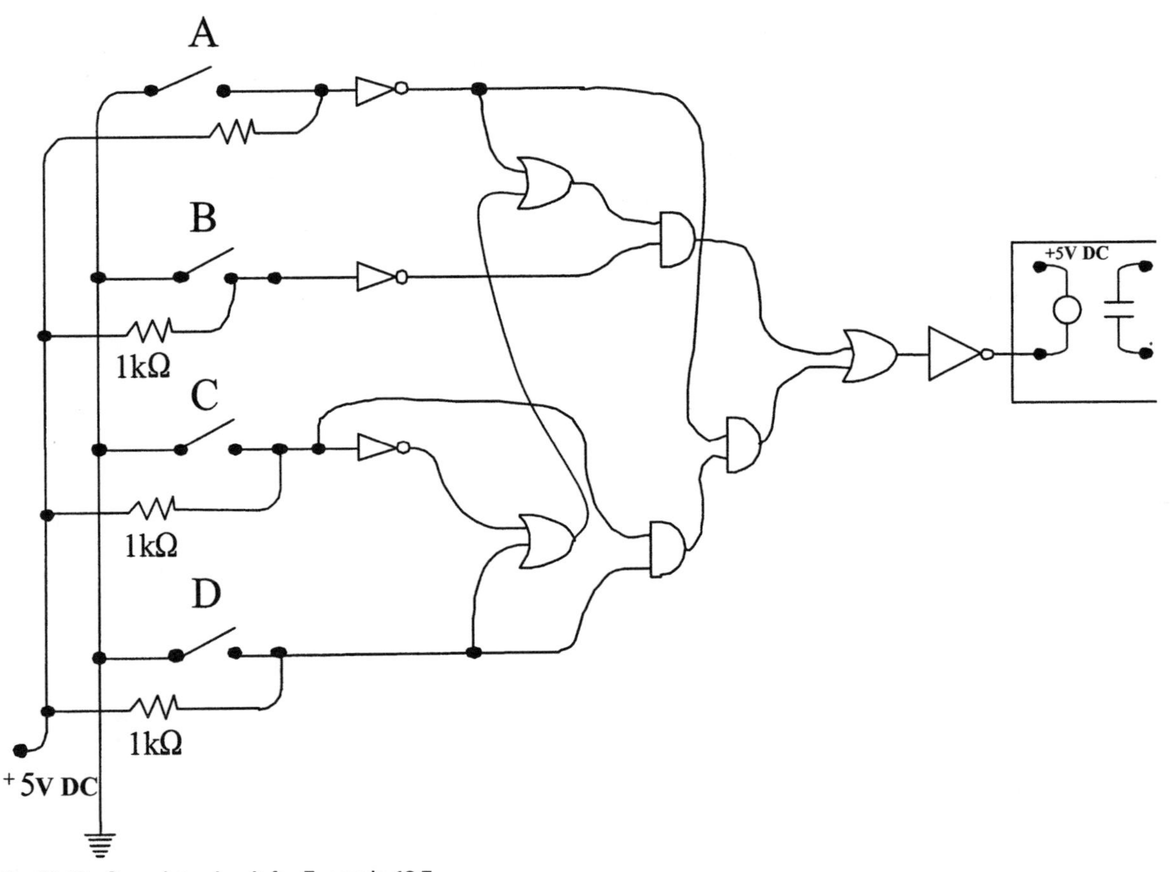

Fig.13.18. Complete circuit for Example 13.7.

Maximum current draw is 10 mA. The contacts can switch up to 6 A at up to 140 V (Elec-Trol, Inc., circa 1975).

The circuit for Example 13.7 is designed so that the motor is to come on when Q is at logic 1. Common TTL gates cannot supply the 10 mA required to turn on the solid-state relay with the current coming from a gate output at logic 1. Therefore, the output logic is reversed (by insertion of an inverter) and the solid-state relay coil is connected to a 5-V source. This is shown in Figure 13.18. Now when Q is at logic 0, the relay coil is energized and the load comes on.

References

Elec-Trol, Inc. Circa 1975. Catalog 113, Solid state relays. Elec-Trol, Inc. Sangus, CA.

Holdsworth, B. 1982. Digital logic design. Butterworth and Co., London.

Mimms III, F. M. 1986. Engineer's mini-notebook—Digital logic circuits. Tandy Corporation, Fort Worth, TX.

Mimms III, F. M. 1982. Archer engineer's notebook II. Radio Shack, Fort Worth, TX.

Problems

13.1. Convert these numbers to decimal:

(a) 100101101_2

(b) 1000_2

(c) 00001_2

(d) 11111111_2

13.2. What is the largest decimal number which can be defined by a 12-bit binary number?

13.3. Convert these numbers to binary:

(a) 254_{10}

(b) 1111_{10}

(c) 2047_{10}

13.4. Write the truth table and equation for a circuit consisting of 2 inverters in series.

13.5. An even number of inverters in series is functionally the same as

__________ __________.

13.6. An odd number of inverters in series is functionally the same as

__________ __________.

13.7. Draw the circuit for a gate having this truth table:

A	B	C	Q
0	0	0	0
0	0	1	0
0	1	0	0
0	1	1	0
1	0	0	0
1	0	1	0
1	1	0	0
1	1	1	1

13.8. Write the truth table for this gate:

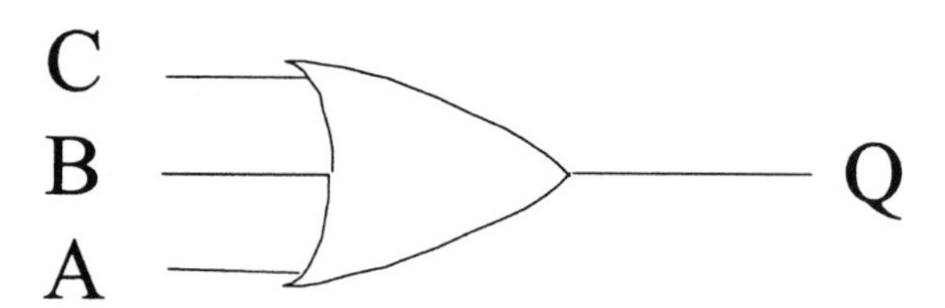

13.9. Can a NAND gate be built by placing inverters in each input line ahead of an AND gate instead of one inverter behind the AND gate? (Answer by constructing truth tables.) What kind of a gate does this form?

13.10. How many possible input combinations are there for a logic circuit having 6 inputs?

13.11. A vent fan is to be turned on if a manual switch is closed or if a thermostat senses that temperature has exceeded a present limit. The preceding sentence infers use of what logic gate?

13.12. Write the equation of Q along with a truth table for this circuit:

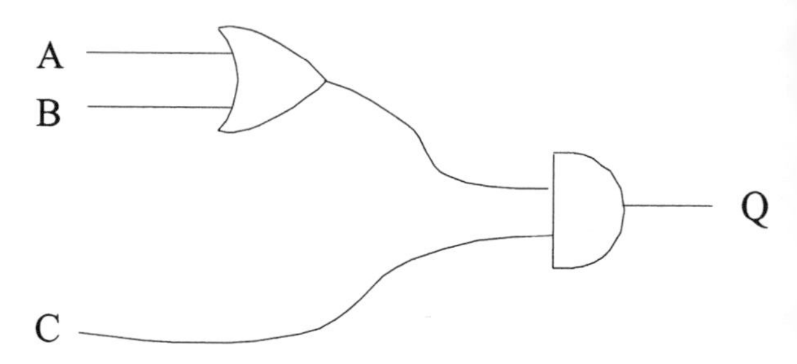

13.13. Draw a circuit and write a truth table for this equation:

$$Q = (\overline{A} + B)(\overline{\overline{C} + \overline{D}})$$

13.14. Draw a circuit which builds an inverter from:

(a) a 2-input NAND gate

(b) a 2-input NOR gate

13.15. Using inverters, ANDs, and ORs, draw a circuit with the same response as an XOR gate.

13.16. What logic function is being performed by this circuit?

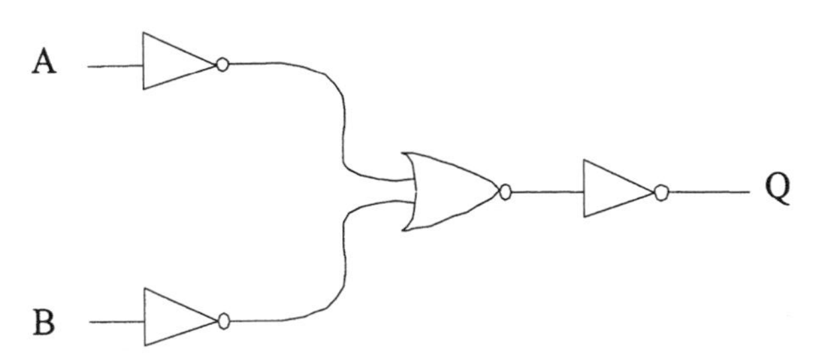

13.17. What logic function is being performed by this circuit?

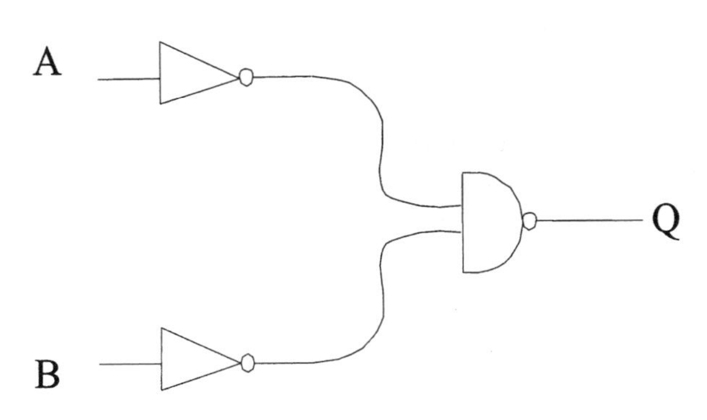

13.18. (a) Draw a circuit satisfying this equation (as is): $Q = A + B + \overline{B}$

(b) Simplify the equation.

(c) Verify the simplification with a truth table.

13.19. (a) Draw a circuit which satisfies this equation (as is): $Q = \overline{A}(\overline{B}(B + A) + C)$

(b) Simplify the equation.

(c) Draw a circuit which satisfies the simplified equation.

(d) Verify by a truth table that the simplified equation is functionally the same as the original.

13.20. Simplify this equation:

$$Q = ABC + AB\overline{C} + \overline{A}BC + \overline{A}B\overline{C}$$

13.21. (a) Write the equation for this circuit.

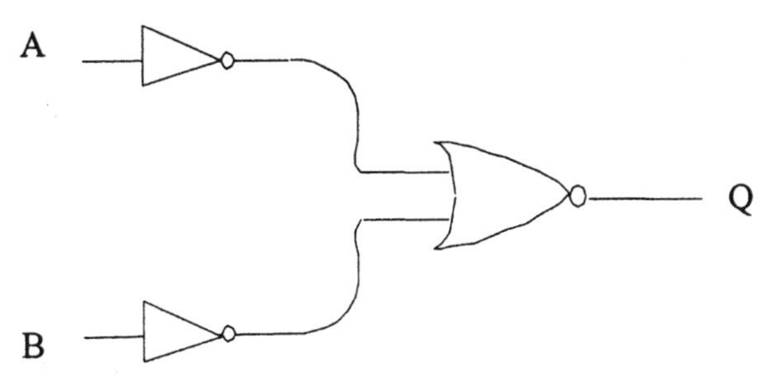

(b) Change to an AND form.

13.22. Simplify this logic equation:

$$Q = \overline{(A + \overline{B} + \overline{C})}\,\overline{(A + \overline{C})}$$

13.23. Simplify this equation:

$$Q = \overline{(A\overline{B}\overline{C} + BC)}\,(A\overline{B})$$

13.24. $Q = \overline{A}\overline{B}C\overline{D} + \overline{C}\overline{D}\overline{A}\overline{B} + \overline{C}\overline{B}\overline{D}A + A\overline{B}\overline{C}D + A\overline{B}CD + \overline{B}AC\overline{D} + CDAB$

(a) Simplify by Karnaugh mapping.

(b) Draw a circuit to satisfy the simplified equation, using inverters and 2-input ANDs and ORs.

(c) Construct the truth table.

13.25. Design a control system to control the 230-V, single-phase fan on a natural-air drying system. The control system inputs include:

A Ambient temperature (thermostat)
$1 = \text{temp} > 32°F$
$0 = \text{temp} \leq 32°F$

B Ambient relative humidity (humidistat)
$1 = RH < 60\%$
$0 = RH \geq 60\%$

C Grain temperature (differential thermostat)
$1 = $ temp 10°F or more over ambient
$0 = $ temp $< 10°F$ over ambient

D Ambient relative humidity (humidistat)
$1 = RH \geq 90\%$
$0 = RH < 90\%$

The fan is to operate (QF = 1) if ambient RH < 60% and ambient temp ≤ 32°F regardless of grain temperature. If ambient relative humidity exceeds 90%, the fan is to run whenever grain temperature exceeds 10°F above ambient. If ambient relative humidity is at least 60% but less than 90%, the fan is to run when ambient temperature is over 32°F regardless of grain temperature. A warning lamp is to operate (QW = 1) if a humidistat malfunction causes one humidistat to indicate humidity is less than 60% while the other humidistat indicates humidity is over 90%.

Procedure:

1. Write the truth table including a column for QF and a column for QW.
2. Write equations for QF and QW, and simplify them.
3. Write the truth table for your simplified equations.
4. Design a circuit, having QF and QW as outputs, which satisfies the simplified equations. Use 2-input gates.
5. Verify your circuits with a truth table.
6. Draw a ladder diagram in which 2 light bulbs in series simulate the 230-V fan. The bulbs are controlled by a magnetic motor starter which is controlled by a solid-state relay driven by your logic circuit. The warning condition (QW = 1) is indicated by a 120-V light bulb being turned on. The bulb is controlled by a second solid-state relay which is driven by your circuit.

Chapter 14

Stray Voltage

Stray voltage can be a major problem in dairy and other livestock buildings. Decreases in milk production and reduced weight gain can be attributed to the effects of current flow through the animals caused by stray voltage. This low voltage is known by terms such as "tingle voltage," "transient voltage," and more technically as *neutral-to-earth* (N–E) voltage.

The common terms used to describe this condition imply mysterious appearances of voltage that can be here today, gone tomorrow. It is true that the problem can vary from day to day because an N–E voltage can be caused by many factors such as frayed insulation, corrosion, heavily loaded power lines, undersize neutral wires, and even loads operating on a neighboring farm. A neutral-to-earth voltage can exist even with a properly wired system with no ground faults. In general, a significant N–E voltage is possible whenever there is current flowing in the grounded neutral system.

Figure 14.1 illustrates how an animal is subjected to stray voltage in a typical barn. The cow standing on wet concrete is in contact with the earth (true ground), and also with the grounded neutral network that includes barn metal and water pipes. When there is current in the neutral conductor due to unbalanced loading of the service entrance panel (or other reasons), there is a voltage between the neutral buss in the panel and the earth. Voltage as low as 0.5 V can be sensed by animals, and potentially cause problems with production (MWPS 1992).

What Is "Stray Voltage"?

As you are reading this text, there are two general principles of electric distribution to keep in mind:

1. Electric current leaving a power source, such as a generator or transformer, must return to the power source.
2. Electric current does *not* take the "path of least resistance" but takes *all* available paths, regardless of the resistance. It is true that much of the current returns to the source by low-resistance paths. However, some current flows through all available paths such as the earth, water pipes, buildings, or animals as the current returns to the source. The magnitude of the current depends on the resistance of the pathway.

Stray voltage is a small difference in electrical potential that exists between two points that an animal can touch, sometimes referred to as "cow contact points." When an animal comes in contact with these two points, current flows through its body as defined by Ohm's Law. If the voltage is large enough, the resultant current flow can cause stress for the animal.

Anyone who has received an electrical shock has experienced firsthand the result of contact between two points of different potential. These shocks can be disturbing

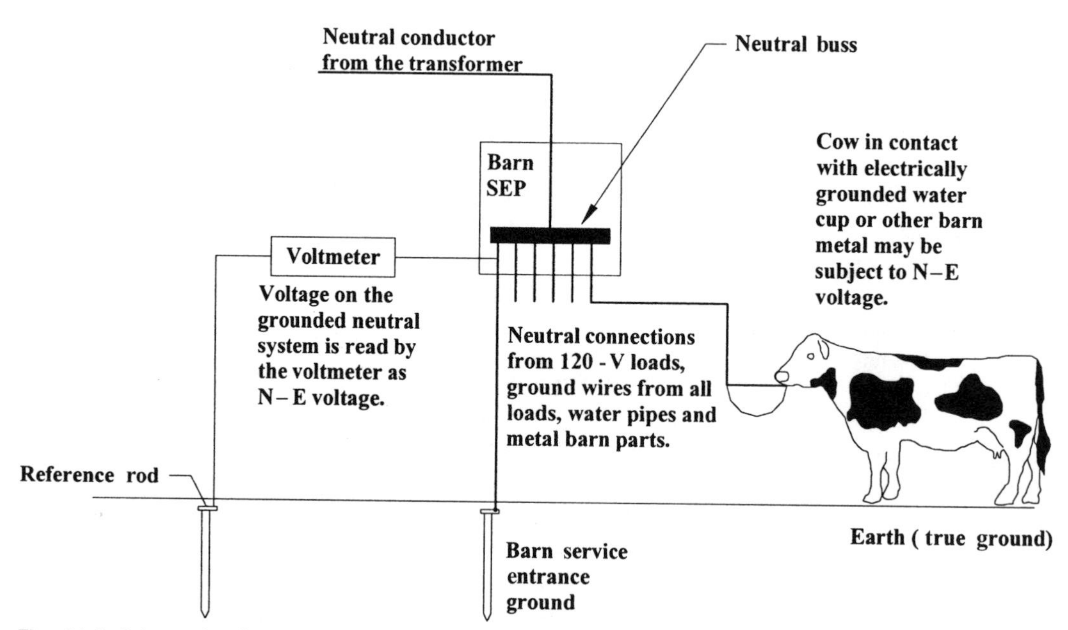

Fig. 14.1. Schematic of a cow subjected to a neutral to earth (N–E) voltage.

or fatal because the potential difference (voltage) is great enough to cause a relatively large current flow through the body (see Table 3.2). Current flow due to stray voltage is the same principle, but the potential difference, and resulting current, is usually much smaller.

A source of confusion is how a voltage potential can exist between the neutral buss and the earth (Figure 14.1) since there is a direct connection between the two through the ground rod, and much of the barn metal is in direct contact with the earth. Current flow in the neutral conductor, ground rods, and barn metal occurs because there is a potential difference between two points. Relatively simple circuits representing the grounded neutral system and application of Ohm's Law can explain the N–E voltage.

The neutral and grounding network can be a complicated electrical circuit. Every part of the system, including the conductors, the connections, the earth, and contact between grounding rods and the earth has some electrical resistance. To better understand the role of these resistances in a circuit, consider the simplified schematic of a single 120-V load (Figure 14.2). The ground rods (R_{Ground}) are installed in accordance with the National Electrical Code (MWPS 1992).

In chapter 5, this circuit (without the ground rods) was presented as three loads in series where voltage drops across each resistance and current could be easily calculated using Ohm's Law (see Figure 5.2). Although addition of the two ground rods does not greatly affect the current and voltage drops in this circuit, the presence of ground rods can play a major role in stray voltage. The following example illustrates the effect of these ground rods in the circuit.

Example 14.1

This example examines the effect of ground rods in a simplified model of 120/240-V single-phase electrical service typical of farm buildings. In part A, electrical parameters are calculated as if ground rods are not present. In part B, the ground rods are incorporated into the analysis for comparison purposes.

A) Consider the circuit illustrated in Figure 14.2 with the ground rods removed from the circuit. Calculate the total resistance, total current, and voltage drops across each resistance.

Assume:

$E_T = 120$ V (voltage available at the transformer)

$R_{Hot} = 0.024\ \Omega$ (equivalent to 300 ft of AWG-0000 aluminum conductor)

$R_{Neutral} = 0.024\ \Omega$ (equivalent to 300 ft of AWG-0000 aluminum conductor)

$R_{Load} = 10\ \Omega$

Solution:

$R_T = 0.024\ \Omega + 10\ \Omega + 0.024\ \Omega = 10.048\ \Omega$

$I_T = (120\ V)/10.048\ \Omega = 11.94$ A (current flowing in the circuit)

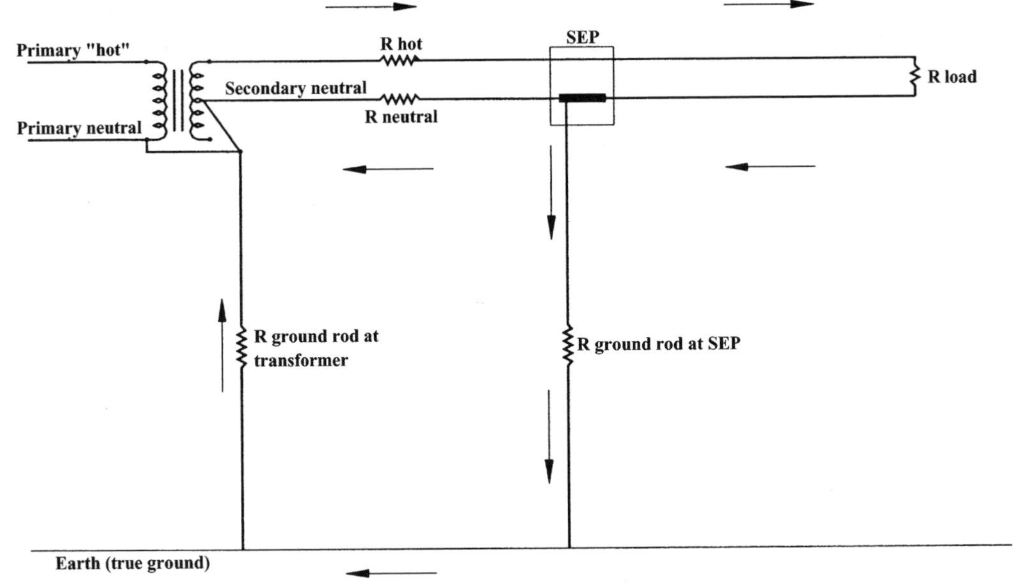

Fig. 14.2. Schematic of the single 120-V load circuit used in Example 14.1.

$E_{Load} = (11.94\ A)(10\ \Omega) = 119.4\ V$ (voltage across the load)

$E_{Hot} = E_{Neutral} = (11.94\ A)(0.024\ \Omega)$
$= 0.287\ V$ (voltage drop across both the hot and neutral conductors)

B) Now consider the same circuit with the ground rods included in the circuit. Calculate the total current and voltage drops across each resistance. Compare results with part A.

Assume the resistance of a single ground rod:
$R_{Ground} = 5\ \Omega$ (discussed later in the text)

Solution:

Total resistance is a combination series/parallel circuit with the ground rods in parallel with the neutral conductor. First, calculate equivalent resistance (R_{eq}) of the ground rods (series) in parallel with the neutral conductor.

$1/R_{eq} = 1/(5\ \Omega + 5\ \Omega) + 1/0.024\ \Omega$
$R_{eq} \approx 0.024\ \Omega$

Calculate total resistance of the circuit:

$R_T = 0.024\ \Omega + 10\ \Omega + 0.024\ \Omega = 10.048\ \Omega$

Note the addition of the ground rods does not appreciably change the total resistance of the circuit. Therefore, voltage drops throughout the circuit are essentially the same as in part A. However, since the ground rods are in parallel with the neutral conductor, there are two current paths between the neutral buss and the center tap of the transformer:

$I_{Neutral} = 0.287\ V / 0.024\ \Omega = 11.94\ A$
$I_{Ground} = 0.287\ V / 10\ \Omega = 0.03\ A$

Voltage drops and total current in the circuit are essentially the same with or without the ground rods present in the circuit. Errors in total current are due to roundoff errors.

The key point of this example is that a small voltage drop does exist between the transformer and the neutral buss in the service entrance panel and between the neutral buss and the earth whenever current is flowing in the neutral conductor. The presence of this voltage drop is the basis behind understanding stray voltage.

The relationship between voltage drop across the neutral conductor and N–E voltage can be understood by studying Figures 14.1 and 14.2. In Figure 14.1 the N–E voltage is measured directly by measuring the voltage between the neutral buss and a reference ground rod that is at the potential of the earth. In Figure 14.2, the N–E voltage can be calculated as the voltage drop across R_{Ground} connected to the neutral buss. Recall that R_{Ground} is the total resistance of the grounding conductor, grounding rod, connections, and the ground rod/soil contact resistance.

The N–E voltage in Example 14.1 is approximately equal to 0.14 V (half of $E_{Neutral}$ because equal R_{Ground} was assumed). Since this voltage is less than 0.5 V, this circuit as illustrated should not pose a significant stray voltage hazard. However, note the circuit in this example has stray voltage, although the magnitude is small and the circuit is constructed according to the National Electrical Code. In other words, if electrical service is installed similar to Figure 14.2, an N–E voltage will exist if there is current in the neutral conductor.

The magnitude of N–E voltage will depend on resistance of the wires, resistance of the ground rods, and current in the circuit. It is left to the reader to verify that doubling the current in Example 14.1 (B) would result in an N–E voltage of 0.28 V and that doubling (or tripling) the resistance of the ground rods would have very little effect on the N–E voltage.

But consider a slight change in the resistance of the neutral conductor. The resistance of the conductor is a function of wire size and type as discussed in chapter 5. This resistance does not consider resistance of connections, splices, or corrosion in the system. Example 14.2 illustrates the effect of increasing neutral conductor resistance on N–E voltage.

Example 14.2

Consider the circuit used in Example 14.1 (B). For illustration purposes, assume the resistance of the neutral conductor is increased by 0.5 Ω. This increase could easily be caused by a loose connection in the service entrance panel, or some corrosion on the connection. Therefore, in Figure 14.2, $R_{Neutral} = 0.524\ \Omega$ and all other resistance is as illustrated. Calculate the N–E voltage for the circuit.

Solution:

Calculate the equivalent resistance of the neutral conductor and ground rods.

$1/R_{eq} = 1/0.524 + 1/(5 + 5)$

$R_{eq} = 0.498\ \Omega$

$R_T = 10.52\ \Omega$

$I_T = 120/10.52 = 11.4$ A

$E_{Neutral} = (11.4\ A)(0.498\ \Omega) = 5.7$ V

$E_{N\text{–}E} = 0.5(5.7\ V) = 2.8$ V (because resistance of ground rods assumed equal)

The total current in this example was reduced by approximately 0.6 A when compared to Example 14.1 (B). More importantly, the N–E voltage was increased to nearly 3 V, which is great enough to cause stress for the animal.

Ground Rod Resistance

The resistance of the ground rods was assumed to be 5 Ω in Examples 14.1 and 14.2. The resistance of a good ground installation is often 3 Ω or less (MWPS, 1992). If the ground rod installation is greater than 25 Ω, more rods should be installed as per the National Electrical Code (NEC 250-81) to reduce the effective resistance of the ground rod installation. This maximum resistance allowed by the NEC could conceivably result in a total ground resistance of 50 Ω in Example 14.1.

Soderholm (1982) states the resistance of individual ground rods may range from a few ohms to several hundred ohms, depending on soil and moisture conditions. Reines and Cook (1999) found an average resistance of 78 Ω, determined from measuring over 3,000 ground rods installed along distribution lines in Wisconsin. These studies show that actual resistance of individual ground rods could be much higher than anticipated.

Regardless of the actual resistance of individual ground rods, Example 14.1 shows the voltage drop across the neutral conductor is only slightly affected because of low resistance of the neutral conductor. However, the N–E voltage is dependent on the actual resistance of the ground rods.

Grounded Neutral Network

In Examples 14.1 and 14.2, the grounded neutral network was simplified by considering only the neutral conductor in parallel with the ground rods. In reality there are many connections to the neutral buss such as water pipes, barn metal, and all 120-V neutral and ground connections.

A schematic of a grounded neutral network in the form of a ladder diagram is illustrated in Figure 14.3. The vertical lines represent potential of the neutral center tap at the transformer, potential of the earth, and potential of the neutral buss. The potential of the earth can be considered a common electrical point, regardless of the physical location of the ground rods or barn metal. When current flows from the transformer to the neutral buss, some current flows down the ground rod at the transformer to earth, splits to each of the ground connections, and flows up to the neutral buss. This illustrates the principle that current flows through all available paths.

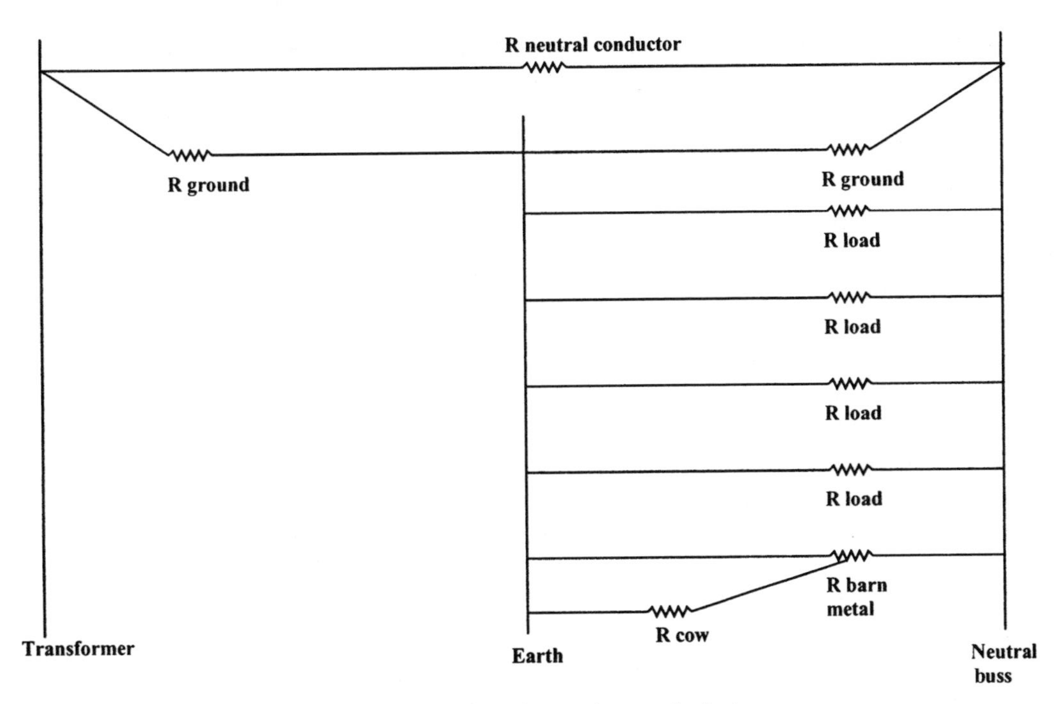

Fig. 14.3. Ladder diagram illustrating the grounded neutral network of a barn.

Figure 14.3 also illustrates how an animal can be exposed to stray voltage. The connection of water pipes to the neutral buss places a point on the water pipe at or near the potential of the neutral buss while another point somewhere in the water pipe is at the potential of the earth, thus current will flow through the water pipe. The water cup is electrically located somewhere between the potential of the neutral buss and the earth. When an animal drinks from the cup, there is a potential difference between the mouth and hooves, and current will flow through the animal.

Causes of Stray Voltage

Although there are many causes of an N–E voltage, keep in mind that the voltage exists if and only if there is current in the secondary neutral system. Also keep in mind that the resistance of each component of the system determines the magnitude of the problem.

Figure 14.4 illustrates possible current paths of the single 120-V load connected to a 120/240-V single-phase service considered in Example 14.1. In previous chapters, current leaving the transformer was

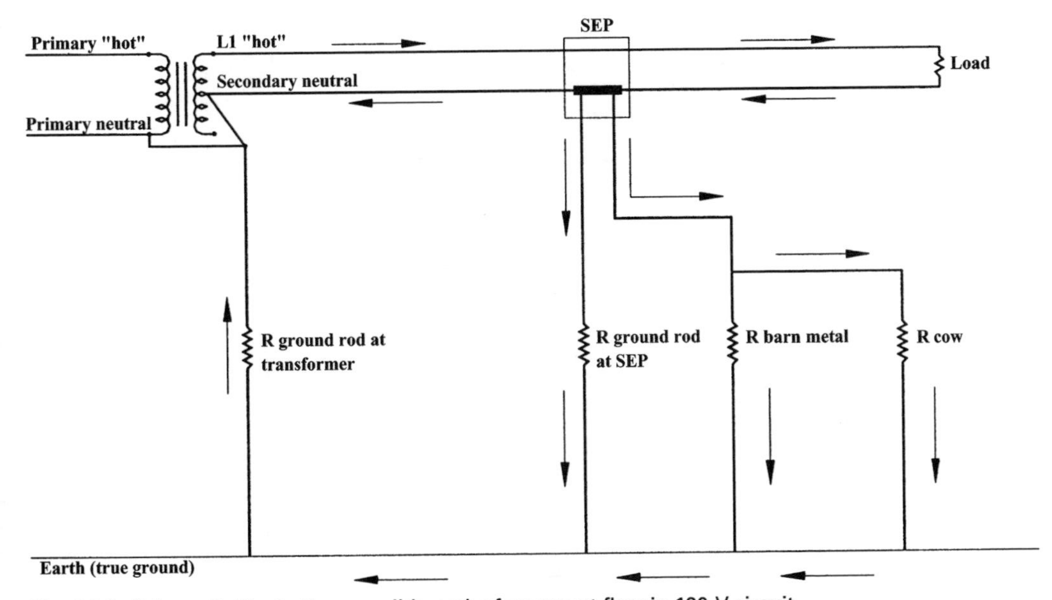

Fig. 14.4. Schematic illustrating possible paths for current flow in 120-V circuit.

assumed to move in a loop through the load and back to the center tap of the transformer and the ground at the transformer was not considered to be part of the circuit (see Figure 1.25). However, note that the ground rod at the transformer, the earth, the ground rod at the barn SEP, the metal barn parts, and the animal simply provide alternate pathways for the current back to the center tap of the transformer.

The following sections discuss possible sources of N–E voltage, but realize that in the real world any combination of these conditions can contribute to the problem (Cloud et al. 1987, MWPS 1992). The source of an N–E voltage can be determined and corrected if there is a clear understanding of how the current flows in both the primary and secondary distribution systems.

Primary Neutral Current External to the Farm

In order to understand how loads operating off the farm can contribute to stray voltage on the farm, it is necessary to have a basic understanding of the distribution (primary) circuit. Similar to 120/240-V circuits, a simplified primary circuit can be considered as a loop between the power source (generator or regional facility called a substation) at one end of the loop and the transformer located near the farmstead at the other end of the loop. Therefore current flowing from the substation "hot" must return to the substation through the primary neutral or through the earth via ground rods installed at the transformer and at points along the distribution line (Figure 14.5).

Figure 14.5 illustrates a single farmstead connected to the primary system, whereas it is likely that several farms will be connected to the primary system. Current flowing in the primary neutral conductor can cause stray voltage on any farm connected to the primary system.

The primary neutral conductor and the secondary neutral conductor are bonded together at the transformer serving a farm (Figure 14.6). A primary–secondary neutral bond is established to protect the customer in case there is damage to the power lines or transformer that results in a conducting path from the primary hot to some point on the secondary circuit (MWPS 1992).

Because there is a direct connection between the primary and secondary neutral wires at the transformer, the secondary neutral system on the farm is an alternate pathway for primary neutral current to earth. As loads are operated on neighboring farms, some primary neutral current will flow through the primary–secondary neutral bond and through the farm secondary neutral system. In this case, an N–E voltage can exist even with the main farm disconnect open.

Soderholm (1982) illustrated a significant stray voltage on the farm is possible due to primary neutral current. The primary neutral system was considered to be a single resistance (R_p) and the farm neutral system was considered to be a single resistance (R_f) in parallel with the primary system (Figure 14.7). By applying Ohm's Law between the neutral connection on the transformer and the earth, it was demonstrated that a significant N–E voltage could occur under normal operating conditions of a rural electric distribution system.

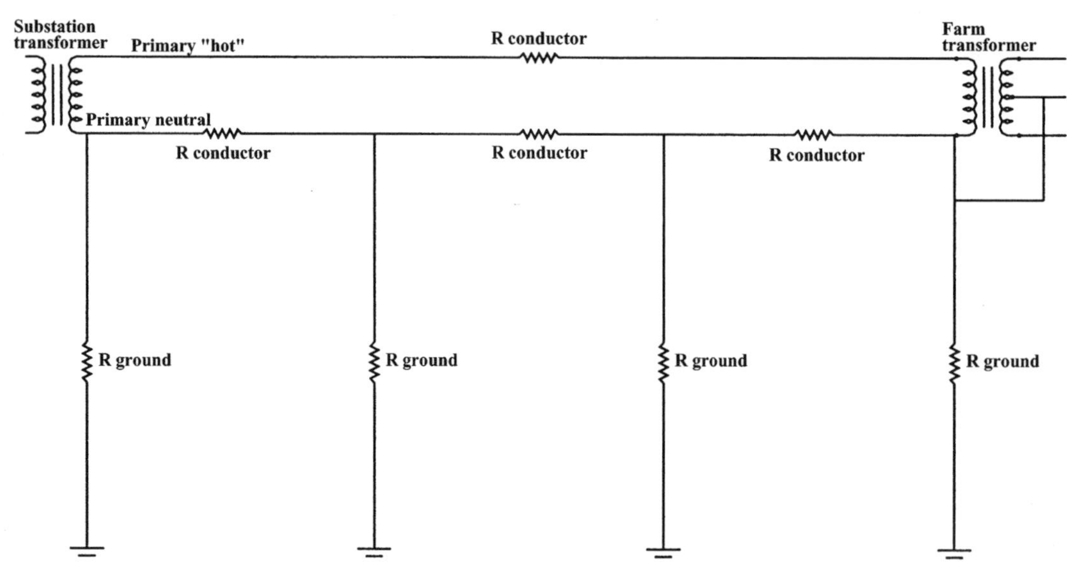

Fig. 14.5. Illustration of primary electrical distribution system.

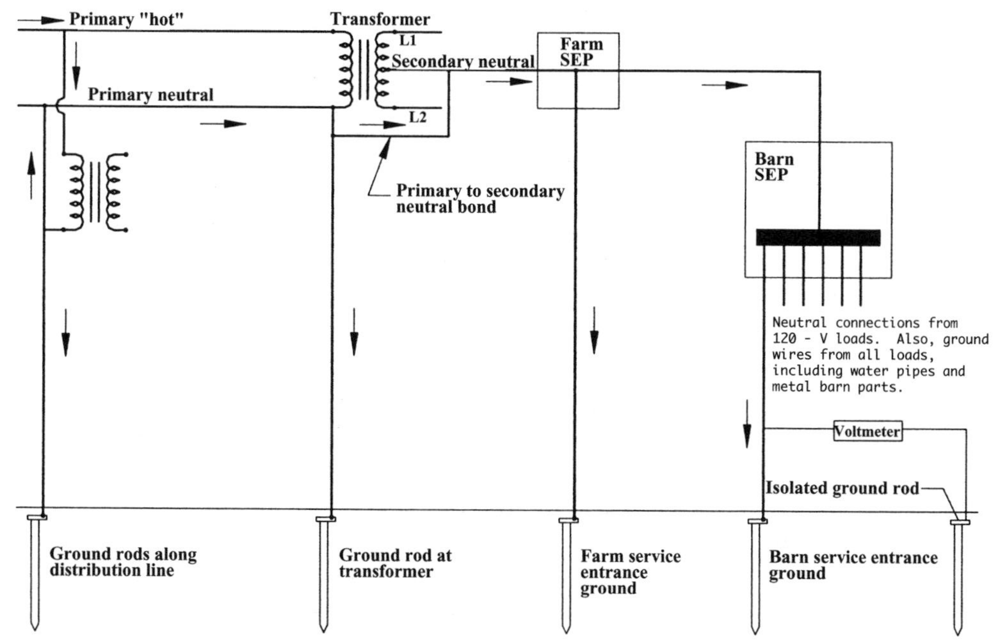

Fig. 14.6. Neutral-to-earth voltage caused by primary neutral current external to the farm.

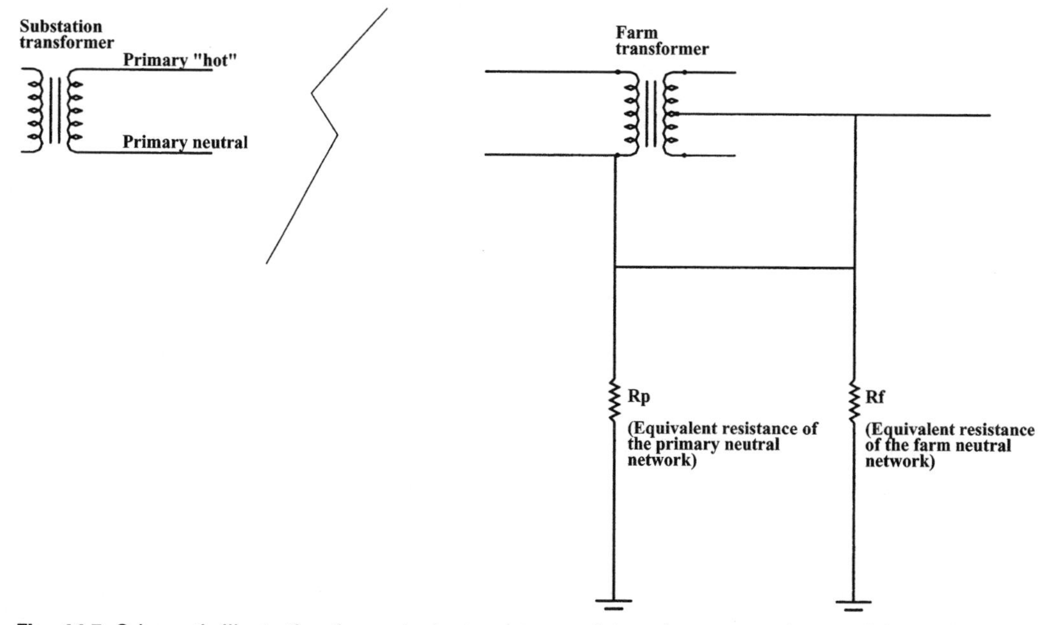

Fig. 14.7. Schematic illustrating the equivalent resistance of the primary neutral network in parallel with the equivalent resistance of the farm neutral network.

In a report to the Minnesota Public Utilities Commission (1998), it was reported that primary neutral impedance (R_p in Figure 14.7) ranged from 0.2 Ω to 1.0 Ω with a mean resistance of 0.48 Ω and farm grounding impedance (R_f) ranged from 0.2 Ω to 2.6 Ω with a mean resistance of 1.1 Ω. These values were determined from a comprehensive analysis of 19 dairy farms in Minnesota and Wisconsin.

Example 14.3 illustrates how typical resistance of the primary neutral system and resistance of the farm neutral system can result in a significant N–E voltage under normal operating conditions. The example is simplified by assuming purely resistive loads.

Example 14.3

Calculate the voltage between the neutral connection on the transformer and the earth for the following three conditions. Note this voltage is proportional to N–E voltage in the barn.

Assume: Current in primary neutral = 10 A
$R_p = 0.48\ \Omega$ and $R_f = 1.1\ \Omega$ (mean resistance values)
$R_p = 0.2\ \Omega$ and $R_f = 2.6\ \Omega$ (low primary resistance, high farm resistance)
$R_p = 1.0\ \Omega$ and $R_f = 0.2\ \Omega$ (high primary resistance, low farm resistance)

Solution:

Use the simplified circuit in Figure 14.7 and apply Ohm's Law.

Calculate the equivalent resistance for each case:
$R_T = (0.48)(1.1)/(0.48 + 1.1) = 0.334\ \Omega$
$R_T = 0.186\ \Omega$
$R_T = 0.167\ \Omega$

Calculate the voltage between the transformer and the earth:
$E_T = I_T R_T = (10\ A)(0.334\ \Omega) = 3.34\ V$
$E_T = 1.86\ V$
$E_T = 1.67\ V$

This example illustrates that current flowing in the primary neutral system results in a voltage between the transformer and the earth. The N–E voltage in the barn would depend on many factors such as the resistance of the secondary neutral conductor and the quality of grounding connections throughout the barn.

Primary Neutral Current from On-Farm Loads

The previous discussion illustrated that primary neutral current can cause an N–E voltage on the farm. It follows that operation of loads on the farm will also increase primary neutral current. In this case, operation of on-farm loads will generally increase N–E voltage.

A common misconception is that connecting all on-farm loads to 240 V or perfectly balancing the farm service entrance panel will eliminate stray voltage problems. As discussed in chapter 1, balancing the service entrance panel results in zero current flow in the neutral wire. However, operation of these 240-V loads will increase the current in the primary neutral wire and can be reflected back to the farm through the primary–secondary neutral bond.

In some cases, operation of on-farm loads can result in a lower N–E voltage. For example, if the farm is supplied by one leg of an unbalanced three-phase system, operation of on-farm loads can improve the balance of the primary system and reduce the primary N–E voltage (Cloud et al. 1987).

Secondary Neutral Current from On-Farm Loads

In chapter 1, it was stated that balancing the 3-wire service entrance panel in the barn was desirable because a balanced 3-wire circuit results in zero current flow in the neutral wire. This is desirable because if current in the neutral is zero, the N–E voltage is also zero. However, it is usually impossible to completely balance the panel due to the different number of loads required at various times of the day.

Balancing the SEP can reduce the N–E voltage if secondary neutral current is the only source. However, an unbalanced SEP can either increase or decrease the N–E voltage if primary neutral current is contributing to the problem.

Current in the secondary neutral can either be in phase with the primary neutral, or 180° out of phase with the primary neutral. If current in the primary and secondary neutral conductors are in phase, an unbalanced SEP will increase the N–E voltage, whereas if the currents are 180° out of phase, an unbalanced SEP will decrease the N–E voltage (Cloud et al. 1987).

The three previously discussed causes of N–E voltage are present on all farms where electrical service is installed similar to Figure 14.6. N–E voltages are an inherent characteristic of these types of systems. Installing the system "up to code" can usually control the magnitude of N–E voltage.

The following cases of N–E voltage are caused by equipment failure or improper use of the equipment. These conditions can be due to deterioration of electrical connections, improper installation of electrical service, or seemingly unrelated practices common in barns.

Ground Fault Currents

A potentially dangerous source of N–E voltage can result from a ground fault. Recall from earlier chapters that a ground wire installed in parallel with the neutral wire is necessary to provide a safe path for current to flow in the event of a ground fault (see Figure 6.2). In cases where the ground fault current is not large enough to open the circuit breaker, an N–E voltage exists between the faulty equipment and the earth. A lethal potential may exist if the corrosive environment of the livestock facility has compromised the low-resistance ground circuit. Intermittent ground faults can be difficult to locate and correct. For example, in dusty conditions, a ground fault can be more serious during humid days since the moisture will decrease resistance of the dust that is part of the circuit.

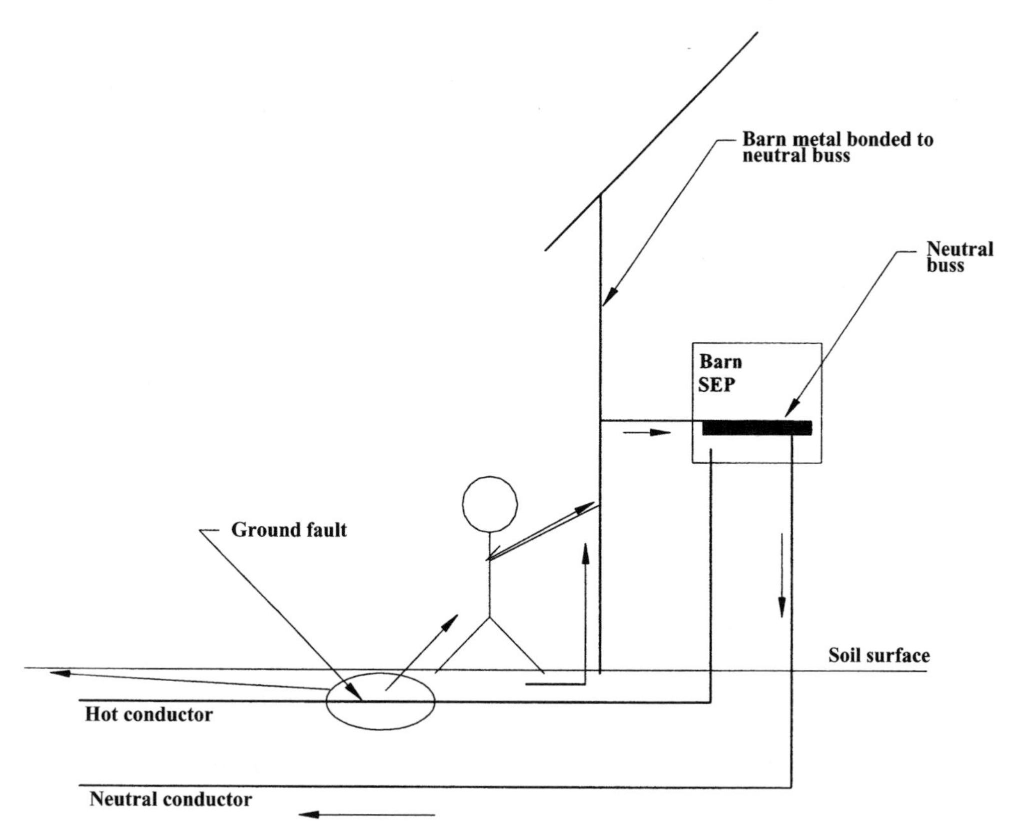

Fig. 14.8. Current flow through a person standing near a ground fault in contact with barn metal.

Ground fault currents are often associated with faulty equipment. Any current "leaking" from an energized conductor to earth can cause potentially dangerous N–E voltages. For example, consider current flow from a ground fault if the insulation fails on the service entrance cable (Figure 14.8). If the insulation on the buried service entrance cable is damaged during installation or by any other means, the "leaking" current can return to the center tap of the transformer through the earth and ground rod(s). However, a person or animal may become part of the circuit when standing near the ground fault and touching barn metal or anything else connected to the secondary neutral system (MWPS 1992).

Improper Use of Neutral and Grounding Conductors

Properly wired electrical systems consist of a completely separate neutral conductor and ground conductor bonded together only at the building service entrance panel. The ground conductor is designed to carry current only in the event of a ground fault. But if the neutral conductor is used as the ground or is interconnected with the ground conductor at the load or any place other than the SEP, current flows through both the neutral and grounding conductors under normal operating conditions. This can result in a high-spike N–E voltage due to the high current resulting from start-up of motors. Interconnection of the neutral and ground conductors at locations other than the SEP is a violation of the National Electric Code as well as a safety hazard and contributing factor to stray voltage problems (Cloud et al. 1987, MWPS 1992).

Induced Voltages

It is possible for electrically isolated metal equipment such as water pipes, milk pipelines, and vacuum lines to become electrically charged. The isolated equipment acts as a capacitor and a short duration shock can occur when the animal shorts it to earth, similar to a static electricity shock. Common sources of induced voltages are high-voltage cow trainers used in dairy barns, or even extension cords wrapped around the isolated equipment. The current produced in this case is usually small but can produce a shock large enough to trouble the animals. These induced voltages can be eliminated by properly grounding the metal equipment (MWPS 1992).

Measurement and Diagnostic Procedures

The procedure to determine the source of stray voltage requires careful recording and interpretation of voltages observed during tests. Detailed instructions for conducting these tests are published elsewhere (Cloud et al. 1987, Surbrook and Reese 1981, Reines and Cook 1999).

The following steps can aid in identifying the source(s) of stray voltage and taking proper action to reduce or eliminate the problem. The reader is cautioned to conduct tests using advice and services from the local power company engineer and an electrician, as some tests require connecting and disconnecting electrical circuits.

A high-quality voltmeter is necessary to determine the magnitude and source of stray voltage. Inexpensive VOM meters can measure DC voltages on the AC setting and can result in erroneous conclusions. Digital readout meters capable of reading 0.1 to 0.01 V are generally suitable for measuring neutral-to-earth voltage. If possible, two or three voltmeters should be available to simultaneously measure voltage between different points. A clamp-on ammeter should also be used to measure current in various tests.

Secondly, it is helpful to install an isolated ground rod to provide a common reference point. A 4- to 8-ft copper or copper clad ground rod installed 25 ft or more from the barn and other metallic objects such as water pipes should be used. Unless grounding conditions are poor, a 4-ft rod installed in moist soil should be sufficient. Next, connect a suitable length of conductor to the isolated ground rod to facilitate voltage measurements in the barn. The size of the conductor is not important as only voltage measurements between the isolated ground rod and various points in the barn are necessary (Cloud et al. 1987).

Initial Voltage Measurements

Connect a voltmeter between the isolated ground rod and the barn service-grounding conductor, and connect a second voltmeter between the isolated ground rod and the transformer-grounding conductor (Figure 14.9). Measurements exceeding 0.5–1.0 V indicate a possible stray voltage problem. The N–E voltage measured between the isolated ground rod and the barn neutral buss is generally the maximum expected voltage between cow contact points anywhere in the barn. Either voltage may be higher depending on the phase relationship between the primary and secondary current. Measurements should

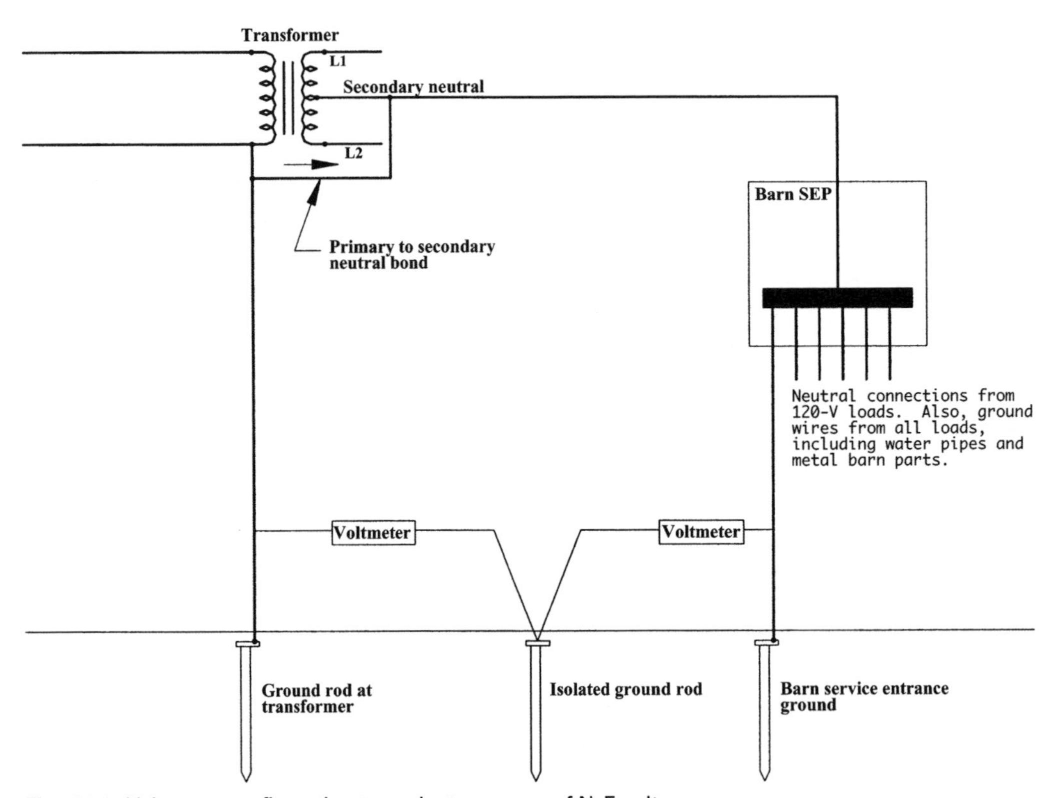

Fig. 14.9. Voltmeter configuration to evaluate sources of N–E voltage.

be recorded at various times because they may change with loads operating on or off the farm.

The difference between the two readings is due to a voltage drop somewhere in the secondary neutral network. A significant difference between the two (0.5 V or more) may be due to an undersized secondary neutral wire.

Primary Neutral Test

For this test, use the same voltmeter configuration described above to determine if primary neutral current is contributing to the N–E voltage. As a preliminary test, open the main farm disconnect to eliminate N–E voltage due to on-farm conditions. Primary neutral current from external loads is contributing to the problem if any N–E voltage is observed.

The next step is to determine if increasing primary neutral current from on-farm loads is contributing to the problem. Select three or four 240-V loads, such as the electric water heater, or large (3-hp or greater) motors to be used in the test. Using 240-V loads will increase the primary neutral current while not causing any secondary neutral current. If it is not possible to eliminate all 120-V loads during this test, try to balance the 120-V loads to keep the current in the secondary neutral as close to zero as possible.

In this test, record the N–E voltages and current in the barn conductors with no loads added, and incrementally add the selected loads while observing changes in N–E voltage and current in the conductors. Then, remove the loads in reverse order until the test loads are disconnected. If the base N–E voltage has changed after the selected loads are removed, the test should be repeated because it is possible that loads off the farm or loads elsewhere on the farm were activated during the test.

An increase in N–E voltage as each load is added can be caused by an increase in primary neutral current or by a ground fault current from the circuit. If "clean" 240-V loads are used in the test, the N–E voltage will increase in direct proportion to each load. Primary neutral current from on-farm loads is contributing to the problem if adding 240-V loads causes a proportional increase in N–E voltage.

Secondary Neutral Test

This test involves observing N–E voltages and circuit currents as 120-V loads are connected. Differences in the N–E voltages are due to a voltage drop somewhere on the secondary neutral.

With no other loads on the service, measure N–E voltages and current in conductors as known 120-V loads are connected and disconnected to each side of the SEP. This can be accomplished with existing loads or by plugging a load into identified receptacles. The primary neutral current and the N–E voltage measured between the isolated ground rod and the transformer-grounding conductor will increase as loads are added and decrease as loads are disconnected. However, the N–E voltage measured between the isolated ground rod and the barn service-grounding conductor may increase or decrease as loads are added depending on the phase relationship between primary and secondary current. A significant difference between the transformer and barn N–E voltages may be due to corroded connections, an unbalanced SEP, or an undersized secondary neutral conductor.

Locating Electrical Problems in the Barn

Circuits should be visually inspected for corrosion, dirt, or improper wiring practices such as interconnected neutral and ground conductors. This test should be conducted by measuring voltages at cow contact points while operating one circuit at a time. Comparing N–E voltage at the barn SEP and voltage measured between the isolated ground rod and various cow contact points can be used to determine locations of ground faults and improper wiring practices overlooked during the visual inspection.

During this test, it may be helpful to measure current in the hot, neutral, and ground conductors. When a known load, such as a 120-V 10-A load, is connected to a circuit, current in the hot and neutral conductors should be 10 A. Any current measured in the ground conductor indicates a ground fault somewhere on the circuit or an improper wiring practice.

If the voltage for a contact point (such as a motor casing) is greater than the barn N–E voltage, a wiring or ground fault is likely to exist on that circuit. If the voltage is significantly less, then it is likely the object is not effectively grounded.

Continuous Monitoring

If the previous tests do not conclusively identify the source(s) of the problem, it may be necessary to continuously monitor voltages during milking time or over a longer period of time. Short-term peak voltages may be present in the system due to starting and stopping of loads that were not observed during previous tests.

Isolated Neutral Test

This test must be conducted with cooperation of the local utility by having them remove the primary–secondary neutral bond and removing any other connection that may affect the isolation. Removing this bond isolates the farm from the effects of primary neutral current. This will also test the effectiveness of isolating neutrals as a possible solution to the problem on the farm.

When the farm is isolated, an N–E voltage at the barn will be caused by on-farm problems. Repeat the previous tests to determine the on-farm source of the N–E voltage. The N–E voltage should not change with connection of a 240-V load. A change indicates a ground fault in the 240-V load or that there is poor isolation of the primary system from the secondary system.

Standby Power Supply

If the testing procedures indicate that primary neutral current from on-farm loads is a principal source of stray voltage on the farm, a standby power supply may be used to determine if neutral isolation is a solution to the problem. Using a standby power supply eliminates off-farm sources but does not change on-farm conditions. If N–E voltages measured during critical operation times were at acceptable levels, then isolation of the secondary neutral system from the primary neutral system may be an effective solution to the problem.

Solutions to the Problem

Once the source(s) of stray voltage have been identified, steps can be taken to minimize the problem. After suitable solutions are found, measured N–E voltages should be reduced to acceptable levels.

There are three basic solutions (Gustafson et al. 1984):

1. Eliminate or reduce the voltage causing the problem.
2. Gradient control by the installation of equipotential planes.
3. Isolation of the grounded neutral system from the animals.

Voltage Reduction

Elimination of voltages caused by ground fault current and improper wiring practices can be easily carried out. In cases where voltage drop in the secondary neutral system is contributing to the problem, an option is to use a 4-wire service to the barn that will eliminate N–E voltage due to secondary neutral drop by completely separating the neutral and grounding systems (MWPS 1992).

This solution requires installation of a 4-wire cable to the barn and a service entrance panel in the barn where the grounding buss is completely isolated from the neutral buss (Figure 14.10). The ground rod at the barn is eliminated and the grounding conductor (fourth wire in the cable) is connected to the neutral at the source, either the farm SEP or the transformer. Since the grounding circuit is isolated from the neutral circuit, current is forced to return through the neutral conductor, eliminating pathways through the barn metal and animals. N–E voltage will remain from sources due to primary neutral current because a pathway exists through the primary-secondary neutral bond.

It is imperative that the grounding circuit is completely isolated from the neutral circuit, except at the origin of the 4-wire service. If there is *any* connection between the two, an N–E voltage is possible. For example, if the water pipe in the house is connected to the neutral buss in the house, and connected to the ground buss in the barn, there is a "bypass" in the isolated grounding system.

Installation of a 4-wire system will eliminate on-farm sources of N–E voltage during normal operation. However, future ground-fault currents will produce an N–E voltage in the barn. These ground faults should be identified and eliminated as quickly as possible.

If faulty wiring is not the cause of stray voltage, there are devices to eliminate N–E voltages in other ways (MWPS 1992). One possible solution is an electronic device that measures N–E voltage as input to an amplifier and current to a remote grounding electrode is adjusted to offset the N–E voltage. This system is commercially available.

Gradient Control

Gradient control by installation of equipotential planes during construction is an effective way to negate the effects of N–E voltage regardless of the source by reducing the potential between cow contact points. Since stray voltage is a problem because a potential exists between contact points such as a watering cup and the floor, it follows that if all contact points are at the same potential then current will not flow.

Generally, installing equipotential planes is accomplished by bonding reinforcement metal in the concrete floor and all metal in the area to the neutral buss in the service entrance panel. This practice

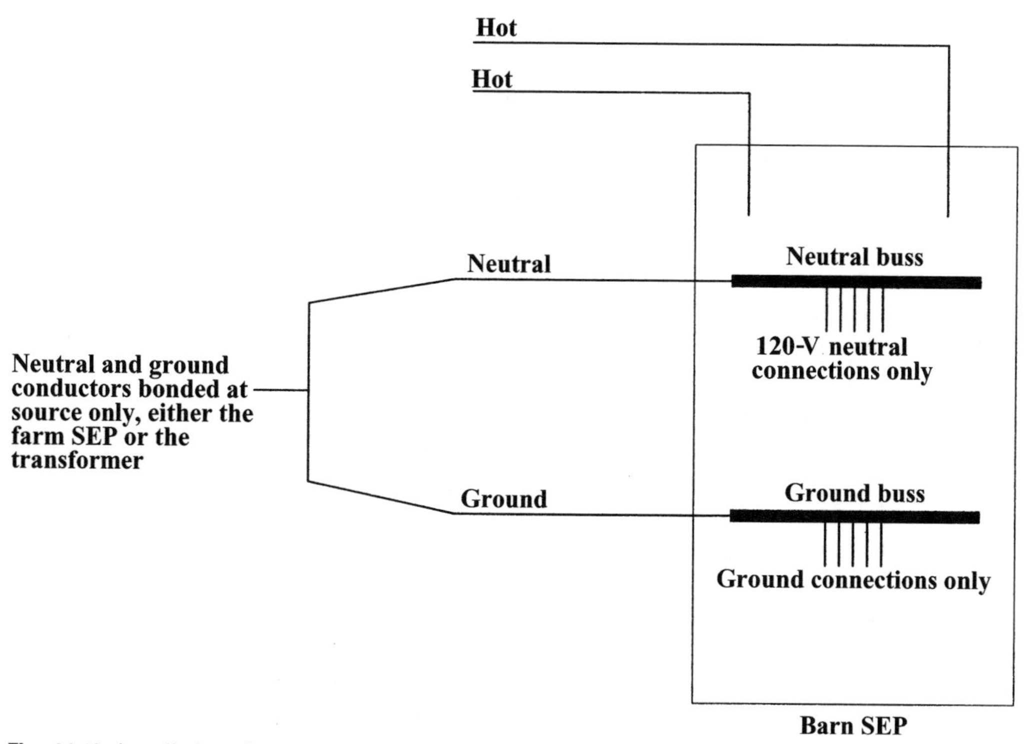

Fig. 14.10. Installation of a 4-wire service entrance cable in the barn service entrance panel.

greatly reduces a potential difference between the barn metal and the floor by raising the potential of the floor to near the potential of the neutral buss.

Installing an equipotential plane eliminates the potential difference between contact points in the barn. However, a potential difference may still exist between true earth and the barn floor and animals will be subject to a voltage as they enter or leave the barn. Therefore, use of a transition ramp is required at the entrance. A transition ramp consists of bonding metal bars to the reinforcement metal in the barn floor and installing them at an angle into the earth away from the entrance to the barn. This ramp slowly changes the potential from true earth to the potential of the barn floor thus reducing the voltage to which the animal is subjected (MWPS 1992).

Specific requirements for installation of equipotential planes can be obtained from sources such as "Design Criteria for Equipotential Planes" (Kammel et al. 1986). Publications such as this should be consulted as they address local and state electrical codes to ensure proper installation of equipotential planes for a specific location.

Isolation

If tests show that primary neutral current is a major source, it is possible to isolate the primary and secondary neutral systems. One option is for the power supplier to disconnect the conductor bond between the two systems. However, some suppliers may not allow this due to safety concerns. Other options include installation of isolation transformers to eliminate effects from the primary system.

Conclusion

Stray voltage can be a major problem for livestock producers. This chapter briefly discussed possible sources of stray voltage in a barn. In reality, several of these conditions may contribute to stray voltage. Careful measurements and interpretation of these measurements are necessary to successfully solve the problem.

In general, if there is current in the neutral system, there is an N–E voltage. The magnitude of the voltage depends on the resistance of every component of the system including wires, connections, dust, and the grounding rod/soil interface.

Carefully measuring and interpreting voltage measurements at various points in the system can determine the source of N–E voltage. Measurements may be necessary under controlled conditions, during normal operation, and/or over extended periods of time to identify the source(s) of the N–E voltage.

Experience shows that problems on the farm can be common sources of stray voltage. Of the 19 farms considered in a study (MPUC 1998), six of the farms had supply lines that were severely overheating. These farms were probably older systems and simply were not designed to carry the present electrical demand. Every farm in this study had one or more pieces of electrical equipment not grounded. Thirteen of the farms did not have a grounding electrode installed as required by the National Electrical Code.

The study concluded that wiring conditions on many farms could be a primary cause of high N–E voltage. It was stated that primary reasons for code violations included:

1. At least some wiring on the farm was installed by the owner or untrained persons.
2. Many code violations were present in work done by electrical contractors.
3. Electrical inspectors failed to note and/or enforce code regulations.
4. Wiring was not maintained or replaced when defective.

Unfortunately, the cause(s) and effects of stray voltage on animals and humans can be very difficult to identify. Lawsuits have been filed against power suppliers where stray voltage has been cited as the cause of farms going out of business and even human health problems (Hardie 2000). In most cases, utilities have conducted many tests on the farms and were unable to locate sources of stray voltage, or if stray voltage was present. Regardless of the presence or absence of stray voltage, multimillion-dollar settlements have been awarded to farmers in these lawsuits (Wiff 1999).

Stray voltage problems can usually be corrected or prevented by strict adherence to the National Electrical Code. Careful evaluation of existing systems by qualified personnel should be conducted to determine if stray voltage is present on a farm, and appropriate steps taken to eliminate or reduce the voltage. Electricians specifically trained to install electrical systems in new facilities should be employed.

If stray voltage is suspected on a farm, it is recommended to contact the rural electric cooperative or power supplier and a certified electrician. Many cooperatives have equipment and personnel trained for stray voltage analysis and sometimes offer stray voltage measurement as a service to their customers.

References

Cloud, H. A., R. D. Appleman, and R. J. Gustafson. 1987. Stray voltage problems with dairy cows. North Central Regional Extension Publication 125. Minnesota Extension Service, University of Minnesota, St. Paul, MN.

Gustafson, R. J., H. A. Cloud, and V. D. Albertson. 1984. Techniques for coping with stray voltages. Agricultural Engineering 65(12):11–15.

Hardie, C. 2000. Farmers bear burden of stray voltage. LaCrosse Tribune, LaCrosse, WI. July 9.

Kammel, D. W., L. A. Brooks, B. Jones, and R. Hau. 1986. Design criteria for equipotential planes. Fiche 86-3021. American Society of Agricultural Engineers, St. Joseph, MI.

Minnesota Public Utilities Commission. 1998. Final report of the science advisors to the Minnesota Public Utilities Commission: Research findings and recommendations regarding claims of possible effects of current in the earth on dairy cow health and milk production. MPUC, St. Paul, MN.

MWPS, 1992. Farm buildings wiring handbook, 2nd ed. Midwest Plan Service, Ames, IA.

Reines, R. S. and M. A. Cook. 1999. PSC staff report: The Phase II Stray Voltage Testing Protocol. Rural Electric Power Services Public Service Commission of Wisconsin, Madison, WI.

Soderholm, L. H. 1982. Stray-voltage problems in dairy milking parlors. Trans ASAE 25(6): 1763–1767, 1774.

Surbrook, T. C., and N. D. Reese. 1981. Stray voltage on farms. ASAE Paper No. 81-3512. American Society of Agricultural Engineers, St. Joseph, MI.

Wiff, J. 1999. St. Croix verdict orders NSP to pay nearly $4 million on farm stray voltage case. River Falls Journal, River Falls, WI. December 3.

Problems

14.1. Calculate N–E voltage at the barn service entrance panel if the neutral conductor extending to the SEP is an AWG 00 copper conductor, the unbalanced SEP causes 15 A in the neutral system, and the resistance of the ground rod at the barn SEP is 2 Ω. Assume the length of the neutral conductor is 200 ft, including connections.

14.2. What is the effect of very good ground rod installation at the barn (resistance less than 1 Ω) on N–E voltage?

14.3. What is the N–E voltage at the barn in Example 14.1 (B) if the neutral conductor between the transformer and the SEP is accidentally disconnected?

Chapter 15

Electrical Power Use Patterns

We are accustomed to always having electrical loads come on whenever we turn on a switch. It's easy to take for granted what is required to give us this convenience. For a light bulb to come on at our command, a complete electrical system must be in place with every component from spinning generator to light bulb filament ready to supply and carry the necessary flow of electrons. This service has to be ready to meet the needs of anyone in the country or on the continent at any time. In this chapter, we will explore some issues related to possessing this capability and appropriately charging electrical customers for it.

Electrical Demand

Electrical service billing is always done on the basis of energy (kWh) use during the billing period, which is usually one month. In chapter 1, a residential service billing procedure was described. In addition to the energy charge, for large agribusiness and commercial and industrial customers, the electrical bill can include a demand charge. The demand charge reimburses the electric power company for having the capability in generation, transmission, and distribution equipment to supply the quantity of power (kW)—in other words, the demand—needed by the customer at any time. A customer who uses a low, constant kW level costs the electric company less in ownership cost than another customer who uses the same energy (kWh) but takes a large kW level for a short period, and then has long periods of low power usage. An example illustrates these principles.

Example 15.1

Load profiles for Customer 1 and Customer 2 are graphed in Figure 15.1 for a 24-hour period. Both customers use the same energy during the 24-hour period shown: Customer 1: (1 kW) (24 h) = 24 kWh. Customer 2: (6 kW) (4h) = 24 kWh. However Customer 2's demand was 6 kW for a 4-hour period while Customer 1's demand was 1 kW for the entire 24-hour period.

The power company can supply Customer 1 at a lower cost than Customer 2 because Customer 1 never uses more than 1 kW, whereas Customer 2 uses 6 kW for part of the day. The power company must buy a generator and all the other equipment to deliver 6 kW, even though it is used only four hours during the day. The power company can pass along this cost by billing for a demand charge in addition to the energy charge. When the power company has hundreds or thousands of customers like Customer 2, the effect is lessened because the times of the peak loads will likely be distributed over the day. However, most power companies do have systems in place to charge customers for demand load.

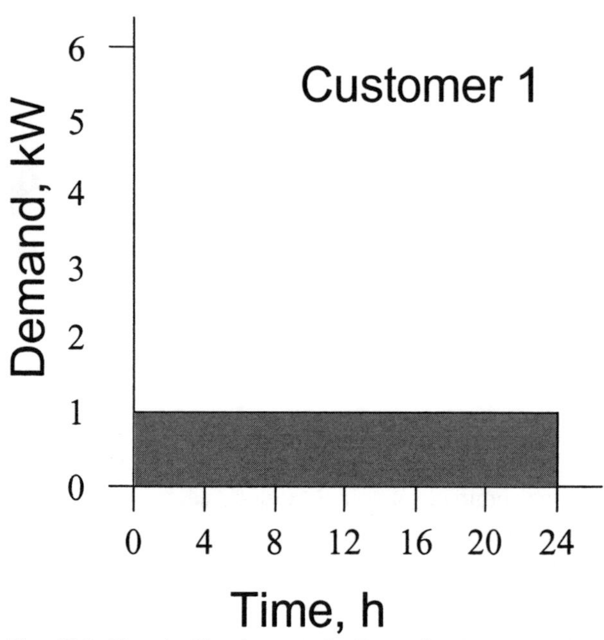

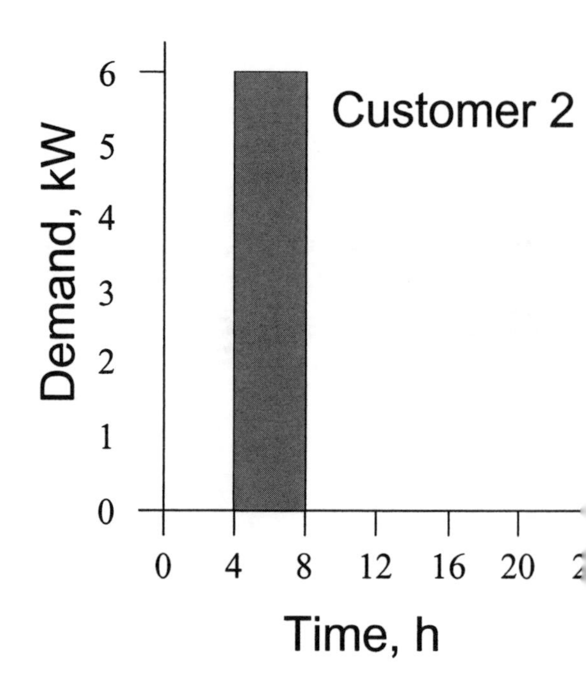

Fig. 15.1. Electrical load pattern for Example 15.1.

Demand Metering

Watthour meters installed for large farm, commercial, and industrial customers often have the capability to measure and register both energy (kWh) and demand (kW) for the usual monthly billing period. These demand meters read out the highest average demand during a selected demand interval between 15 and 30 min during the month. For example, if the demand interval selected is 20 min, when the meter is read, it will display the highest 20-min demand in kW which has occurred since the last meter reading. The meter reader records this and then resets the meter to zero kW demand. If the meter is computerized, these operations are done automatically. Demand peaks occurring for times less than 15 min do not tend to be detrimental to electrical equipment because the high current does not flow long enough to overheat equipment. For example, incandescent lamps and induction motors draw high currents for milliseconds and a few seconds, respectively, when they start. But the times are so short that these current spikes do not increase required generator size or heat up equipment and conductors. The city of Ames, Iowa, for example, uses a 15-min demand interval.

Example 15.2

Figure 15.2 illustrates the electrical demand pattern of a customer. Let's compute the maximum 15-minute demand. If the demand meter is a mechanical or thermal type, the demand intervals are fixed by the time at which the meter was placed into service. Let's say the meter is one of these types and was placed in service at time = 0.

Demand registered for each period would be calculated as follows:

$$\text{Period 1: } \frac{0\text{ kW} + 0\text{ kW} + 2\text{ kW}}{3} = 0.67\text{ kW}$$

$$\text{Period 2: } \frac{5\text{ kW} + 3\text{ kW} + 0\text{ kW}}{3} = 2.67\text{ kW}$$

$$\text{Period 3: } \frac{1\text{ kW} + 3\text{ kW} + 2\text{ kW}}{3} = 2\text{ kW}$$

$$\text{Period 4: } \frac{2\text{ kW} + 2\text{ kW} + 0\text{ kW}}{3} = 1.33\text{ kW}$$

$$\text{Period 5: } \frac{3\text{ kW} + 5\text{ kW} + 5\text{ kW}}{3} = 4.33\text{ kW}$$

$$\text{Period 6: } \frac{4\text{ kW} + 1\text{ kW} + 1\text{ kW}}{3} = 2\text{ kW}$$

For this 90-min time of consideration, the highest 15-min demand was 4.33 kW, which occurred during Period 5. A watthour meter fitted with a demand metering system is able to display the highest demand for the month.

Rolling Demand

If the electric power company uses computer equipment to record demand data for each customer,

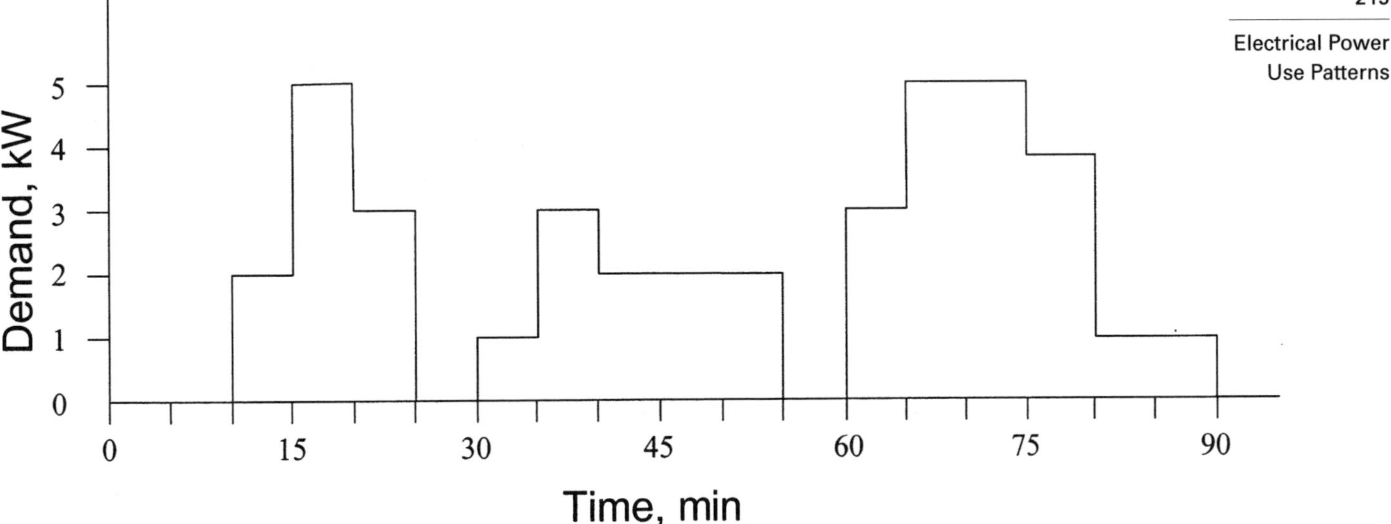

Fig. 15.2. Load pattern for Example 15.2.

then the computer software can check every possible 15-min interval, rather than only successive demand intervals beginning at time = 0. Demands computed by this procedure, which is called "rolling demand," will never be smaller, and may be larger than demands computed using successive demand intervals.

Looking at Figure 15.2, we see that a demand period starting at 65 min and ending at 80 min yields a 15-min demand of $(5 + 5 + 4)/3 = 4.67$ kW, which is larger than the previous high demand value of 4.33 kW.

In the example, demand was assumed to be recorded every 5 min. In practice, computerized equipment will record demand several times per second.

Calculating Demand Charges

Electric power companies calculate monthly charges for large power users by computational procedures which use monthly demand and energy data for each customer, and algorithms developed to bring in appropriate income for the company. Table 15.1 is the electrical rate schedule which the city of Ames uses for large power customers. All non-residential customers who use over 10,000 kWh per month of energy are billed as large power customers. An example illustrates use of this schedule.

Example 15.3

An electric power customer of the city of Ames has this use data for a one-year period:

Month	Energy, kWh	Demand, kW
J	20,000	130
F	21,000	60
M	20,000	61
A	18,000	62
M	18,000	59
J	21,000	61
J	26,000	65
A	25,000	67
S	21,000	77
O	20,000	58
N	18,000	54
O	20,000	62

Compute the electric bill for November. Since this customer uses over 10,000 kWh/month, the large power rate is appropriate. From Table 15.1 we first compute the billing demand, which is the greatest of

a. current month's demand — 54 kW
b. 75% of the high demand from the previous 4 summer months: (77 kW)(.75) = 58 kW
c. 60% of the high demand from the previous 11 months: (130 kW)(.6) = 78 kW

The billing demand is 78 kW.

Compute demand charge
First 50 kW @ \$3.05/kW — \$152.50
Next 28 kW @ \$2.65/kW — 74.20

Energy charge (based on billing demand)
First 200 h @ \$0.047/kWh:
(200 h)(78 kW)(\$0.047/kWh) = 733.20

TABLE 15.1 City of Ames, Iowa, electric large power rates

Summer—Bills mailed on or between July 1 and October 31.

Net—per meter per month

Demand Charge: Based on billing demand

First 50 kW	$4.25 per kW
Next 100 kW	$3.85 per kW
Over 150 kW	$3.45 per kW

Energy Charge: Based on billing demand

First 200 hours	5.5¢ per kWh
Next 200 hours	3.8¢ per kWh
Over 400 hours	3.5¢ per kWh

Customer Charge: $40.00 per month

Minimum bill shall be the customer charge plus current demand charge.

Winter—Bills mailed on or between November 1 and June 30.

Net—per meter per month

Demand Charge: Based on billing demand

First 50 kW	$3.05 per kW
Next 100 kW	$2.65 per kW
Over 150 kW	$2.25 per kW

Energy Charge: Based on billing demand

First 200 hours	4.7¢ per kWh
Next 200 hours	3.7¢ per kWh
Over 400 hours	3.3¢ per kWh

Customer Charge: $40.00 per month

Minimum bill shall be the customer charge plus current demand charge.

The "billing demand" shall be for the greater of:

a. the current month's demand;
b. 75% of the high demand in the previous 4 summer months; or
c. 60% of the high demand in the previous 11 months;
d. provided that the demand used for billing shall in no case be less than 15 kW after discounts.

Source: City of Ames (2000); reprinted with permission.

(200 h)(78 kW) = 15,600 kWh
18,000 − 15,600 = 2400
2400/78 = 30.8
next 30.8 h @ $0.037/kWh
(30.8 h)(78 kW)($0.037/kWh) = 88.89

Customer charge 40.00

Total $1088.79

Discussion

The hypothetical data for this problem was derived with a high 1-month demand in order to show the effect of such an occurrence on the electric bill. The Ames rate schedule is written with a "ratchet clause" which allows a single month's demand value to influence charges for an entire year, if it is large enough (billing demand item c). The 130-kW demand in January will determine the billing demand for the entire following year.

Notice that the Ames rate schedule allows billing demand to influence energy cost, in addition to demand cost, since a high demand causes more of the energy to be purchased at the higher rates.

In the example, reducing the January demand to below 96 kW (with no other changes in demand or energy use) will reduce the energy bill for November, and for the year, by nearly 9%. This will be about $1,100 for the year.

Reducing Demand

For large power users, demand reduction can result in large energy cost reductions. For example, consider the customer in Example 15.3. The November energy use was 18,000 kWh. Considering that there are 720 h in the month this energy could theoretically be moved to the customer with a constant demand of 25 kW (720 h × 25 kW = 18,000 kWh). If the billing demand is 25 kW, the electric bill goes down to $760.25. Reductions of this magnitude are probably never possible in practice, but significant improvements are usually possible.

Scheduling use of electrical loads, with an objective of reducing demand, can be an effective way of reducing demands. Some power companies offer lower rates for electrical energy used during times of the day when their demand is low. This can influence customers to schedule load use in ways which reduce demand.

Demand Controllers

Demand controllers are programmable devices installed on an electric power customer's system to keep demand from exceeding some selected level. The controller continuously monitors demand, and has the capability of turning off and turning on selected loads on the customer's system. Loads like air conditioners, freezers, refrigerators, and electric water heaters are good candidates for being shed because they can be shut off for a few minutes without causing problems. Energy use is not reduced by such control, since load use is delayed, rather than being prevented.

The demand controller is programmed to shed loads in a selected sequence for a selected time, in order to prevent demand from exceeding a selected level. The decision to purchase a demand controller can be made, based on an analysis of its costs versus its savings. Some demand-type watthour meters have a limited demand control capability.

References

City of Ames. 2001. City of Ames utility rates. City of Ames, Ames, IA.

Problems

15.1. An electrical service has these kW demands during 12 successive 5-minute periods: 1:9, 2:4, 3:1, 4:0, 5:12, 6:10, 7:3, 8:2, 9:2, 10:1, 11:4, 12:6.

(a) Sketch a graph of this demand record with demand on the vertical (Y) axis and time on the horizontal (X) axis.

(b) Compute the energy use during the one-hour period in kWh.

(c) Compute the highest 10-minute demand in kW, during the one-hour period of record, assuming successive 10-minute demand periods starting at time = 0.

(d) Compute the highest 10-minute demand in kW, during the one-hour period of record, assuming a rolling demand record.

15.2. An electrical service uses 40,000 kWh of energy every month. Electrical demand is 120 kW for every month except January, when it reaches 250 kW.

(a) Compute the January electrical bill.

(b) Compute the February electrical bill.

(c) Compute the annual electrical bill.

(d) What is the theoretical lower limit of demand for this service?

(e) What would the annual electrical bill be if January demand is reduced to 120 kW?

(f) What would the annual electrical bill be if demand is reduced to the theoretical lower limit calculated in (d)?

Appendix A

POWER REQUIREMENTS OF MACHINES

TABLE A.1 Approximate auger capacities (bu/h), dry corn, 450 rev/min

Nominal Auger Diameter (inches)	Angle of Elevation				
	0°	30°	45°	60°	90°
4	280	220	180	150	120
6	1,000	800	700	600	450
7	1,300	1,100	900	800	600
8	1,800	1,500	1,300	1,100	800
10	3,000	2,400	2,100	1,800	1,250

Source: From Meyer (1973).

TABLE A.2 Approximate HP for each 10 ft of auger, dry corn, 450 rev/min

Nominal Auger Diameter (inches)	Angle of Elevation				
	0°	30°	45°	60°	90°
4	.30	.32	.35	.34	.32
6	.43	.63	.75	.68	.56
7	.90	1.10	1.18	1.15	1.12
8	1.24	1.50	1.65	1.60	1.50
10	1.75	2.25	2.50	2.50	2.25

Source: From Meyer (1973).

TABLE A.3 Approximate auger capacities and horsepower needs (6-in. auger, 500 rev/min, corn and soybeans)

	Corn			
	Bu/h		HP for 10 ft	
Slope	Dry	25%	Dry	25%
0°	1,110	640	.43	1.72
30°	970	580	.63	1.89
40°	750	450	.72	1.80
60°	540	320	.68	1.36
90°	440	260	.56	.59

	Soybeans	
Slope	Bu/h	HP for 10 ft
0°	1,000	.62
30°	800	.85
40°	630	.94
60°	460	.86
90°	350	.74

Source: From Meyer (1973).

TABLE A.4 Approximate power requirements of U-trough augers (250 rev/min)

Sizes (inches)		6	8	10	12
Capacity (bu/h)		600	1600	2400	3000
		HP			
Auger Length (ft)	10	¾	1	1	2
	20	1	1	2	3
	30	1	1	3	5
	40	1½	1½	3	7½
	50	1½	2	5	10
	60	2	3	5	10
	70	2	3	7½	
	80	3	5	7½	
	90	3	5	10	
	100	5	7½	10	

Source: From Meyer (1973).

TABLE A.5 Typical bucket elevator capacity and power

Bucket Size (in.)	Bucket Spacing (in.)	Belt Speed (ft/min)	Capacity (bu/h)	Power Requirements (hp/10 ft height)
4 × 3	8	240	200	0.10
	6	270	300	0.125
6 × 4	4 ¼	270	550	0.20
	4 ¼	335	700	0.25
7 × 5	8	335	900	0.30
	6	335	1,200	0.33
9 × 5	7	265	1,600	0.5
	6	300	1,800	0.5
9 × 6	12	385	1,500	0.625
	6	385	3,000	1.25
12 × 7	10	565	5,000	2
15 × 7	9	565	7,500	3
14 × 8	10	650	10,000	4

Source: MWPS (1987); reprinted with permission.

TABLE A.7 Approximate power requirements of flight elevators

Length (ft)	(35°–40°) Approximate Capacity (bu/h)	Motor Size (hp)
30–36	900–1,200	2
38–50	900–1,200	3

Source: From Meyer (1973).

TABLE A.8 Approximate power requirements of 12-in. horizontal belt conveyor (100 ft/min), 4 tons/h capacity

Conveyor Length (ft)	Approximate Horsepower
20	½
30	¾
40	¾
50	1
60	1
80	1½
100	1½

Source: From Meyer (1973).

TABLE A.9 Approximate power requirements of silo unloaders

Top Silo Unloaders		
Silo Diameter (ft)	Motor size (hp)	Minutes to Unload 1 Ton
12–18	5	12
16–24	7½	8
20–30	10	6
24–40 Heavy Duty	15–20	3 to 4
Bottom Silo Unloaders		
Silo Diameter (ft)		Motor Size (hp)
20–26		7½–Auger ½–Sweep arm 2–Exit auger

Source: From Meyer (1973).

TABLE A.6. Approximate power requirements of bin unloading augers

Bin Diameter (ft)		Sweep Auger (hp) 4 in.	Sweep Auger (hp) 5 in.	Unloading Auger (hp)
18	800–1,000 bu/hr	⅓	½	1
21		½	½	1
24		½	¾	1½
27	1,000–1,300 bu/hr	½	1	1½
30		¾	1	2
36		¾	1½	2

Source: From Meyer (1973).

TABLE A.10 Approximate power requirements of grinders

Hammer Mill		Roller Mill	
Capacity (tons/h)	Motor Size (hp)	Capacity (tons/h)	Motor Size (hp)
½ to 1	2	¾ to 1½	2
¾ to 1½	3	1¼ to 2¼	3
1½ to 3	5	2¼ to 4½	5
2 to 4	7½	3 to 6	7½

Source: From Meyer (1973).

References

Meyer, V. M. 1973. Tables for estimating power needs. Iowa State University Cooperative Extension Service Publication A. E.–1061. Iowa State University, Ames.

MWPS. 1987. Grain drying, handling, and storage handbook, MWPS 13. Midwest Plan Service, Iowa State University, Ames.

TABLE A.11 Approximate power requirements of axial ventilation fans

Fan Diameter (in.)	Horsepower of Motor (approximate)	Capacity (at 1/8-in. water column static pressure) (ft^3/min)
12	1/30	450
12	1/20	900
12	1/12	1100
16	1/6	1900
16	¼	2400
16	⅓	2800
20	¼	3000
20	⅓	3500
20	½	4500

Source: From Meyer (1973).

Appendix B

ELECTRICAL EQUIPMENT DATA

TABLE B.1. Electrical equipment data

Equipment	Usual Voltage Rating (V)[1]	Current for Typical Watts (A)	Typical (Range of) Power Use (W)	Typical Energy Use (kWh/mo)[2]
Air conditioner				
Window mount				
6,000 Btu/h	115	6.2	618	
12,000 Btu/h	115	12	1,330	
Central				
3-ton	230	26	4,500	840
5-ton	230	43	7,500	
Blanket, electric	115	1.5	175	21
Blender	115	5.6	385	0.1
Broiler, portable	115	9.9	1,140	7
Can opener	115	1.4	100	
Clipper, hair	115	0.1	10	
Clock	115	0.03	2	1.5
Coffee maker	115	7.4	850	9
Cooker, egg	115	4.5	516	1.1
Computer, personal	115 or 230			33
Corn popper	115	5.0	575	
Crusher, ice	115	1.0	120	
Curling iron	115	0.3	40	
Dehumidifier	115	2.1	240	64
Dishwasher	115	15	1,200	30
Disposer, garbage	115	3.7	420	2–3
Drill, 3/8-inch	115	3.5	240	
Dryer, clothes				
Electric	230	24	5,000	83
Gas	115	4.6	400	
Dryer, hair				
Hand-held	115		1,200(900–1800)	3–4
Electrostatic cleaner	115	0.4	50	
Fan				
20-in box	115	0.64	46	
Attic, 1/3-hp	115	4.8	330	24
Circulating	115		88	4
Kitchen exhaust	115	2.3	160	
Furnace, 1/3-hp fan	115	4.8	330	79
Freezer, 15-ft^3	115	4.9	341	100
Frost-free, 15-ft^3	115	5.5	440	147
Ice cream	115	1.9	130	
Fry pan	115	10	1,150	8
Fryer, deepfat	115	13	1,500	7
Furnace				
Gas or oil, 1/3-hp fan	115	4.8	330	79
Electric	230	6.5	15,000	
Grinder, food	115	2.2	150	
Heater				
Home electric heating systems	230	87	20,000 (500-25,000)	1,850
Portable	115	11.3	1,300	15
Water, quick recovery	230	20	4,500	398
Waterbed	115	3.1	360	100
Heat lamp	115	2.2	250	1
Heat pad	115	0.57	65	

(Continued)

Table B.1. (Continued)

Equipment	Usual Voltage Rating (V)[1]	Current for Typical Watts (A)	Typical (Range of) Power Use (W)	Typical Energy Use (kWh/mo)[2]
Heat pump				
2-ton	230	18.6	3,200	
5-ton	230	39	6,800	
Hi-Fi/stereo	115	0.87	100	
Hot plate	115	10.9	1,250	8
Humidifier	115	2.5	175	7
Iron, fabric	115	8.7	1,000	12
Iron, soldering	115	3.5	400 (30–450)	
Juicer	115	1.3	90	
Knife, electric	115	1.3	92	0.7
Knife sharpener	115	1.6	110	
Lamps (see Table 12.1)				
Fluorescent	115		15–60	
Incandescent	115		10–1000	
Yard light, Hg vapor	115		175	73
Microwave oven	115		1,500	16
Mixer				
Hand	115		100	0.2
Standing	115		150	
Motors[3]				
Under 0.5-hp	115/230, 115 or 230		1,200/hp	
0.5-hp and over	115/230, 115 or 230		1,000/hp	
Opener, can	115	1.4	100	
Printer, laser	115			25
Pump, sump, 1/3-hp	115	4.8	330	
Radio	115		71	7
Range				
6-inch heating coil	115	7.8	900	
8-inch heating coil	115	10	1,200	
4-coil surface unit	230	29	6,700	
Built-in single oven	230	20	4,700	
Complete	230	53	12,200	99
Refrigerator-freezer				
2-ft^3, thermoelectric	115	0.7	60	
14-ft^3	115	4.1	330	95
14-ft^3, frost-free	115	8.9	615	153
20-ft^3, frostless	115			230
Dorm size	115	1.0	70	24
Roaster	115	12	1,350	5
Sewing machine	115	1.1	75	0.9
Shaver	115	0.2	14	0.2
Shaving cream dispenser	115	0.9	60	
Television, color	115		200	17–36
Toaster	115	10.4	1,200	3
Toothbrush	115	0.10	7	0.1
Trash compactor	115	5.8	400	4
Trimmer, hedge	115	3.1	265	
Vacuum cleaner	115	7.5	650	4
Vaporizer				
Cool mist	115	0.87	60	
Steam	115	4.2	480	
Waffle iron	115	10.4	1,200	1.6
Warmer, baby food	115	1.4	165	
Warming tray	115	1.2	140	
Washer, automatic	115	7.2	500	9
Waterbed	115	3.5	400	85
Waterer, cattle	115	5.2	600	
Waterer, hog	115	1.7	200	
Waxer-cleaner, floor	115	5.1	350	
Welder, 230-amp arc	230	47	7,100	

[1]115 or 230 means item is available with 115-V rating or with 230-V rating.

115/230 means item as purchased can be operated in either 115 V or 230 V, with appropriate changes in wiring.

[2]If seasonal (like an air conditioner), only for months when used.

[3]Motor wattages per horsepower assume a motor efficiency of 74.6% for motors of 0.5 hp and above, and 62.2% for motors under 0.5 hp. Running wattages are listed. Starting wattages are four to five times higher (Table 9.2). Motors are assumed to be delivering rated horsepower. Assume PF = 0.75 (lagging) for 0.5 hp and above, 0.6 (lagging) for under 0.5 hp.

Appendix C

Answers to Selected Problems

Chapter 1

1.1 24,000, 24,000 cycles/min
1.2 12 Ω, 3 W
1.3 (a) 3 A; (b) 360 W; (c) 108 kWh
1.4 (a) 0.111 A; (b) 81 Ω
1.5 8.7 A, 1326 kWh/yr, $92.82/yr
1.6 (a) 1.30 A, 88.5 Ω; (b) 0.652 A, 353 Ω; (c) 0.652 A, 176 Ω; (d) 0.326 A, 705 Ω
1.7 61.2 W
1.8 $P = I^2R$, $P = E^2/R$
1.9 1.3 A, 88.2 Ω
1.10 (a) 1.0 kWh; (b) 336 kWh
1.11 51.8 A
1.12 18.5 A, $0.22/h
1.13 $126.34, $0.0735/kWh
1.14 15 Ω, 2 A, 20 V, 10 V
1.15 200 Ω
1.16 25 Ω, 37.5 Ω
1.17 (a) 0.6 A; (b) 19.17 Ω; (c) 11.5 V; (d) 0.667 A, 12.8 V, 8.54 W; (e) 76.8 W
1.18 (a) 12 Ω; (b) 4 A, 6 A; (c) 1200 W
1.19 (a) 24 Ω; (b) 4 Ω; (c) 30 A; (d) 3 A, 12 A, 10 A, 5 A
1.20 $R_T = R_1R_2/(R_1+R_2)$
1.21 60 Ω, 2 A, 20 V, 100 V, 0.67 A, 1.33 A, 2 A
1.22 12 Ω, 10 Ω, 3 Ω, 2 Ω, 24 Ω, 5 Ω, 20 V, 15 V, 15 V, 1.67 A, 25 V
1.23 8 A, 5 Ω, or 4 A, 20 Ω
1.24 (a) 192 V, 48 V; (b) 96 W, 24 W
1.25 32 Ω, 128 Ω, 128 Ω
1.26 180 Ω, 1036.8 W
1.27 (a) 69.12 Ω; (b) 833.33 W, 66.67 W, 100 W
1.28 (a) 15,000 turns; (b) 3.47 A, 104 A
1.29 10 Ω, 36 A, 24 A, 12 A
1.30 3 A, 3 A, 5 A, 2 A, 5 A, 2 A, 2 A, 7 A, 9 A, 12 A, 3 A, 17 A, 12 A, 5 A
1.31 68.3 A, 72.5 A, 4.17 A

Chapter 2

2.3 11.69
2.4 35.21
2.5 39.47°

2.6 44.72°
2.7 25.24
2.8 35.199
2.9 70°
2.10 32°
2.11 41.23 ∠ 75.96
2.12 103.92 − j60
2.13 (c) 0 + j0; (d) 240 + j0
2.14 15.095 ∠ 10.798°, 14.828 + j2.828, 15.095 ∠ −34.2°, 12.485 − j8.485
2.15 5.76 + j0, 5.76 ∠ 0°, 41.667 ∠ 0°, 41.667 + j0, 240 ∠ 0°, 240 + j0
2.16 29.25 Ω, 0 + j29.25, 29.25 ∠ 90°
2.17 400 Hz
2.18 0.562 ∠ 0°, 0.562 + j0, 120 ∠ 62.05, 56.24 + j106.0, 213.39 ∠ 62.05, 100 + j188.5, 67.44 VA, 31.609 W, 59.57 VAR
2.19 109.33 ∠ −23.85°, 100 − j44.21, 1.098 ∠ 0°, 1.098 + j0, 0.915, 120.6 W
2.20 104.75 ∠ 17.32°, 100 + j31.19, 1.146 ∠ 0°, 1.146 + j0, 0.955, 131.24 W
2.21 (b) 13.0 ∠ 22.62°; (c) 1440 W; (d) 9.231 ∠ −22.61°
2.22 (a) 14.4 Ω, 21.7 Ω; (b) 122 μF, 8.33 A, 1000 W; (c) 0.554, 15.04 A, 1000 W; (d) 4.42 Ω, 399 μF
2.23 (a) 18.38 Ω, 15.42 Ω, 31.32 Ω, 37.33 Ω; (b) 1838.5 W, 1838.5 W; (c) 1838.5 W, 1838.5 W
2.24 (a) 51.32°; (b) 15 ∠ 51.32°; (c) 9.375 Ω, 11.71 Ω; (d) 24 Ω 19.22 Ω; (e) 140 μF; (f) 5.0 A
2.25 6 Ω, 10 Ω
2.26 383 μF, 166 μF
2.27 42 A, 100 A, 129.6 A

Chapter 3

3.2 (a) 519 mA; (b) 259 V
3.3 1 in 2857

Chapter 4

4.1 0 + j0
4.2 (a) 96.23 A; (b) 55.56 A; (c) 4.32 Ω
4.3 (a) 720.5 W; (b) 60 Ω; (c) 180 Ω
4.4 (a) half; (b) half
4.5 15 A, 25.98 A, 240 V, 240 V, 16 Ω
4.6 30 A, 30 A, 120 V, 207.9 V, 4 Ω
4.7 $R_1 = R_A R_B/(R_A + R_B + R_C)$, $R_2 = R_A R_C/(R_A + R_B + R_C)$, $R_3 = R_B R_C/(R_A + R_B + R_C)$
4.8 3.47 Ω, 3.06 Ω
4.9 4.426 + j0, − 4.536 − j1.215, − 4.536 + j1.215, 4.646 + j0, 1470 W
4.10 92.59 A, 166.7 A, 44.44 A, 75.24 A
4.11 $I_a = I_b = I_c = 23$ A, $I_x = I_y = I_z = 39.84$ A, 15.87 kW
4.12 6.0 A
4.13 6.06 ∠ 102.76°
4.14 30.0 A, 10.81 kW

Chapter 5

5.1 0.316 in.
5.2 2625 V
5.3 (a) 114.94 V; (b) 4.22%; (c) 120 V
5.4 UF2 W/G, AWG-8, CU, 2%, 242 ft
5.5 66 ft, NM 2 W/G, CU, 2%, AWG-10, AWG-6, AWG-6
5.6 2%, 495 ft, USE, AL, AWG-0, 226.15 V
5.7 (a) NM 2 WG; (b) Black, white, green or bare; (c) AWG-12; (d) AWG-10; (e) AWG-10; (f) 240 V; (g) 236.14 V
5.8 44.20°, 234.17 V
5.9 132.6 W, 7.96 kWh/mo

Chapter 6

6.4 (a) 0.336 s; (b) 0.0372 s; (c) 0.030 s
6.5 (a) yes; (b) 23 s
6.6 (a) > 1000 s; (b) 0.014 s

Chapter 7

7.1 228.9 A, 169.4 A
7.2 (a) 230, 125; (b) 400 A

Chapter 9

9.1 $55.56/kWh
9.2 48¢/kWh
9.3 60.5 Hz, 1815 r/min
9.4 35.0 kW, 99.2 kW, 17.7 kW, 41.7 kW
9.5 (a) 21.2 kW, 21.2 kW; (b) 36.1 kW, 21.2 kW; (c) 48.6 kW, 21.2 kW
9.6 7, 46.0 kW, 32.5 kW

Chapter 10

10.2 (a) 5; (b) 720 r/min; (c) 4800 r/min; (d) 24,000 r/min
10.3 (a) 6; (b) 50 r/min; (c) 4.2%
10.4 66.7 Hz
10.5 (a) 75 r/min; (b) 4.17%
10.7 (a) 152 lb·ft; (b) 319.2 lb·ft; (c) 407.4 lb·ft
10.8 5.3 hp
10.9 $T_{FL} = 0.7614$ (HP)(P)
10.10 10.0 hp

10.13 (a) 3.04 lb·ft; (b) 12.03 lb·ft
10.16 (a) NM, CU; (b) AWG-8; (c) 38.7 A; (d) 113.8 V; (e) 109.3 V; (f) 0.54
10.17 (a) 18 r/min; (b) 145 A; (c) 2.28 lb·ft (d) 3.43 lb·ft; (e) 1750 r/min; (f) 50.4%; (g) 130°C
10.19 (a) 2-value cap, 5 hp, 230 V, 4 pole, TEFC, BALL, SOLID; (b) 1. THHN, AWG-4, 264 ft; 2. 2 pole, 35 A, 50 A; 3. 5 hp, 2 pole; 4. 32.2 A; 5. 2 pole, 5 hp
10.20 AWG4, MAX: 125A, MIN: 60A, NEMA 3, 10 hp rating
10.21 (a) 2, 50 A, 80 A; (b) 2, 7.5 hp, 46 A; (c) 2, 2; (d) 50 A; (e) USE, AL, AWG-6, AWG-3, AWG-3, 132 ft
10.22 2-value cap, 2 hp, 1725 r/min, solid, ball, TEFC.
10.23 (a) 5 hp, 3-phase squirrel cage, ODP, solid, 3450 r/min, ball; (b) cap start

Chapter 12

12.1 1 fc
12.2 10 ft
12.3 78, 540 lm
12.4 127 cd
12.6 34%
12.8 (a) 2982 h, 13.18 lm/W, 1120 lm, 85 W; (b) 215 h, 19.1 lm/W, 2208 lm, 116 W

Chapter 13

13.1 (a) 301; (b) 8; (c) 1; (d) 255
13.2 4095
13.3 (a) 11111110; (b) 10001010111; (c) 11111111111
13.10 64

Chapter 14

14.1 0.14 V
14.2 little effect
14.3 29.96 V

Chapter 15

15.1 (b) 4.5 kWh; (c) 6 kW; (d) 11 kW
15.2 (a) $3032.50; (b) $2237.50; (c) $27,645; (d) 55.56 kW; (e) $25,176; (f) $20,727

INDEX

Page numbers in *italics* refer to tables.

D

P

R

S

T

U

V

W